Lecture Notes in Artificial Intelligence   1360

Subseries of Lecture Notes in Computer Science
Edited by J. G. Carbonell and J. Siekmann

Lecture Notes in Computer Science

Edited by G. Goos, J. Hartmanis and J. van Leeuwen

T0223204

Springer
Berlin
Heidelberg
New York
Barcelona
Budapest
Hong Kong
London
Milan
Paris
Santa Clara
Singapore
Tokyo

Dongming Wang (Ed.)

Automated Deduction in Geometry

International Workshop
on Automated Deduction in Geometry
Toulouse, France, September 27-29, 1996
Selected Papers

 Springer

Volume Editors

Dongming Wang
Laboratoire LEIBNIZ, Institut IMAG
46, avenue Félix Viallet, F-38031 Grenoble Cedex, France
E-mail: Dongming.Wang@imag.fr

in cooperation with

Ricardo Caferra
Laboratoire LEIBNIZ, Institut IMAG
46, avenue Félix Viallet, F-38031 Grenoble Cedex, France
E-mail: ricardo.caferra@imag.fr

Luis Fariñas del Cerro
Institut de Recherche en Informatique deToulouse, Université Paul Sabatier
118, route de Narbonne, F-31062 Toulouse Cedex, France
E-mail: farinas@irit.fr

He Shi
Institute of Systems Science, Academia Sinica
Beijing 100080, China
E-mail: hshi@mmrc.iss.ac.cn

Cataloging-in-Publication Data applied for

Die Deutsche Bibliothek - CIP-Einheitsaufnahme

Automated deduction in geometry : selcted papers / International
Workshop on Automated Deduction in Geometry, Toulouse, France,
September 27 - 29, 1996. Dongming Wang (ed.). - Berlin ;
Heidelberg ; New York ; Barcelona ; Budapest ; Hong Kong ;
London ; Milan ; Paris ; Santa Clara ; Singapore ; Tokyo : Springer,
1998
    (Lecture notes in computer science ; Vol. 1360 : Lecture notes in
    artificial intelligence)
    ISBN 3-540-64297-8

CR Subject Classification (1991): I.2.3, F.4.1, I.3.5, G.2

ISBN 3-540-64297-8 Springer-Verlag Berlin Heidelberg New York

© Springer-Verlag Berlin Heidelberg 1997
Printed in Germany

Typesetting: Camera ready by author
SPIN 10631780   06/3142 – 5 4 3 2 1 0   Printed on acid-free paper

Preface

This volume contains selected papers presented at the Workshop on Automated Deduction in Geometry held September 27–29, 1996 in Toulouse, France. The workshop, organized by Ricardo Caferra, Luis Fariñas del Cerro, He Shi, and Dongming Wang, and sponsored by Programme de Recherches Avancées de Coopérations Franco-Chinoises (PRA M94-1), Université Paul Sabatier de Toulouse, and PRC-GDR AMI du CNRS, brought together 20 researchers from Asia, Europe and North America. Two invited talks were given by Deepak Kapur and Volker Weispfenning.

In addition to those presented at the workshop, a few papers whose authors were invited to attend the workshop but could not come have also been considered for inclusion in the volume. All the submitted papers underwent a refereeing process at the usual conference standard; finally 11 papers have been accepted.

Automated deduction in geometry is one of the classical research subjects in artificial intelligence. Remarkable success has been achieved since the invention of Wu's method in the later 1970s. Research practice continues to demonstrate the high power and capability of advanced methods for automating geometric problem-solving with modern computing technologies. This collection of state-of-the-art contributions from leading experts and active researchers presents recent advances and new trends on the subject: existing methods are extended, implemented and applied, new elimination and coordinate-free techniques are introduced and developed, and integration of different approaches is attempted. We hope that this book will not only serve as an up-to-date reference but also motivate further developments on automated geometric deduction.

We thank the authors, the referees, and those who contributed to the preparation of this volume.

February 1998   The Editors

Contents

Automated Geometric Reasoning: Dixon Resultants, Gröbner Bases, and Characteristic Sets

Deepak Kapur*

Institute for Programming and Logics
Department of Computer Science
State University of New York
Albany, NY 12222
kapur@cs.albany.edu

Abstract. Three different methods for automated geometry theorem proving—a generalized version of Dixon resultants, Gröbner bases and characteristic sets—are reviewed. The main focus is, however, on the use of the generalized Dixon resultant formulation for solving geometric problems and determining geometric quantities.

1 Introduction

Developing computer programs to prove geometry theorems attracted a great deal of attention in Artificial Intelligence research in its early days. The influential work of Gelernter, in particular, led to a number of key concepts and ideas in automated reasoning which are still used in many artificial intelligence systems employing reasoning. For geometry theorem proving, Gelernter's approach, however, had limited success as it could prove only simple plane geometry theorems whose statements involved lines and angles, and relations among lines and angles [14]. It is difficult to single out any theorem proved by Gelernter's program which could not be easily proved by humans.

It has been well-known that geometry theorem proving falls within the theory of real closed fields for which Tarski gave a decision procedure [55]. But algebraic methods were not successful in proving geometry theorems even with an implementation of a more efficient decision procedure due to Collins [1] for Tarski's geometry. The situation changed dramatically in the late 70's and early 80's when Wu Wen-tsün [62, 63] proposed an algebraic method based on Ritt's characteristic set computation [50] for proving a restricted set of geometry theorems.

For geometry problems involving incidence, congruence and parallelism relations (and not involving the betweenness relations), Wu showed that if a geometric configuration is specified as a finite set of hypotheses translated into polynomials, then conclusions about these relations also expressed as polynomials can be decided. Hypotheses can be transformed into a triangular form, which is then used to check whether the conclusions reduce to 0 using the triangular

---

* Partially supported by the National Science Foundation Grant no. CCR-9622860.

form. This method can be easily implemented, and can be used for proving many geometry theorems including many nontrivial theorems which even humans do not find easy to prove. Wu's approach and its variations have been experimented by many researchers including Wu himself [63, 64], his students Wang and Gao [61], Chou [4, 5], and Ko and Hussain [41]. A particular mention must be made of Chou's work and his system which has proved nearly 500 geometry theorems [5].

Impressive results achieved using Wu's method sparked a great deal of interest in automatic geometry theorem proving. In particular, Wu's work encouraged researchers to investigate algebraic methods other than the ones based on characteristic set computation for geometry theorem proving. Kapur [28], Chou and Schelter [13], and Kutzler and Stifter [42] independently demonstrated the use of Gröbner basis algorithms for geometry theorem proving.

Wu observed that a good heuristic for studying many Euclidean geometry problems not involving the betweenness relation is to consider these problems over an algebraically closed fields such as the complex numbers, instead of over the reals.[2] Consequently, algorithms for analyzing the zeros of a polynomial set over algebraically closed fields developed in algebraic geometry could be fruitfully used. Wu also demonstrated that Ritt's characteristic set algorithm could be used to prove nontrivial theorems including Morley's theorem and Steiner's theorem [62].

Because of this remarkable but simple insight of Wu, the field of geometry theorem proving has been flourishing with researchers, particularly Chou, Gao and Zhang, developing and experimenting with different heuristics and methods for proving hundreds of theorems in plane and solid Euclidean geometries, as well as in other geometries. Powerful theorem provers have been developed over the last 10 years which have been shown to automatically find proofs of theorems considered difficult by humans as well as discover new geometry theorems. Further, some of these provers can produce succinct, elegant and understandable proofs.

This paper is an attempt to give an overview of my group's research in geometry theorem proving since 1984. The use of Gröbner basis, characteristic set and Dixon resultant methods for algebraic and geometric reasoning as investigated by us is reviewed. In the next section, two commonly-used and related formulations of geometry conjectures are discussed. This is followed by a discussion of two different approaches towards automated geometry theorem proving – the direct approach proposed by Wu and the refutational approach proposed by me. Section 4 reviews three different algebraic methods we have used for geometry theorem proving. Our most recent work using the generalized Dixon resultant formulation is first discussed. The proof by example approach of Hong, Zhang, Yang and Deng is discussed using the Dixon resultants, which can serve as a good heuristic. This is followed by a brief mention of the use of the Gröbner basis method, the first method tried by our group in 1985 lead-

---

[2] For a geometry theorem involving incidence and parallelism which is true for the reals but not in the algebraically closed field, the reader may consult [16].

ing to the development of the theorem prover *GEometer*. Finally, the use of the characteristic set method investigated in my former student Hoi Wan's M.S. thesis is reviewed. The discussion in this section is focussed mostly on my group's research; other researchers' work is mentioned only briefly, mostly for completeness and comparison purposes. Section 5 discusses the use of Dixon resultants for computing geometric quantities. Section 6 is a brief review of other geometric reasoning problems including implicitization and computing invariants. Section 7 is a mention of a few new promising approaches towards geometry theorem proving which have been successfully investigated by other researchers. Section 8 includes a few comments on techniques for deducing subsidiary conditions and additional hypotheses in case a given geometric conjecture is not valid. These techniques are especially useful for discovering new theorems.

2 Formulating Geometry Problems

Different kinds of geometry problems arise in various application domains. This ranges from proving geometry theorems, computing formulas and quantities, determining whether a geometric construction can be done with a ruler and compass, implicitization of curves and surfaces, parameterization, invariants of geometric configuration with respect to a group of transformation, etc. In this paper, we will briefly review a few such problems, discussing how our research group has attempted to solve these problems over the last 10 years or so. Our experience suggests that no single method is uniformly good for different problems. Instead, different methods must be used for different problems, which calls for a need for developing a geometry reasoning system that supports a variety of elimination algorithms including the characteristic set, Gröbner basis and resultant methods [30].

In most of the recent research for geometry theorem proving, two different but related formulations for geometry statements have been considered. In both formulations, a coordinate system is associated with a geometry configuration, and geometry statements are usually studied in an algebraic closed field, in particular the field of complex numbers.[3]

2.1 Generic Formulation

In the first formulation, henceforth called the *generic formulation*, which was used in Wu's original paper [62, 63], a geometry statement is specified as a finite set of hypotheses translated into polynomials of a finite set of *dependent* and *independent* variables (independent variables are also called *parameters*). A conclusion, which is also a polynomial, is given separately. The objective is to decide whether the conclusion follows from the hypotheses *generically*. If the

---

[3] In that sense, these approaches are restrictive and serve only as a heuristic for deciding formulas in the quantifier-free subset of Tarski's geometry. For a way to get around this, and an approach for geometry theorem proving over the real closed field, see [20].

answer is yes, then determine also the degenerate cases, if any, which must be ruled out. If the answer is no, then decide a weaker statement, i.e., whether the conclusion is valid for at least one (or more) of the generic configurations specified by the hypotheses, and if so, determine the configurations for which the conclusion is valid as well as list the associated degenerate cases which have to be ruled out. In case the conclusion does not follow at all (it is not valid for any geometric configuration specified by the hypotheses), then declare the conclusion to be false.

Nondegenerate (also called *subsidiary*) conditions such as points being distinct, three distinct points being not collinear, line segments of nonzero lengths, circles of nonzero radii, etc., are left out from this formulation. Instead, they are discovered in case the conclusion follows generically from the hypotheses.

2.2 Logical Formulation

The second formulation, henceforth called the *logical formulation*, is based on a traditional approach taken in developing decision procedures for theories. In this formulation, a geometry statement, which is usually a finite set of hypotheses implying a conclusion, is given and the objective is to decide whether the statement as it is (i.e., without adding any other conditions) is valid. In contrast to the generic formulation above, the hypotheses in this formulation usually include relations ruling out degenerate cases (such as distinctness, noncollinearity, etc.) Variables do not have to be classified into two disjoint categories of parameters (independent) and dependent variables. This formulation was first used by me in [28].

Besides the above two formulations of geometry theorems, many other different geometry problems arise in many application domains. Many of these problems can be formulated as elimination problems. For example, as we shall see later, the problem of determining an implicit representation of a surface from a parameterization representation can be specified as an elimination problem. The problems of determining certain geometric quantities, e.g. Heron's formula – the area of a triangle in terms of its three sides, Brahmagupta's formula on the area of a cyclic quadrilateral in terms of its sides, the maximum volume of a tetrahedron, can also be formulated as elimination problems.

In the first part of the paper, we discuss theorem proving in Euclidean geometry using the generic and logical formulations. Later, we briefly review other geometric reasoning problems.

3 Geometry Theorem Proving: Direct and Refutational Approaches

There have been quite a few approaches proposed for automating geometry theorem proving using the algebraic formulation of geometry problems, prominent about them being the *direct* approach proposed by Wu [62, 64], and the *refutational* approach proposed by Kapur [28]. Below, we briefly review the direct and refutational approaches.

3.1 Parallelogram Example

Let $\mathbb{Q}[u_1, \cdots, u_k, x_1, \cdots, x_l]$ be a polynomial ring over the rationals with variables (indeterminates) $u_1, \cdots, u_k, x_1, \cdots, x_l$. Whenever needed, $u_1, \cdots, u_k$ are considered as independent variables (parameters) and $x_1, \cdots, x_l$ as dependent variables; further, we assume a total ordering $x_l > \cdots > x_1 > u_k > \cdots > u_1$.

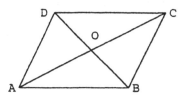

Fig. 1. Midpoint of diagonals in a parallelogram.

As a running example, consider a simple geometry theorem about parallelograms. This example is simple enough that most calculations can even be done by hand to get insight into various methods, and at the same time, the example is quite revealing. Let $ABCD$ be a parallelogram shown in Figure 1; prove that the intersection of the diagonals is the midpoint of a diagonal. The coordinate system chosen is: $A = (0,0), B = (u_1, 0), C = (u_2, u_3), D = (x_2, x_1), O = (x_3, x_4)$. The hypotheses in polynomial form are:

1. $x_1 - u_3 = 0$	AB is parallel to CD.
2. $x_1(u_2 - u_1) - x_2 u_3 = 0$	BC is parallel to AD.
3. $x_4(x_2 - u_1) - x_1(x_3 - u_1) = 0$	B, O, and D are collinear.
4. $x_3 u_3 - x_4 u_2 = 0$	A, O, and C are collinear.
5. $u_1 \neq 0$	A and B are distinct points.
6. $u_3 \neq 0$	A, B, and C are not collinear.

The conclusion is: 7. $x_3^2 + x_4^2 = (u_2 - x_3)^2 + (u_3 - x_4)^2$, i.e., $|AO|^2 = |OC|^2$.

In the above specification, u_1, u_2, u_3 are independent variables (or parameters), and x_1, x_2, x_3, x_4 are dependent variables.

3.2 Direct Approach

The direct approach for geometry theorem proving, originally in Wu's papers [62, 63, 64, 66], consists of two parts. The hypotheses consisting of polynomial equations are first processed and brought in a *standard* form. Wu and others following his approach used a triangular form, called a *characteristic* set, a concept due to Ritt [50], as well as a finite family of characteristic sets. Chou and Schelter as well as Kutzler and Stifter used a Gröbner basis as a standard form. In the second step, using this standard form, it is decided whether the conclusion follows from the hypotheses. While making this check, degenerate conditions, if any, that must be ruled out, are determined.

For instance, in the parallelogram example, hypotheses 1, 2, 3 and 4 are brought into a standard form. Hypotheses 5 and 6 are not included while attempting a proof of the conclusion 7 using the direct approach; instead, they can be derived.

For computing a characteristic set or a Gröbner basis from the hypotheses, variables appearing in the hypotheses must be classified into independent and dependent variables. This is so because computations are done over the field of rational functions with independent variables as transcendentals. For checking whether a conclusion follows from the hypotheses, any polynomials in independent variables appearing in the denominators of coefficients or as initials of polynomials used in simplification of the conclusion are used as expressing degenerate conditions since those polynomials becoming 0 must be ruled out. Some of these conditions may, however, be unnecessary for the conclusion to follow from the hypotheses.

The main idea is to ensure that the zero set of the conclusion contains the zero set of the hypotheses. Since the conclusion is only required to be *generically* true for the hypotheses, some of the zeros from the zero set of the hypotheses can be excluded insofar as they do not change the generic nature of the geometric problem. This can be ensured by disallowing certain values of independent parameters. For the parallelogram example, it can be assumed that the points A, B, C are not collinear, even if the hypotheses do not explicitly state that; the zeros of the hypotheses corresponding to the case when A, B, C are indeed collinear, do not have to be considered.

Many heuristics can be employed to perform a check for containment of the zero set of the hypotheses in the zero set of a conclusion. The variety defined by the hypotheses can be decomposed into components; the nature of the desired decomposition determines the effort involved. Full decomposition into irreducible components can be, in general, computationally quite expensive. See [56, 39, 5, 58] for such a discussion.

For the parallelogram example, a triangular form can be easily obtained using a characteristic set construction algorithm or a Gröbner basis algorithm (since the hypotheses are linear in the dependent variables). (See below a brief review of Gröbner basis and characteristic sets.) And, it can be easily checked that the conclusion indeed reduces to 0 with respect to the triangular form of the hypotheses.

3.3 Refutational Approach

The second approach discussed in [28] is *refutational*, similar to the resolution method for first-order theorem proving and other refutational methods for reasoning. The conclusion of a conjecture is negated, and the inconsistency (or unsatisfiability) of the negated conclusion and the hypotheses is decided over an algebraically closed field.

If a hypothesis is a polynomial inequation (e.g., 5 and 6 in the parallelogram example), it can be converted into a polynomial equation by introducing a new variable as demonstrated in [28]. By the following theorem proved in [28], the validity of a geometry conjecture is equivalent to checking whether a finite set of polynomial equations generated from the conjecture has a common solution in the field of complex numbers.

Theorem 1 ([28]): Given a quantifier-free formula

$$f = ((p_1 = 0 \wedge \cdots \wedge p_i = 0 \wedge s_1 \neq 0 \wedge \cdots \wedge s_j \neq 0) \Rightarrow c),$$

where $p_1, \cdots, p_i, s_1, \cdots, s_j, c \in \mathbb{Q}[u_1, \cdots, u_k, x_1, \cdots, x_l]$. The conjecture f is valid over an algebraically closed field containing $\mathbb{Q}$ if and only if $\{p_1 = 0, \cdots, p_i = 0, s_1 z_1 - 1 = 0, \cdots, s_j z_j - 1 = 0, cz - 1 = 0\}$ has no solution over the algebraically closed field, where $z, z_1, \cdots, z_j$ are new variables.

For the parallelogram example, the last two hypotheses above can be converted into equations using new variables z_1 and z_2 to give:

$$5'. \ u_1 z_1 - 1 = 0, \quad \text{and} \quad 6'. \ u_3 z_2 - 1 = 0,$$

respectively. The negated conclusion can also be translated to

$$7'. \ z((x_3^2 + x_4^2) - ((u_2 - x_3)^2 + (u_3 - x_4)^2)) - 1 = 0.$$

Checking whether a set of polynomial equations has no common solution in an algebraically closed field can be done by the Gröbner basis or the characteristic set construction, or even using other elimination methods including Dixon resultants.

The number of new variables in the above translation can be reduced to a single variable by combining all the subsidiary conditions together with the negated conclusion. For the above quantifier-free formula f, it thus suffices to require that $\{p_1 = 0, \cdots, p_i = 0, s_1 \cdots s_j cz - 1 = 0\}$ has no solution. Even though the complexity of all the elimination methods grows with the number of variables to be eliminated, we have often found that the Gröbner basis method as well as the characteristic set method work better for many examples when subsidiary conditions and conclusion are separated in the translation using new variables. The Dixon resultant method, of course, work much better with fewer variables. However, the optimized translation using a single extra variable does not seem to help since some of the subsidiary conditions reappear as extraneous factors in a projection operator computed by the Dixon resultant method. Unless these extraneous factors can be explicitly identified, it becomes difficult to discover subsidiary conditions or to check whether the needed subsidiary conditions have been identified.

Generation of Subsidiary Conditions: The refutational approach can also be used for deducing subsidiary conditions when a geometry theorem proving problem is given in generic formulation, as shown in [28]. The hypotheses and the negated conclusion are transformed together to a standard form to see whether the conclusion logically follows from the hypotheses. If yes, then the method terminates affirmatively. If not, then information about additional subsidiary conditions needed, under which the hypotheses logically imply the conclusion, can be extracted from the standard form. The following theorem in [28] serves as the theoretical basis of the method.

Theorem 2 ([28]): Given a quantifier-free formula

$$f = ((p_1 = 0 \land \cdots \land p_i = 0) \Rightarrow c),$$

where $p_1, \cdots, p_i, c \in \mathbb{Q}[u_1, \cdots, u_k, x_1, \cdots, x_l]$ such that f is not valid over an algebraically closed field containing $\mathbb{Q}$. There is a nonzero polynomial s in the parameters $u_1, \cdots u_k$, such that $f = ((p_1 = 0 \land \cdots \land p_i = 0 \land s \neq 0) \Rightarrow c)$ is valid over an algebraically closed field containing $\mathbb{Q}$ if and only if the elimination ideal of $(p_1, \cdots p_i, cz - 1)$ in the parameters $u_1, \cdots, u_k$, i.e., $(p_1, \cdots p_i, cz - 1) \cap \mathbb{Q}[u_1, \cdots, u_k]$, is different from 0.

The selection of a polynomial from the elimination ideal to serve as a subsidiary condition is an interesting research problem. If the elimination ideal is principal, its unique generator can be a good candidate. Otherwise, a polynomial(s) with a geometric meaning has to be selected. Factoring polynomials in the elimination ideal can lead to identifying simpler subsidiary conditions corresponding to some of the factors. In a later section, we discuss this issue in more detail.

4 Dixon Resultants, Gröbner Bases, and Characteristic Sets

In this section, the generalized Dixon resultant formulation as developed in [35, 38] is first reviewed. It is shown how Dixon resultants can be used to implement the refutational approach to geometry theorem proving. This is followed by a brief review of Gröbner basis and characteristic set methods to implement the direct as well as refutational approaches for geometry theorem proving. For more details, about Dixon resultants, Gröbner basis and characteristic set methods, the reader may consult [31, 30]. A detailed example illustrating the use of Dixon resultants is given later in a subsection on implicitization. It is recommended that the reader read the next subsection along with the subsection on implicitization.

4.1 Dixon Resultants

Let $X = \{x_1, \ldots, x_n\}$ be a set of n variables. Let $P = \{p_1, \cdots, p_{n+1}\}$ be a set of $n + 1$ polynomials in X. Unless otherwise stated, by a polynomial system, we will mean this set of $n + 1$ polynomials in n variables, X. The coefficients of the polynomials in P are assumed to be polynomials from $\mathbb{Q}[U]$, where U is a set of parameters, say $u_1, \cdots, u_k$.

Let I be the ideal generated by P in $\mathbb{Q}[U, X]$. All polynomials in the ideal $J = I \cap \mathbb{Q}[U]$, the elimination ideal of I over U, are known as the **projection**

operators of P with respect to X. Clearly, projection operators of P vanish at all those specializations of A from the algebraic closure, $\bar{\mathbb{Q}}$, of $\mathbb{Q}$, for which P has a common solution in $\bar{\mathbb{Q}}^n$. Note that 0 is trivially always a projection operator of any P. The **resultant** of P is defined to be the unique (up to a scalar multiple) generator of the ideal J. Note that the resultant of P is 0 *if and only if* P has no nonzero projection operator. Also, a nonzero resultant of P must divide all projection operators of P. Finally, any factor ($\in \mathbb{Q}[U]$) of a projection operator of P which is not the resultant of P, is called an **extraneous factor** in that projection operator.

In [19], Dixon proposed a generalization based on Cayley's reformulation of Bezout's resultant for eliminating a single variable from two polynomials.[4] Dixon used it for simultaneously eliminating two variables from three generic bi-degree polynomials, and commented that the generalization would work for a generic n-degree polynomial system as well. A polynomial system P is called **generic** n-**degree** if there exist nonnegative integers $d_1, \ldots, d_n$,

$$p_j = \sum_{i_1=0}^{d_1} \cdots \sum_{i_n=0}^{d_n} a_{j,i_1,\ldots,i_n} x_1^{i_1} \cdots x_n^{i_n} \text{ for } 1 \leq j \leq n+1,$$

where each coefficient $a_{j,i_1,\ldots,i_n}$ is a distinct indeterminate. $(d_1, \ldots, d_n)$ is known as their n-degree.

Dixon's method, as discussed in [19], does not work for an arbitrary polynomial system. In fact, it almost always fails if the polynomial system is not generic and even if generic, but of not n-degree.

In [35], we developed a generalization of Dixon's resultant formulation so that it works for a family of polynomial systems that is much larger than systems of generic n-degree polynomials. This method has been experimentally found to be superior in performance on a wide variety of examples, in comparison with other elimination methods including Macaulay resultants, sparse resultants [21], the characteristic set construction, and the Gröbner basis construction. The method takes less time, less space, as well as the extraneous factors seem to be fewer (except in the case of the Gröbner basis method which gives the exact resultant) [38, 37]. Recently, we have also been able to show that if the polynomial system consists of polynomials with the same set of terms (an **unmixed** system), the Dixon formulation, in fact, implicitly exploits the sparse structure of the polynomial system, i.e., its computational complexity is governed by the Newton polytope of the unmixed system, not by the Bezout bound [36].

Below, we first briefly review the generalized Dixon formulation for computing a projection operator of P. From now on, by a projection operator, we mean the projection operator as computed by the extended Dixon resultant formulation. Later, we discuss its application to geometric reasoning.

Let $\bar{X}$ be the set of n new variables $\{\bar{x}_1, \ldots, \bar{x}_n\}$. The **Cancellation matrix**

---

[4] See [30] for more historical details.

C_P of P is defined to be the following $(n+1) \times (n+1)$ matrix:

$$C_P = \begin{bmatrix} p_1(x_1, x_2, \ldots, x_n) & \cdots & p_{n+1}(x_1, x_2, \ldots, x_n) \\ p_1(\bar{x}_1, x_2, \ldots, x_n) & \cdots & p_{n+1}(\bar{x}_1, x_2, \ldots, x_n) \\ p_1(\bar{x}_1, \bar{x}_2, \ldots, x_n) & \cdots & p_{n+1}(\bar{x}_1, \bar{x}_2, \ldots, x_n) \\ \vdots & \cdots & \vdots \\ p_1(\bar{x}_1, \bar{x}_2, \ldots, \bar{x}_n) & \cdots & p_{n+1}(\bar{x}_1, \bar{x}_2, \ldots, \bar{x}_n) \end{bmatrix},$$

where $p_i(\bar{x}_1, \cdots, \bar{x}_k, x_{k+1}, \ldots, x_n)$ stands for uniformly replacing x_j by $\bar{x}_j$ for all $1 \leq j \leq k \leq n$ in p_i.

Since for all $1 \leq i \leq n$, $(x_i - \bar{x}_i)$ is a zero of $|C_P|$, $\prod_{i=1}^{n} (x_i - \bar{x}_i)$ divides $|C_P|$. The **Dixon polynomial** δ_P of P is defined as:

$$\delta_P = \frac{|C_P|}{\prod_{i=1}^{n} (x_i - \bar{x}_i)}.$$

Dixon Matrix and Rank Submatrix Computation: Let V be a column vector of all monomials in X which appear in δ_P, when δ_P is viewed as a polynomial in X. Similarly, let W be a row vector of all monomials in $\bar{X}$ which appear in δ_P, when δ_P is viewed as a polynomial in $\bar{X}$. The **Dixon matrix**, D_P, of P is defined to be the matrix for which $\delta_P = V D_P W$. The entries of the Dixon matrix of P are polynomials in U.

If the Dixon matrix is square and non-singular, then its determinant is a projection operator of P. The vanishing of the projection operator of P is a necessary condition for the system P to have a common solution.

It is possible for the Dixon matrix to be rectangular, or even if it is square, to be singular. In that case, the *rank submatrix construction, RSC*, described in [35] can be used to extract identically non-zero projection operators from D_P. The construction works as follows: if the removal of the column labeled with the monomial 1 reduces the rank of the Dixon matrix, then any maximal minor of the Dixon matrix gives a projection operator. Often, there can be constraints on the variables being eliminated, e.g. variables being nonzero. Then, if the removal of any column from the Dixon matrix, that is labeled by a monomial constrained to be nonzero, leads to a smaller rank matrix, then any maximal minor of the Dixon matrix produces a projection operator.

This method has been successfully used to compute projection operators for elimination problems arising in many different application domains; the reader may consult [35, 38, 52] for details.

Below, we first discuss how the generalized Dixon resultant formulation can be used for geometry theorem proving. Later, we discuss the application of the method for other geometric reasoning problems.

Refutational Approach: In [34], we reported on our preliminary investigation on the use of the generalized Dixon method to implement the refutational approach for geometry theorem proving. This method has been further developed and experimented with. Below, we discuss this method.

As discussed earlier, in the generic formulation of a geometry conjecture, there are as many hypotheses as dependent variables. The negated conclusion introduces an extra variable, say z, which must be nonzero. The generalized Dixon formulation is used to eliminate the dependent variables from the hypotheses and the negated conclusion. The resulting projection operator includes factors expressing conditions under which the hypotheses and the negated conclusion have common zeros.

Theorem 3: Given a quantifier-free formula $f = ((p_1 = 0 \land \cdots \land p_l = 0) \Rightarrow c)$, where $p_1, \cdots, p_l, c \in \mathbb{Q}[u_1, \cdots, u_k, x_1, \cdots, x_l]$, and $u_1, \cdots, u_k$ are independent parameters. If the projection operator as computed by the generalized Dixon resultant of $\{p_1, \cdots, p_l, cz - 1\}$ obtained by eliminating $x_1, \cdots, x_l$ is of the form cz^k, where c is a rational number, then f is valid over an algebraically closed field containing $\mathbb{Q}$.

A projection operator being cz^k, where c is a number and z is assumed to be nonzero, implies that the hypotheses and the negated conclusion do not share any common zero, so the conclusion follows from the hypotheses.

In general, a projection operator may have additional factors, some of which may even be extraneous. We have a more general version of the above theorem:

Theorem 4: Given a quantifier-free formula $f = ((p_1 = 0 \land \cdots \land p_l = 0) \Rightarrow c)$, where $p_1, \cdots, p_l, c \in \mathbb{Q}[u_1, \cdots, u_k, x_1, \cdots, x_l]$, and $u_1, \cdots, u_k$ are independent parameters. If the projection operator as computed by the generalized Dixon resultant of $\{p_1, \cdots, p_l, cz - 1\}$ obtained by eliminating $x_1, \cdots, x_l$ is of the form $cz^k s_1^{k_1} \cdots s_j^{k_j}$, where $s_1, \cdots, s_j$ are polynomials in $u_1, \cdots, u_k$, then f is generically valid over an algebraically closed field containing $\mathbb{Q}$.

In addition to cz^k, which is nonzero, if the projection operator includes only polynomials in the independent parameters as the other factors, this means that the negated conclusion shares common zeros with the hypotheses only for certain values of the independent parameters. If those cases can be ruled out as subsidiary conditions, then the conclusion follows from the hypotheses under those subsidiary conditions. In other words, the conclusion generically follows from the hypotheses, much like the direct approach used in Wu's method. A factor in the independent parameters corresponds to the subsidiary conditions that must be ruled out.

For the parallelogram example, when the Dixon resultant method is tried on polynomials $1, 2, 3, 4$, and the negated conclusion $7'$ and variables x_1, x_2, x_3, x_4 are eliminated, a projection operator is:

$$8u_1 u_3^4 (u_2^2 + u_3^2)^2 z^2.$$

It can be thus declared that the conclusion generically follows from the hypotheses. The above polynomial must vanish on every common zero, if any, of the hypotheses and negated conclusion. It is clear that for the parallelogram to be non-degenerate, both u_1 and u_3 be nonzero. Further, $u_2^2 + u_3^2$ should be also nonzero, implying that the point C be distinct from A. (Note that u_3 being nonzero does not guarantee that $u_2^2 + u_3^2$ is nonzero over the field of complex numbers.)

The above method has been successfully tried on almost all examples discussed in [28, 39], including Simson's theorem, Pappus' theorem, Desargues' theorem, Gauss' theorem, nine point circle theorem. Our experiments suggested that the Dixon method compared much more favorably with the Gröbner basis method for deducing subsidiary conditions, but it is slower than the characteristic set method on problems that can be formulated using linear relations. Again like the characteristic set method, the Dixon method produced more subsidiary conditions than necessary. Some of these conditions may even correspond to extraneous factors present in a projection operator.

Logical Formulation: The generalized Dixon method was also tried on geometry statements expressed using the logical formulation in which the subsidiary conditions are included in the problem statement. For every subsidiary condition, a new variable is used to convert an inequation into an equation, and each of these new variables is assumed to be nonzero. All dependent variables and some of the independent parameters constrained by the subsidiary conditions are eliminated from the hypotheses, subsidiary conditions and the negated conclusion. If the projection operator is a polynomial in the new nonzero variables used to express negated conclusion and subsidiary conditions (z, z_1, z_2 for the parallelogram example, for instance), such that if these variables take nonzero values, then the projection operator is also nonzero, then the conjecture is valid; otherwise the conjecture is not valid. That would be the case if a projection operator is of the form $cz^k z_1^{k_1} \cdots z_j^{k_j}$, where $z, z_1, \cdots, z_j$ are the new nonzero variables.

Let us illustrate this on the parallelogram example again. When variables $x_1, x_2, x_3, x_4, u_1, u_3$ are eliminated from polynomials $1, 2, 3, 4, 5', 6'$ and $7'$, the generalized Dixon resultant gives:

$$32u_2^6(z_2^2 u_2^2 + 1)z_1^7 z_2^4 z^7$$

as the projection operator. So, it cannot be said that the conclusion follows the hypotheses and the two subsidiary conditions. If we add an additional subsidiary condition

$$8'. \quad u_2 * z_3 - 1 = 0,$$

to stand for $u_2 \neq 0$, and then eliminate u_2 as well, the resulting projection operator is:

$$128(z_2^2 + z_3^2)z_1^{10} z_2^7 z_3^9 z^{10},$$

implying that the conclusion still does not follow from the hypotheses under the subsidiary conditions of $u_1, u_2, u_3 \neq 0$. However if instead of $8'$, the subsidiary condition:

$$8''. \quad (u_2^2 + u_3^2) * z_3 - 1 = 0,$$

the projection operator produced by the generalized Dixon resultant is:

$$65536 z_1^{14} z_2^{12} z_3^{14} z^{14},$$

thus implying that the modified conjecture is valid.

As the above example illustrates, even for a geometry conjecture given in logical formulation that does not turn out to be valid, it may be possible to extract additional degenerate conditions from the projection operator which must be ruled out to make the conjecture true.

Our experiments on plane Euclidean geometry theorems including Simson's theorem, Butterfly theorem, Ptolemy's theorem, suggest that the Dixon method works reasonably well on geometry conjecture in the logical formulation, but it is much slower than the Gröbner basis method as well as the characteristic set method for detecting inconsistency. This is so because more variables (including both dependent and some of the independent variables) must be eliminated if a logical formulation is used. The complexity of the Dixon resultant method is determined by the volume of the Newton polytope, which is dependent upon the number of variables to be eliminated.

A main limitation in the use of the generalized Dixon resultants for geometry theorem proving is due to the extraneous factors generated when the projection operator is computed. It becomes necessary to distinguish between extraneous factors and the subsidiary conditions relevant for the configuration. That is why the statements of the above theorems are only in one direction. Extraneous factors are, however, not peculiar to Dixon resultants. They arise in all other elimination methods based on multivariate resultants - successive applications of Sylvester resultants, Macaulay resultants or sparse resultants. In fact, we have experimentally observed that the generalized Dixon formulation produces the least extraneous factors. For more details on extraneous factors in the projection operator computed by the generalized Dixon resultant formulation, refer to [37].

For the parallelogram example, instead of $5', 6', 8''$ and $7'$, subsidiary conditions can be combined together with the negated conclusion to have a single polynomial:

$7''.$ $u_1 u_3 (u_2^2 + u_3^2)((x_3^2 + x_4^2) - ((u_2 - x_3)^2 + (u_3 - x_4)^2))z - 1 = 0.$

If the Dixon resultant method is attempted on $1, 2, 3, 4$ and $7''$, the resulting projection operator is:

$$8 u_3^6 u_1^3 (u_2^2 + u3^2)^4 z^2,$$

still including the subsidiary conditions as extraneous factors.

Proving by Example: In [34], we also reported the use of the Dixon resultants on a generic formulation of a geometry conjecture, using a method of theorem proving by example. The Dixon resultant formulation performed very well. Geometry conjectures could be proved in a few milliseconds since the Dixon matrix entries are mostly numbers when the method of theorem proving by example is used. And, the rank submatrix computation can be performed very quickly in that case.

To our knowledge, Zhang, Yang and Deng were the first one to develop a geometry prover based on the idea of theorem proving by example [72] using the characteristic set construction. Their approach is based on an idea proposed by Hong [26]; see also [57]. For a generic formulation of a geometry conjecture, randomly select a specific geometric configuration by choosing particular values of independent parameters, and use the configuration for proving a theorem. If the conclusion indeed follows from the hypotheses, then argue based on genericity and randomness, that the theorem is generically true. Many instances of the conjecture can be tried in parallel. In [70, 72], Zhang et. al. showed how many interesting and nontrivial geometric theorems could be easily proved even on a 386 personal computer in a few seconds, e.g., Thebault's theorem, Morley trisector theorem, etc.

Using this approach of proving by example, Rege [48] has implemented a geometry prover employing Sylvester resultants and Wu's characteristic set method.

4.2 Gröbner Basis Construction

A Gröbner basis of a set of polynomials is a special basis of their ideal which has the property that (i) every polynomial in the ideal *reduces* to 0 with respect to the basis, and (ii) every polynomial has a unique normal form (canonical form) with respect to the basis. The concept of a Gröbner basis of an ideal was introduced by Buchberger in [3], where he also gave an algorithm for computing a Gröbner basis of a polynomial ideal from its finite basis in the case when the coefficients of polynomials are from a field such as the rationals. Since then, Gröbner bases and algorithms for computing them have been extensively studied, analyzed and generalized [2].

There are two key ideas used in the construction of a Gröbner basis of an ideal specified by a finite set of polynomials: (i) reduction of a polynomial by another polynomial such that the reduction process always terminates, and (ii) inclusion of new polynomials from the ideal into its basis from a pair of polynomials already in the basis using *S-polynomial* construction (also called *critical pair* construction in the rewriting literature). First, an admissible term ordering on terms is selected; some commonly used orderings are degree, lexicographic and reverse lexicographic, as well as block orderings. Terms in a polynomial are sorted with respect to the term ordering. Simplification of a polynomial by another polynomial is done based on whether the first polynomial has terms that are multiple of the highest (leading) term of the second polynomial. If yes, the appropriate multiple of the second polynomial is subtracted from the first polynomial. If not, the first polynomial cannot be reduced by the second polynomial. For S-polynomial computation, the smallest term, say t, that is a multiple of the highest terms of the two polynomials (but not a product) is used to determine the terms with which each polynomial must be multiplied; the two resulting polynomials are subtracted to get a new polynomial whose leading term is smaller than t.

This process of simplification of a polynomial by another polynomial as well as generation of new polynomials from the ideal is continued until it is no more possible to generate new polynomials in the basis. And, this process is guaranteed to stop because of Hilbert's basis theorem. The result is a Gröbner basis of the input ideal.

Refutational Approach: In [28, 29], a Gröbner basis algorithm was used to implement the refutational approach for proving geometry theorems. To determine whether a set of polynomial equations has a common solution, the ideal of the polynomials must be nontrivial, i.e., different from 1, and this can be checked by computing a Gröbner basis of the ideal.

For the parallelogram example, the Gröbner basis algorithm on $1, 2, 3, 4, 5', 6'$, and $7'$ results in 1 as the Gröbner basis implying that the conjecture is true.

In these papers, it was also shown that in case the hypotheses do not imply the conclusion, from a Gröbner basis of the hypotheses and the negated conclusion, subsidiary conditions under which the hypotheses imply the conclusion, can be extracted.

For both problems, it is not necessary to classify variables into parameters and dependent variables. For finding subsidiary conditions, it is however useful to specify the variables in which such conditions must be expressed. In that case, as stated above, a Gröbner basis of the elimination ideal in the parameters can be computed from the hypotheses and the negated conclusion. The smallest polynomial in a Gröbner basis of an elimination ideal is often a good candidate for expressing the simplest subsidiary conditions. This method of deducing subsidiary conditions appears to yield the weakest subsidiary condition under which the hypotheses imply the conclusion, in contrast to the characteristic set method and the resultant method.

For the parallelogram example, the Gröbner basis method, when attempted on $1, 2, 3, 4$ and $7'$, gives $u_1 \neq 0, u_3 \neq 0$ as the subsidiary conditions, whereas the Dixon resultant method as well as the characteristic set method discussed below produce subsidiary conditions in addition to these two conditions which are sufficient.

In [28, 29], many examples are discussed and details about implementing the refutational approach are provided. A theorem prover, *GEometer* based on these ideas is discussed in [18, 15]

Direct Approach: In [13] as well as in [42], the direct approach is employed using a Gröbner basis algorithm. A Gröbner basis of the hypotheses polynomials is first computed without taking into account the conclusion. This computation is done viewing polynomials as multivariate polynomials in the dependent variables with their coefficients being polynomials in the independent parameters. In other words, the computation is performed over the field of rational functions expressed in terms of parameters. In the method used by [13, 42], it is thus essential to classify the variables in the geometry problem formulation into dependent variables and independent parameters.

As stated before, once a Gröbner basis is computed from the hypotheses, the conclusion is simplified with respect to the Gröbner basis. If the conclusion simplifies to 0 (i.e., the conclusion polynomial is in the ideal of the hypotheses polynomials), then it follows from the hypotheses under the assumption that the leading coefficients of polynomials of the Gröbner basis used in the simplification are non-zero. Those leading coefficients, which are polynomials in the independent variables, not being zero are the subsidiary conditions.

Strictly speaking, the membership of the conclusion polynomial should be checked in the radical ideal of the hypotheses polynomials. However, for most geometry problems, checking whether the conclusion polynomial is in the ideal of the hypotheses turns out to be a fairly effective heuristic. For more details, the reader can refer to [13, 42].

The subsidiary conditions produced by this method often turn out to be more complicated than the ones found using the refutational approach discussed in the previous subsection. Further, some of these subsidiary conditions can be unnecessary.

4.3 Characteristic Set Construction

The key primitive operation used in characteristic set computation is that of *pseudo-division* of a multivariate polynomial by another multivariate polynomial (see [40] for more details about pseudo-division). Pseudo-division is different from the reduction used in a Gröbner basis computation. Firstly, the recursive representation of polynomials is used: a multivariate polynomial is considered as a univariate polynomial in its highest variable. Division is performed between two univariate polynomials, but the initial of the divisor, which is its leading coefficient when the divisor is viewed as a univariate polynomial, can be used to multiply the dividend. Reduction is defined using pseudo-division.

A polynomial p is *reduced with respect to* another polynomial q if (a) the highest variable, say x_i, of p is $<$ the highest variable, say x_j, of q, or (b) the degree of x_j in q is $>$ the degree of x_j in p. If p is not reduced with respect to q, then p *reduces* to r by pseudo-dividing by q, i.e., by multiplying p with the *lowest* power, say e, of the initial I_q of q and dividing the result by q such that

$$I_q^e p = sq + r,$$

where r is reduced with respect to q.

It is easy to see that (i) the pseudo-remainder r is in the ideal of p and q, (ii) the common zeros of p and q are also the zeros of remainder r, and (iii) if $r = 0$, then the common zeros of p and q are the same as the zeros of q insofar as they are not the zeros of I, the initial of q. These properties are crucial in the proof of Ritt's theorem below.

The following definition and theorem from [64] are attributed to Ritt and serves as the basis of the characteristic set construction:

Definition: Given a finite set Σ of polynomials in $\mathbb{Q}[u_1, \cdots, u_k, x_1, \cdots, x_l]$, a *characteristic set* Φ of Σ is defined to be either (i) $\{g_1\}$, where g_1 is a polynomial in $\mathbb{Q}[u_1, \cdots, u_k]$ or (ii) $\{g_1, \cdots, g_l\}$, where g_1 is a polynomial in $\mathbb{Q}[u_1, \cdots, u_k, x_1]$ with initial I_1, g_2 is a polynomial in $\mathbb{Q}[u_1, \cdots, u_k, x_1, x_2]$, with initial I_2, $\cdots$, g_l is a polynomial in $Q[u_1, \cdots, u_k, x_1, \cdots, x_l]$ with initial I_l, such that (a) any zero of Σ is a zero of Φ, and (b) any zero of Φ that is not a zero of any of the initials I_i is a zero of Σ.

In the above definition, $u_1, \cdots, u_k$ are the independent parameters, and $x_1, \cdots, x_l$ are the dependent variables. In contrast to a Gröbner basis, a characteristic set includes at most one polynomial for every dependent variable.

Theorem 5 (Ritt): Given a finite set Σ of polynomials in $x_l, \cdots, x_1, u_1, \cdots, u_k$, there is an algorithm which computes a *characteristic set* Φ of Σ.

An algorithm for computing a characteristic set is given in [64]; this algorithm is based on the description of how a characteristic set can be constructed in Ritt's book. Since then, many modifications and heuristics to improve the performance of this algorithm have been proposed; an interested reader may consult [64, 5, 6, 39].

Even though the characteristic set of Σ is small, intermediate computations can swell, leading to large polynomial sets. The worst case complexity of computing a characteristic set is doubly exponential in the degree of the input polynomials.

Wu had a remarkable insight into the theory of characteristic sets. Wu relaxed a main requirement from Ritt's definition of a characteristic set, particularly, that every polynomial in an ideal need not have to be pseudo-divided to 0 with respect to its characteristic set. Wu demonstrated that without this property, a characteristic set can be computed much more easily, since there is no need to require a characteristic set to be **irreducible**. The modified definition works as a good heuristic for geometry theorem proving. For more details about the differences between Ritt's and Wu's definition of a characteristic set, the reader may consult [30].

Direct Approach: A characteristic set is first constructed from the hypotheses. Then the conclusion is pseudo-divided using the polynomials in the characteristic set. If the conclusion pseudo-divides to 0, then the conclusion generically follows from the hypotheses. The initials of the polynomials in the characteristic set used to pseudo-divide the conclusion to 0 correspond to the degenerate conditions which must be ruled out.

For the above parallelogram example, a characteristic set can be easily constructed from the hypotheses. Pseudo-division of hypothesis 3 by hypothesis 4 gives a remainder which serves as the polynomial introducing x_3, whereas hypotheses 1, 2 and 4 serve as introducing x_1, x_2, x_4 respectively. Using these polynomials. the conclusion pseudo-divides to 0, implying that the conclusion generically follows from the hypotheses.

In case a conclusion does not pseudo-divide to 0, then it cannot be declared that the conclusion does not generically follow from the hypotheses since Wu had relaxed one condition on his definition of a characteristic set. Instead, additional work must be done. The characteristic set must be checked for *irreducibility*; see [62] for a definition. It must be checked whether each polynomial included in the characteristic set factors over the extension field defined by the smaller polynomials in the characteristic set; this is an expensive computation. If yes, then the polynomial must be factored, and each factor must be considered separately, thus giving rise to a **family** of characteristic sets. Such decomposition corresponds to a possible ambiguity in the formulation of a geometric problem, giving rise to many geometric configurations, e.g., corresponding to in-circles and out-circles of a triangle, interior angles and exterior angles between two lines, etc [66].

The conclusion is pseudo-divided with respect to each characteristic set in a family of irreducible characteristic sets generated from the input hypotheses (which corresponds to all possible geometric configurations implied by the hypotheses). If the conclusion pseudo-divides to 0 in all cases, then again, the conclusion generically follows from the hypotheses. If the conclusion pseudo-divides to 0 for a subset of characteristic sets, then the conclusion follows from the hypotheses only for the corresponding geometric configurations. If the conclusion does not pseudo-divide to 0 in any case, then the conclusion does not follow from the hypotheses at all for any case.

The direct approach has been successfully used by Wu, Chou, Gao, and Wang to prove hundreds of geometry theorems, including many nontrivial theorems in Euclidean geometry such as Steiner's theorem, Morley's theorem, Thebault's theorem, as well as discover new theorems in Euclidean geometry. Chou's first book is an excellent source for the information about all the theorems proved using this method [5]. As the reader would notice, for proving most geometric theorems, factoring over extension fields is not needed; the heuristic of computing a characteristic set (a la Wu) from the hypotheses works well. In [5], an algorithm for factoring polynomials, in which the degree of variables does not exceed 2, over extension fields is developed. This algorithm has been used by Chou and his colleagues to prove nontrivial theorems including Thebault's theorem. An interested reader may also consult [69] where a different method that attempts to avoid factoring over extension fields is given; see also [59].

The direct approach has also been shown to work for other geometries including solid geometry, non-Euclidean geometry, as well as differential geometry.

Refutational Approach: In [56, 39], we showed how the characteristic set construction can also be employed to implement the refutational approach. It was shown that for many geometry problems, this method turns out to be superior in performance to the Gröbner basis method for the refutational approach. The following theorem serves as the basis of the refutational method for geometry theorem proving using the characteristic set construction.

Theorem 6: Given a finite set Σ of polynomials, if its characteristic set includes a constant, then Σ does not have a common zero.

The converse of the theorem does not hold; however, the following holds:

Theorem 7: Given a finite set Σ of polynomials, if its characteristic set Φ is irreducible and does not include a constant, then Σ has a common zero.

Much like the refutational approach implemented using the Gröbner basis method, to check the validity of a geometry conjecture, a characteristic set is computed from the hypotheses and the negated conclusion. If the characteristic set includes a constant, then the hypotheses and the negated conclusion do not have a common zero, implying that the conclusion indeed follows from the hypotheses. The method works on the generic as well as the logical formulations.

For the parallelogram problem, the characteristic set of $\{1, 2, 3, 4, 5', 6', 7'\}$ includes 1, implying the validity of the conjecture.

If the computed characteristic set does not include a constant, it does not mean that the hypotheses and the negated conclusion have common zeros. For example, consider $\{(x - 1)^2 = 0, (x - 1) \neq 0\}$ which clearly does not have a common zero. If we replace the inequation by an equation, we get

$$\Sigma = \{(x^2 - 2x + 1) = 0, (x - 1)z - 1 = 0\}.$$

Under the ordering $z > x$, Σ is also a characteristic set. (However, under the ordering $x > z$, the characteristic set of Σ includes 1.)

There are a number of possibilities if a characteristic set computed from the hypotheses polynomials and the negated conclusion does not include a constant. (i) Φ is reducible and each of the irreducible characteristic sets of Φ includes a constant. (ii) Φ includes a polynomial in $u_1, \cdots, u_k$, which can serve as an additional subsidiary condition to modify the conjecture to make it valid, or (iii) the geometry conjecture is false.

In [39], an implementation of the above approach is discussed which is based on detecting a constant being included in a characteristic set. However, if case (ii) above is identified in which a polynomial in independent variables (whenever such variables are identified) is in a characteristic set, then it suggests that a subsidiary condition in the hypotheses may be missing. The original geometry statement is not a theorem but it can be modified to include the new subsidiary condition to obtain a geometry theorem.

For decomposing a reducible characteristic set, factorization over extension fields has to be performed. With reducibility check by factorization over extension field, the refutational method using characteristic set computation is complete just like the refutational method using the Gröbner basis method [62, 28].

For the refutational approach, the characteristic set method compares favorably with a method using Gröbner basis algorithm. This method was also tried for deducing subsidiary conditions. When no subsidiary conditions are given in a problem formulation, the method turned out to be quite slow. Moreover, the subsidiary conditions found were usually too complicated. In that sense, the method using the Gröbner basis algorithm is superior than this method. However, when an incomplete set of subsidiary conditions were provided, then this method could identify additional subsidiary conditions quickly. In [39], a table comparing different methods on a number of geometry conjectures is given, along with heuristics to improve the performance of the method as well as different algorithms for computing a characteristic set.

5 Computing Geometric Quantities

Many geometry problems involve computing an expression in terms of other parameters, for example, computing the area of a triangle in terms of its sides – Heron's formula, computing the area of a cyclic quadrilateral in terms of its sides, etc. Such problems can be easily formulated as variable elimination problems. For illustration, we discuss how to determine the area of a cyclic quadrilateral in terms of its sides, also known as Brahmagupta's formula.

Let A, B, C, D be four distinct points on a circle, forming a cyclic quadrilateral. Without any loss of generality, assume $A = (0,0), B = (a,0), C = (x_1, x_2), D = (x_3, x_4)$, with $|BC| = b, |CD| = c, |AD| = d$. Let k stand for the area of the cyclic quadrilateral $ABCD$. The objective is to compute k in terms of a, b, c, d, the lengths of the four sides of the cyclic quadrilateral. The hypotheses correspond to:

1. $x_2^2 + x_1^2 - 2ax_1 - b^2 + a^2$, $|BC| = b$.
2. $x_4^2 + x_3^2 + x_2^2 + x_1^2 - 2x_2x_4 - 2x_1x_3 - c^2$, $|CD| = c$.
3. $x_4^2 + x_3^2 - d^2$, $|AD| = d$.
4. $x_2x_4^2 + x_4(-x_2^2 - x_1^2 + ax_1) + x_2x_3^2 - ax_2x_3$, A, B, C, D are on a circle.
5. $x_1x_4 - x_2x_3 + ax_2 - 2k$, k is the area of the quadrilateral.

The area k in terms of a, b, c, d can be computed by eliminating x_1, x_2, x_3, x_4 from the above 5 polynomials. The projection operator computed by the Dixon resultants includes two factors:

$$k^2 - (\frac{p}{2} - a)(\frac{p}{2} - b)(\frac{p}{2} - c)(\frac{p}{2} - d), \text{ and}$$

$$k^2 - \frac{p}{2}(\frac{p}{2} - a - b)(\frac{p}{2} - a - c)(\frac{p}{2} - a - d),$$

where $p = a + b + c + d$. These two factors correspond to the two possible areas depending upon where the fourth point D is in relation to the other three points of the cyclic quadrilateral.

In [65], Wu discussed the use of the characteristic set method for computing such geometric quantities, including Heron's formula, the volume of a tetrahedron in terms of the six edges, etc. Chou and Gao [7] also used the characteristic set method for deriving such geometric formulas. Using a variation of the characteristic set method, Gao and Wang [22] solved an open problem about determining the three sides of a triangle in terms of its angle bisectors. Their approach, however, took a long time and seemed quite inefficient. They reported that their method took about 7 hours of computation time on a SUN 4/470. Below, we review the use of the generalized Dixon resultant formulation for solving the same problem as well as a few other problems that involve computing geometric formulas.

5.1 Side in Terms of Bisectors

Below, we illustrate how determining the side of a triangle in terms of its bisectors can be done easily and much faster using the generalized Dixon's formulation. Here are the three polynomials relating the lengths of one internal bisector a_i, and two external bisectors a_e, b_e, in terms of the lengths of the three sides a, b, c. The objective is to eliminate b, c from these polynomials to get an expression for a in terms of a_i, a_e, b_e.

$$p_1 = a_i^2(b + c)^2 - cb(c + b - a)(c + b + a),$$
$$p_2 = a_e^2(c - b)^2 - cb(a + b - c)(c - b + a),$$
$$p_3 = b_e^2(c - a)^2 - ac(a + b - c)(c + b - a).$$

This problem can be completely solved using the Dixon resultant formulation in less than 6 minutes on a Sun Sparc10 using our interpolation software for computing determinants of matrices with polynomial entries [38, 52]. Even using Maple's linear algebra package, the problem can be done in less than 17 minutes on SUN 4 and less than 9 minutes on a Sparc10. This shows the advantage of using the generalized Dixon resultant formulation over the characteristic set method.

The resulting expression has 330 terms, the same as reported in [22]. It is a polynomial of degree 20 in a, implying using a result reported in Herstein's book that the triangle cannot be constructed using a compass and a rule from the bisectors.

5.2 Maximum Volume Tetrahedron

Here is another interesting problem for computing a geometric quantity taken from [23]. The objective is to compute the maximum volume that a tetrahedron can have in terms of the areas a, b, c, d of its faces. It has been shown in [23] that if there exist parameters x, y, z and w satisfying the following equations, then the tetrahedron is orthocentric and hence with maximum volume with those face areas.

$$p_1 = yz + zw + wy - a,$$
$$p_2 = zx + xw + wz - b,$$

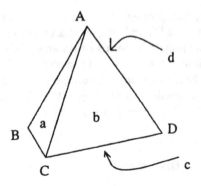

Fig. 2. Orthocentric Tetrahedron.

$$p_3 = wx + xy + yw - c,$$
$$p_4 = xy + yz + zx - d,$$
$$p_5 = 2(xyz + yzw + zwx + wxy) - 9\,T.$$

Polynomial p_5 above relates the square of the volume T to x, y, z, w.

Using the generalized Dixon resultants, the problem can be solved in 76 seconds using interpolation software, and 110 seconds using Maple's linear algebra package. This problem is not easy to do by other methods.

5.3 Gothic Architecture

It is not surprising to know that in the design of cathedrals, geometry was extensively used and there are nice geometric quantities arising there. As an example, consider the design of the top of a catholic church take from [54] (p. 266). There are a collection of arcs drawn in the diagram below: P, Q are centers of the arcs CB, AC, respectively; arcs ED, DF are drawn with R, S as centers, respectively, and AQ as the radius. Further, $AQ = BP = \frac{5}{6}AB$. The goal is to find the radius of the circle O'' tangent to arcs EA, ED, HM, KM as a function of AB.

Only a few of the geometric relations are needed to determine the radius. Assume the origin to be point M so the center of the circle (by symmetry) is on the y-axis. Two tangency points of the inner circle with other arches suffice to completely characterize the inner circle. Let r be the radius of the inner circle and $s = |AB|$. So one point of tangency, say (x_1, y_1), is with arch AHM with A as the center and AM as radius; this is expressed by polynomial q_1, q_3 below. Further this point is collinear with A and O'', expressed by q_2. The second point of tangency, say (x_2, y_2) is with the outer arch AEC with Q as the center and AQ as the radius; this is expressed by q_4, q_6. Further, this point is collinear with Q and O'', expressed by q_5.

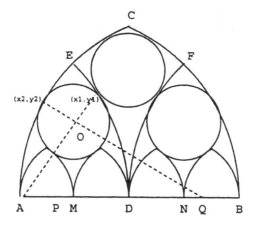

Fig. 3. Top of a Cathedral.

$$q_1 = 2y_1^2 + sx_1 + 2x_1^2,$$
$$q_2 = 4y_3x_1 + sy_3 - sy_1,$$
$$q_3 = r^2 - x_1^2 - y_3^2 - y_1^2 + 2y_1y_3,$$
$$q_4 = 17s^2 - 48y_2^2 + 56sx_2 - 48x_2^2,$$
$$q_5 = 7sy_2 - 7sy_3 + 12x_2y_3,$$
$$q_6 = r^2 - x_2^2 - y_3^2 - y_2^2 + 2y_2y_3.$$

By eliminating x_1, x_2, y_1, y_2, y_3, using the Dixon resultant method, a relationship between r and s can be computed which is: $r = \frac{17}{104}s$, as a factor in the projection operator.

The book [54] is full of such problems from the architectural design, and a great source of entertainment for those interested in geometry; it is a highly recommended reading.

Perhaps the main reason the generalized Dixon resultant formulation is efficient on elimination problems is because it implicitly exploits the sparse structure of the polynomial system. For more details, an interested reader can consult [36] as well as [52].

6 Other Geometric Problems

In addition to geometry theorem proving and computing geometric quantities, algebraic methods including resultants, Gröbner basis and characteristic set constructions, have been successfully used for other problems including implicitization, object recognition and camera calibration in image understanding, object modeling where geometry plays an important role. Below, we briefly review the use of the Dixon resultant formulation for implicitization, invariant computation and an image understanding problem. For further details as well as the use of the Dixon formulation to other related problems, the reader may consult [38, 52].

6.1 Implicitization

We illustrate the use of Dixon resultants for the implicitization problem. Given a variety expressed in a parametric form, e.g.

$$x_1 \; g_1(t_1, \cdots, t_k) = f_1(t_1, \cdots, t_k),$$
$$\vdots$$
$$x_n \; g_n(t_1, \cdots, t_k) = f_n(t_1, \cdots, t_k),$$

the objective is to find an implicitization (or implicit representation) of the variety as a polynomial in $x_1, \cdots, x_n$. This can be formulated as an elimination problem if there are enough equations – one more equation than the number of parameters.

As an example, consider generating an implicit representation of a torus from its parametric representation. The constraints defining the torus are.

1. $x = r \; cos(u) \; cos(t) + R \; cos(t)$,
2. $y = r \; cos(u) \; sin(t) + R \; sin(t)$,
3. $z = r \; sin(u)$.

Quantities r and R are the inner and outer radii, respectively, of a torus. The two trigonometric identities relating sin and cos are:

4. $cos(u)^2 + sin(u)^2 = 1$,
5. $cos(t)^2 + sin(t)^2 = 1$.

The objective is to compute an implicit representation of the torus in terms of x, y, z, r, and R. This can be achieved by eliminating $cos(t)$, $sin(t)$, $sin(u)$ and $cos(u)$ from the above equations.

The Dixon polynomial for this example is given below. Symbols c_u, c_t, s_u, s_t are used instead of $cos(u), cos(t), sin(u), sin(t)$, respectively, and $\bar{c}_u, \bar{c}_t, \bar{s}_u, \bar{s}_t$ are the new variables substituted for c_u, c_t, s_u, s_t, respectively, for formulating the Dixon resultant.

$$\left(r^3 c_t{}^2 - r^3 + r^3 s_t{}^2\right) \bar{c}_u^3 + \left(-r^3 c_u c_t + x r^2 - R c_t r^2\right) \bar{c}_u^2 \bar{c}_t$$
$$+ \left(-r^3 c_u s_t + r^2 y - r^2 R s_t\right) \bar{c}_u^2 \bar{s}_t + \left(-r^3 c_t s_u + r^2 c_t z\right) \bar{c}_u \bar{s}_u \bar{c}_t$$
$$+ \left(-r^3 s_t s_u + r^2 s_t z\right) \bar{c}_u \bar{s}_u \bar{s}_t + \left(x r^2 c_t + r^2 c_t{}^2 R + r^2 y s_t - 2 r^2 R - r^3 c_u + R r^2 s_t{}^2\right) \bar{c}_u^2$$
$$+ \left(-r^3 c_t{}^2 s_u + r^2 s_t{}^2 z + r^2 c_t{}^2 z - r^3 s_t{}^2 s_u\right) \bar{c}_u \bar{s}_u$$
$$+ \left(r x R + r^2 c_t z s_u + x r^2 c_u - r R^2 c_t - r^3 c_t - 2 r^2 c_u c_t R\right) \bar{c}_u \bar{c}_t$$
$$+ \left(-r^3 s_t - r R^2 s_t - 2 r^2 R s_t c_u + R r y + r^2 y c_u + r^2 s_t z s_u\right) \bar{c}_u \bar{s}_t$$
$$+ \left(r c_t R z - r^2 c_t R s_u\right) \bar{s}_u \bar{c}_t + \left(R r s_t z - R r^2 s_t s_u\right) \bar{s}_u \bar{s}_t$$

$$+ \left(-r^3 s_t^2 + r^2 c_t^2 z s_u + r x R c_t + R r y s_t - 2r^2 R c_u - r^3 c_t^2 + z r^2 c_u c_t\right) \bar{c}_u$$
$$+ \left(r^2 y c_u s_t - r R^2 + r^2 s_t^2 z s_u\right) \bar{c}_u + \left(R r s_t^2 z + r c_t^2 R z - r^2 c_t^2 R s_u - R r^2 s_t^2 s_u\right) \bar{s}_u$$
$$+ \left(r c_t R z s_u + r x R c_u - R c_t r^2 - r R^2 c_t c_u\right) \bar{c}_t + \left(-r^2 R s_t + R r y c_u + R r s_t z s_u - r R^2 s_t c_u\right) \bar{s}_t$$
$$+ R r s_t^2 z s_u + R r y c_u s_t - R r^2 s_t^2 + r c_t^2 R z s_u + r x R c_u c_t - r^2 c_t^2 R - r R^2 c_u.$$

The (16×12) Dixon matrix is:

	1	c_u	c_t	s_t	c_t^2	s_t^2	$c_t c_u$	$s_u c_t^2$	$s_u c_t$	$s_t c_u$	$s_u s_t^2$	$s_u s_t$
1	0	$-rR^2$	0	0	$-r^2R$	$-r^2R$	rxR	rRs	0	Rry	rRs	0
c_u	$-rR^2$	$-2r^2R$	rxR	Rry	$-r^3$	$-r^3$	zr^2	r^2s	0	r^2y	r^2s	0
$c_u s_t$	Rry	r^2y	0	$-r^3-rR^2$	0	0	0	0	0	$-2r^2R$	0	r^2s
c_u^2	$-2r^2R$	$-r^3$	zr^2	r^2y	r^2R	r^2R	0	0	0	0	0	0
$c_u c_t$	rxR	zr^2	$-r^3-rR^2$	0	0	0	$-2r^2R$	0	r^2s	0	0	0
s_u	0	0	0	0	0	0	rRs	rRs	0	$-r^2R$	0	$-r^2R$	0
c_t	0	rxR	$-r^2R$	0	0	0	$-rR^2$	0	rRs	0	0	0
s_t	0	Rry	0	$-r^2R$	0	0	0	0	0	$-rR^2$	0	rRs
c_u^2	$-r^3$	0	0	0	r^3	r^3	0	0	0	0	0	0
$s_t s_u^2$	r^2y	0	0	$-r^2R$	0	0	0	0	0	$-r^3$	0	0
$c_t s_u^2$	zr^2	0	$-r^2R$	0	0	0	$-r^3$	0	0	0	0	0
$s_u c_u$	0	0	0	0	r^2s	r^2s	0	$-r^3$	0	0	$-r^3$	0
$c_t s_u c_u$	0	0	r^2s	0	0	0	0	0	$-r^3$	0	0	0
$c_t s_u$	0	0	rRs	0	0	0	0	0	$-r^2R$	0	0	0
$s_t s_u c_u$	0	0	0	r^2s	0	0	0	0	0	0	$-r^3$	0
$s_t s_u$	0	0	0	rRs	0	0	0	0	0	0	0	$-r^2R$

. Its rank is 10, and the projection operator is

$$r^{23} R (-z^2 - R^2 + r^2)(4R^2(r^2 - z^2) - (x^2 + y^2 + z^2 - R^2 - r^2)^2)$$

and the factor $4R^2(r^2 - z^2) - (x^2 + y^2 + z^2 - R^2 - r^2)^2$ is the implicit representation of torus.

6.2 Computing Invariant Relations

Geometric invariants appear to play an important role in object recognition as an aid to building model libraries of objects. Building model-based object recognition system using geometric invariants has recently become an active research area. Invariants of various objects (or their models) can be stored in a model library [46]. Invariants of image features can be used to index the model library to prune the search space for objects being recognized in an image.

Invariant relations are expressions relating model features to image features, and they can be computed using elimination techniques. In [32], the use of the generalized Dixon formulation and the Gröbner basis method are discussed for computing invariant relations, and then invariants, using a separability algorithm.

Object features, e.g., coordinates of points in an object, are related to image features, e.g., coordinates of points in an image through the camera parameters which can be specified using various geometric transformations (depending upon the approximation of the camera viewing used), e.g. affine, scaled orthographic and projective transformations. Given enough object features, transformation parameters can be eliminated from these relations by elimination. From the projection operator thus computed, an invariant relation specifying the relation between object and image features and independent of any transformation parameters can be recovered. In many cases, this invariant relation can be separated, resulting in purely a property of an object, or image(s), which remains invariant under camera viewing.

6.3   Recovering Camera Motion from Two Images

In [21], a camera motion reconstruction, a basic problem in photogrammetry, is formulated as an elimination problem. Given the coordinates of corresponding points in two images under perspective projection on calibrated cameras, an objective is compute the displacement of the camera between two positions Let a_i, a'_i be 3-dimensional vectors corresponding to the images of a point in each image frame. Let $q = <q_1, q_2, q_3>$ and $d = <d_1, d_2, d_3>$ be two 3-dimensional vectors corresponding to the camera displacement. The equations describing the camera motion reconstruction can be specified as:

$$p_1 = (a_1^T q) * (d^T a'_1) + (a_1^T a'_1) + ((a_1 \times q)^T a'_1) + ((a_1 \times q)^T (d \times a'_1)) + (a_1^T (d \times a_1)),$$
$$p_2 = (a_2^T q) * (d^T a'_2) + (a_2^T a'_2) + ((a_2 \times q)^T a'_2) + ((a_2 \times q)^T (d \times a'_2)) + (a_2^T (d \times a_2)),$$
$$p_3 = (a_3^T q) * (d^T a'_3) + (a_3^T a'_3) + ((a_3 \times q)^T a'_3) + ((a_3 \times q)^T (d \times a'_3)) + (a_3^T (d \times a_3)),$$
$$p_4 = (a_4^T q) * (d^T a'_4) + (a_4^T a'_4) + ((a_4 \times q)^T a'_4) + ((a_4 \times q)^T (d \times a'_4)) + (a_4^T (d \times a_4)),$$
$$p_5 = (a_5^T q) * (d^T a'_5) + (a_5^T a'_5) + ((a_5 \times q)^T a'_5) + ((a_5 \times q)^T (d \times a'_5)) + (a_5^T (d \times a_5)),$$
$$p_6 = 1 - (d^T q).$$

The objective is to eliminate d_2, d_3, q_1, q_2, q_3 to get a polynomial in d_1 for a given set of values substituted for a_i and a'_i.

The projection operator is a 10th degree polynomial in d_1, which can be computed in 40 milliseconds on a sparc10; the set up for constructing the Dixon matrix, which has to be done only once, is less than 3 seconds.

It thus appears that the camera motion reconstruction can be done on-line in real time using the generalized Dixon resultants.

6.4   Kinematics: Stewart's Platform

The Stewart platform problem is a well-known benchmark from robotics and kinematics [27, 43, 45, 21]. It is a parallel manipulator with six prismatic joints connecting two platforms, in which the base platform is fixed, while the top platform, or end-effector, is moving in 3-dimensional space, controlled by the lengths of the joints.

The quaternion formulation of the Stewart platform presented below is by Emiris [21]. It contains 7 polynomials in 7 variables. Let $x = [x_0, x_1, x_2, x_3]$ and $q = [1, q_1, q_2, q_3]$ be two unknown quaternions, which are to be determined. Let $q^* = [1, -q_1, -q_2, -q_3]$. Let a_i and b_i, for $i = 2, \ldots, 6$, be ten known quaternions, and let α_i, for $i = 1, \ldots, 6$, be six predetermined scalars. The seven polynomials for solving the Stewart platform problem are:

$$f_1 = x^T x - \alpha_1 q^T q,$$
$$f_i = b_i^T(xq) - a_i^T(qx) - (qb_iq^*)^T a_i - \alpha_i q^T q, \; i = 2, \ldots, 6,$$
$$f_7 = x^T q^*.$$

Out of the 7 variables, $x_0, x_1, x_2, x_3, q_1, q_2, q_3$, any six can be eliminated to compute the resultant as a polynomial in the seventh.

The generalized Dixon resultant formulation can be used to compute a projection operator. Using our implementation of the generalized Dixon method running on a sparc10 station with 64MB memory, we successfully computed the projection operator from the above formulation in total computer time of less than 10 minutes (273.5 seconds to construct the Dixon matrix and 256.79 seconds to interpolate a projection operator from it). The projection operator is a polynomial of degree 40, consistent with similar results reported in [51, 43, 45]. The generalized Dixon resultant formulation is thus able to compute the exact resultant, without any extraneous factors.

To our knowledge, our results are the first successful attempt to compute the resultant for the Stewart platform problem using multivariate elimination methods in which no a priori processing or ad hoc techniques are employed.

The solution to the Stewart platform problem illustrates an interesting aspect of multivariate elimination methods. Even though in a multivariate formulation, variables are eliminated simultaneously, there is an implicit ordering on variables used in different methods, for example, in the Dixon resultants, the order in which original variables to be eliminated are substituted by new variables in the cancellation matrix. This ordering affects significantly the computational performance of the methods as well as extraneous factors generated in the projection operators. These aspects of resultant methods are not understood, and should be investigated. In [37], it is shown that for the Stewart platform problem, for certain orderings, the generalized Dixon method computes the projection operator quite fast, but at the cost of including numerous extraneous factors. On the other hand, we have identified a particular ordering that gives the exact resultant for the Stewart platform problem, without any extraneous factors, but the computation time is much longer. An interesting research issue is to find an ordering on variables that leads to minimal extraneous factors as well as the fastest computation of a projection operator from the Dixon matrix constructed from a polynomial system.

7 New Promising Approaches

The approaches discussed above share one common aspect in the formulation of geometry problems – associating coordinates with points and expressing geometric relations as polynomials. There are, however, other methods currently being investigated, including vector-based approaches discussed in [53, 9], an invariant based approach proposed in [49, 10], the area method discussed in [11] and more recently, the logical approach revived in [12].

A special mention must be made of the *area* method popularized by Prof. Zhang of Academia Sinica. Chou, Gao and Zhang have developed an impressive prover, Euclid, based on this method using which they have shown how hundreds of theorems in plane and solid geometry can be easily proved. An important advantage of this method is that the proofs generated by the prover are short and readable. In the area method, geometric relations are usually expressed in terms of areas of triangles. A proof is typically generated using the heuristic of successively eliminating points. See their excellent book [11] for details. Earlier, Chou, Gao and Zhang had proposed the use of fourteen basic geometry propositions expressing three kinds of geometry quantities: ratios of lengths, areas of triangles and Pythagoras difference for eliminating points [10]

Recently, Chou, Gao and Zhang have been using a classical approach for searching for proofs of Euclidean geometry theorems, based on a data base of useful geometric lemmas, more in the spirit of Gelernter's geometry machine. They have developed an impressive prover based on this idea, which they are planning to use for educational purposes. For details, the reader may consult a forthcoming paper by them [12].

Crapo and Richter-Gebert developed a system based on bracket algebras and invariants for geometry theorem proving in Euclidean, affine and projective geometries. Bracket expressions representing determinants are used to express incidence relations. Proving a theorem amounts to using bracket relations to prove that a conclusion expressed as a bracket expression follows from the hypotheses, also expressed using bracket expressions [49, 17].

Recently, Li and Cheng have developed a modification of Wu's method based on Clifford algebra techniques for proving theorems in Euclidean as well as differential geometries [44]. Clifford algebra techniques are closely related to vector representations of points in Euclidean geometry [53, 9].

8 Subsidiary Conditions and Discovering Geometry Theorems

An interesting byproduct of Wu's approach has been a discussion of the role of degenerate cases of the hypotheses in geometry theorem proving. In [68], Wu said, "almost all theorems in ordinary geometry (i.e., ordinary Euclidean geometry) are only *generically true,* or true only under some non-degenerate conditions usually not explicitly described in the statement of the theorem. This fact is quite fatal to make the usual Euclidean fashion of proving theorems rigorous as one believes to be so, since it is difficult to state clearly the non-degeneracy conditions to be observed and to verify that the previous theorems to be applied in the proof of the theorem in question do not fall into the degeneracy cases under which the previous theorems to be applied might not be true." Wu went on to add that the fact that his approach is able to automatically discover the non-degeneracy conditions is "the crucial point which is responsible for the high efficiency of our method in proving quite non-trivial theorems without the defect pointed above" about theorem proving in the Euclidean fashion.

As discussed in earlier sections, both in the refutational and direct approaches, subsidiary conditions of the form $s \neq 0$ can be derived from the hypotheses and the conclusion of a geometric conjecture using any of the three methods discussed above.

A theoretical justification for this derivation is Theorem 2 in section 3.3 taken from [28, 29] about the elimination ideal of the hypotheses and the polynomial corresponding to the negated conclusion, just in the parameters, being different from 0. In that case, it is possible to find such an s, a polynomial in the parameters whose zeros must be ruled out of the common zeros of the hypotheses for the conclusion to follow from the hypotheses. And, one way to compute the elimination ideal of the hypotheses and the negated conclusion is to use a Gröbner basis algorithm, as discussed in [28, 29].

What if the elimination ideal of the hypotheses and the negated conclusion is 0, i.e., it does not have any polynomials in the parameters? Then it is not possible to find a subsidiary condition $s \neq 0$ (also called *negative* hypothesis below to contrast with *positive* hypotheses of the form $t = 0$) such that the hypotheses along with the subsidiary condition imply a given conclusion. In that case, one can look for an additional positive hypothesis of the form $t = 0$ such that along with this positive additional hypothesis, the conclusion follows (after degenerate cases have been ruled out).

Using this approach, Recio and Velez-Melon [47] have recently developed an educational package for teaching algebraic geometry by mechanically discovering geometry theorems. A geometric conjecture is first tested for validity. In case the conjecture is not true, a subsidiary condition (negative hypothesis) is looked for, possibly ruling out degenerate cases as discussed earlier. If no negative hypothesis can be identified (which will be the case if the elimination ideal is 0), then they propose analyzing the elimination ideal generated from the hypotheses and the conclusion for positive additional hypotheses of the form $t = 0$ in the parameters. Conditions on parameters can be identified in this way which would be true in case the hypotheses imply the conclusion. If such a positive additional hypothesis is identified, the number of parameters is reduced (since the additional hypothesis is a polynomial in the parameters). It is then checked whether the conclusion follows from the augmented set of hypotheses. As before, some inequations may have to be deduced to rule out the degenerate cases of the modified conjecture, or alternatively, the modified conjecture may have to be further strengthened by adding more positive hypotheses.

For instance, consider dropping one of the hypotheses in the parallelogram example, namely, that AB is parallel to CD. Under the condition that two of the sides of a quadrilateral are parallel (i.e., the quadrilateral is a trapezoid), the objective is to find whether the intersection O of the two diagonals AC and BD is the midpoint of AC. In this modified formulation, along with u_1, u_2, u_3, one of x_1 and x_2 (say, x_1) is also an independent parameter. The ideal of polynomials 2, 3, 4, and 7' does not include any polynomial in x_1, u_1, u_2, u_3. This implies that no inequation in x_1, u_1, u_2, u_3 can be found such that the hypotheses and the inequation would imply the conclusion.

But if the elimination ideal of polynomials $2, 3, 4$ and 7 in u_1, u_2, u_3, x_1 is analyzed, it includes among other things, the polynomial:

$$(x_1 - u_3)u_1u_3(u_2^2 + u_3^2).$$

This implies that if the hypotheses and the conclusion are to hold together (i.e., if a quadrilateral with two parallel sides has the intersection point of its diagonals to be the midpoint of one of its diagonal), then the above polynomial in the independent parameters must be 0. The cases of $u_1, u_3, (u_2^2 + u_3^2)$ being zero correspond to the quadrilateral being degenerate. If they are ruled out, the interesting additional constraint on the quadrilateral for the conclusion to hold is:

$$x_1 - u_3,$$

implying that AB is parallel to CD.

Ideas similar to the above have also been successfully tried using the characteristic set method. As stated earlier, Wu [65] discussed how the characteristic set construction can be used for computing geometric formulas. That observation can be viewed as a particular application of the use of a characteristic set construction for multi-variable elimination; for other such uses of the characteristic set method as an elimination technique, the reader may consult [31, 30]. Particularly, the characteristic set method can be used for deducing positive additional hypotheses as well. Chou [4], for instance, used it to discover a generalization of Simson's theorem in this way. A similar idea was also used by Kapur and Mundy [33] to deduce conditions on object features from the properties of images features and vice versa in applying the characteristic set method to image understanding problems. The key idea is summarized below.

Once a characteristic set is computed from the hypotheses, the conclusion is pseudo-divided using the characteristic set. If the conclusion pseudo-divides to 0, then it is said to follow from the hypotheses under certain conditions. If the remainder is nonzero, the remainder can be analyzed to identify an additional positive hypothesis which must be included so that the conclusion follows from the augmented set of hypotheses. Typically the remainder is factored, and if one of the factors is a polynomial in the parameters and has a geometric meaning, then that polynomial can serve as an additional hypothesis.

It is easy to see that the pseudo-remainder of a polynomial p by another polynomial q is in the ideal generated by p and q. Thus, the polynomials in a characteristic set computed without using factorization are in the ideal generated by the hypotheses. The pseudo-remainder obtained by pseudo-dividing the conclusion by the characteristic set generated from the hypotheses is in the ideal of the characteristic set (and hence of the hypotheses, since the ideal generated by the characteristic set is a subideal of the ideal generated by the hypotheses) and the conclusion. We thus get the effect of analyzing certain polynomials in the elimination ideal of the hypotheses and the conclusion.

Analyzing the elimination ideal of the hypotheses and the conclusion in the parameters is equivalent to considering all possible inferences that can be made about a geometric configuration in which the hypotheses and the desired conclusion hold.

A different approach for mechanically generating subsidiary conditions is discussed in [8]. For a family of geometry statements defined to be of constructive type in [8], subsidiary conditions are generated along with the construction of the geometric configuration. It is proved that for a geometry statement of such constructive type to be valid, subsidiary conditions so generated suffice.

9 Concluding Remarks

We have discussed the use of three different algebraic methods for automated geometry theorem proving. Each of these methods has its advantages and disadvantages. If the hypotheses in a geometric problem are linear (or even quadratic) as is typically the case in most Euclidean geometry problems involving lines and points, the characteristic set construction works particularly well. The characteristic set construction is less efficient with hypotheses expressed by quadratics, as is the case with configurations involving distances and circles. For complicated problem formulations, the refutational approach using the Gröbner basis algorithm works particularly well if the conjecture is valid. That is perhaps due to the fact that the zero dimensional ideals can be more easily analyzed using the Gröbner basis and resultant constructions, in contrast to the characteristic set constructions.

A great deal of progress has been made in developing computer programs prove nontrivial geometry theorems, moving towards one of the milestones of the AI research. The main reason perhaps why research in geometry theorem proving has been tremendously successful is because (i) algebraic representations turns out to be an efficient and succinct way to represent all the information about a geometric problem, (ii) algebraic representation can be manipulated efficiently using simple primitive operations which can be implemented well, and (iii) heuristics can be easily incorporated. Of course, the speed of computers have increased significantly since the 50's and 60's, and memory has not been a problem, and these advances in hardware have helped a great deal.

In addition to geometry theorem proving, we have also briefly reviewed other geometric reasoning problems arising in various application domains including solid and geometric modeling, robotics and kinematics, image understanding. We can envision general purpose as well as special purpose software being developed to support geometric reasoning needed in different application domains. As of now, there does not exist a single uniform method that works well for all geometric reasoning problems. Instead, our experience suggests that different methods behave differently on different types of geometric problems.

For elimination problems, the generalized Dixon formulation is especially effective because it implicitly exploits the sparse structure, if any, of the problem. But it becomes necessary to separate extraneous factors in a projection operator from the resultant. On the other hand, the Gröbner basis algorithm, whenever it works, gives the exact resultant. However, for most problems, the Gröbner basis approach is too expensive.

Lots of research and analysis still need to be done. One of our long term goals has been to develop algorithms and their implementations so as to enable off-line and on-line reasoning for object recognition and other image understanding tasks in computer vision. Despite considerable progress, it is still not possible to automatically compute invariants of geometric configurations [46]. When invariant computation is formulated as an elimination problem, it becomes too difficult to manage by currently implemented elimination techniques. With the generalized Dixon resultant formulation, invariants for only simple geometric configurations can be computed automatically [32]. Otherwise, considerable human interaction and guidance are required.

It is still not possible to recover shape from images and camera parameters on-line from the algebraic formulation of shape recovery problems. Elimination techniques must be specialized to such applications, as well as it is necessary to effectively integrate symbolic methods with numerical and approximate methods that are less sensitive to noise and error in data.

Acknowledgments: The generalization of Dixon resultants was developed by the author jointly with Lu Yang and Tushar Saxena. The use of Dixon resultants for geometry theorem proving was explored jointly with Tushar Saxena. Thanks to Dongming Wang for useful comments on an earlier draft of this paper.

References

1. Arnon, D.S., Collins, G.E., and McCallum, S., Cylindrical algebraic decomposition I: The basic algorithm; II: An adjacency algorithm for the plane. *SIAM J. of Computing* 13, 1984, 865-877, 878-889.
2. Becker T., Weispfenning V., in cooperation with Kredel H., *Gröbner Bases, A Computational Approach to Commutative Algebra*. Springer-Verlag, New York, 1993.
3. Buchberger B., Gröbner bases: An algorithmic method in polynomial ideal theory. *Multidimensional Systems Theory* (ed. Bose), D. Reidel Publ. Co., 1985, 184-232.
4. Chou, S.-C., *Proving and Discovering Theorems in Elementary Geometry using Wu's Method*. Ph.D. Thesis, Dept. of Mathematics, University of Texas, Austin, 1985.
5. Chou, S.-C., *Mechanical Geometry Theorem Proving*. D. Reidel Publishing Company, Dordrecht, Netherlands, 1988.
6. Chou, S.-C., and Gao, X.-S., Ritt-Wu's decomposition algorithm and geometry theorem proving. Proc. *10th Intl. Conf. on Automated Deduction (CADE-10)*, Kaiserslautern, LNCS 449, July 1990, 207-220.
7. Chou, S.-C., and Gao, X.-S., Mechanical formula derivation of elementary geometries. Proc. *Intl. Symp. on Symbolic and Algebraic Computation (ISSAC)*, Tokyo, 1990, 265-270.
8. Chou, S.-C., and Gao, X.-S., Proving constructive geometry statements. Proc. *11th Intl. Conf. on Automated Deduction (CADE-11)*, Saratoga Springs, Springer LNAI 607, 1992, 20-34.

9. Chou, S.-C., Gao, X.-S., and Zhang, J.-Z., Automated geometry theorem proving by vector calculations. Proc. *Intl. Symp. on Symbolic and Algebraic Computation (ISSAC)*, Kiev, 1993, 284-291.

10. Chou, S.-C., Gao, X.-S., and Zhang, J.-Z., Automated production of traditional proofs for constructive geometry theorems. Proc. *8th Annual IEEE Symp. on Logic in Computer Science (LICS)*, Montreal, June 1994, 48-56.

11. Chou, S.-C., Gao, X.-S., and Zhang, J.-Z., *Machine Proofs in Geometry: Automated Production of Readable Proofs for Geometry Theorems.* World Scientific Press, 1994.

12. Chou, S.-C., Gao, X.-S., and Zhang, J.-Z., A deductive database approach to automated geometry theorem proving and discovering. Accepted for publication in *J. of Automated Reasoning.*

13. Chou, S.-C., and Schelter, W.F., Proving geometry theorems with rewrite rules. *J. of Automated Reasoning* 2 (3), 1986, 253-273.

14. Coelho, H., and Pereira, L.M., Automated reasoning in geometry theorem proving with PROLOG. *J. of Automated Reasoning* 2 (4), 1986, 329-390.

15. Connolly, C.I., Kapur, D., Mundy, J.L., and Weiss, R., GEometer: A System for modeling and algebraic manipulation, Proc. *DARPA Workshop on Image Understanding*, Palo Alto, May 1989, 797-804.

16. Conti, P., and Traverso, C., A case of automatic theorem proving in Euclidean geometry: The Maclane 8_3 theorem. Proc. *AAECC*, LNCS 948, Paris, 1995, 183-193.

17. Crapo, H., Automatic proofs of geometry theorems: The synthetic approach, Presented at the *Intl. Workshop on Automated Deduction in Geometry*, Toulouse, Sept 1996.

18. Cyrluk, D.A., Harris, R.M., and Kapur, D., GEometer: A theorem prover for algebraic geometry. Proc. *9th Intl. Conf. on Automated Deduction (CADE-9)*, Argonne, IL, LNCS 310, May 1988. 770-771.

19. Dixon, A. L., The eliminant of three quantics in two independent variables. Proc. *London Mathematical Society* 6, 1908, 468-478.

20. Dolzmann, A., Sturm, T., and Weispfenning, V., A new approach for automatic theorem proving in real geometry. Accepted for publication in *J. of Automated Reasoning.*

21. Emiris I., *Sparse Elimination and Applications in Kinematics.* Ph.D. Thesis, Computer Science Division, University of Calif., Berkeley, 1994.

22. Gao, X.-S., and Wang, D.-K., On the automatic derivation of a set of geometric formulae. *J. of Geometry*, 53, 1995, 79-88.

23. Gerber, L., The orthocentric simplex as an extreme simplex. *Pacific J. of Mathematics* 56, 1975, 97-111.

24. Herstein, I. N., *Topics in Algebra - Second Edition.* John Wiley & Sons, Inc., USA, 1975.

25. Heymann, W. Problem der Winkelhalbierenden. *Ztschr. f. Math. and Phys.* 35, 1890.

26. Hong, J., Proving by example and the gap theorems. Proc *27th Annual Symp. on Foundations of Computer Science (FOCS)*, Toronto, Oct. 1986, 107-116.

27. Huang, Y.Z., and Wu, W.D., Kinematic solution of a Stewart platform. Proc. *1992 Intl. Workshop on Mathematics Mechanization* (eds. Wu and Cheng), Intl. Academic Publishers, 1992, 181-188.

28. Kapur, D., Geometry theorem proving using Hilbert's Nullstellensatz. Proc. *1986 Symposium on Symbolic and Algebraic Computation (SYMSAC 86)*, Waterloo, 1986, 202-208.

29. Kapur, D., A refutational approach to geometry theorem proving. *Artificial Intelligence*, 37, 1988, 61-94.

30. Kapur, D., Algorithmic elimination methods. Tutorial Notes, *Intl. Symp. on Symbolic and Algebraic Computation (ISSAC)*, Montreal, July 1995.

31. Kapur, D., and Lakshman, Y., Elimination methods: An introduction. *Symbolic and Numerical Computation for Artificial Intelligence* (eds. Donald, Kapur and Mundy), Academic Press, 1992, 45-87.

32. Kapur, D., Y.N., Lakshman, and Saxena, T., Computing invariants using elimination methods. Proc. *IEEE Intl. Symp. on Computer Vision*, Coral Gables, Florida, November 1995.

33. Kapur, D., and Mundy, J.L., Wu's method and its application to perspective viewing. *Artificial Intelligence*, 37, 1988, 15-26.

34. Kapur, D., and Saxena, T., Automated geometry theorem proving using the Dixon resultants. Invited Talk, *IMACS-95*, Albuquerque, New Mexico, May 1995.

35. Kapur, D., and Saxena, T., Comparison of various multivariate resultant formulations. Proc. *Intl. Symp. on Symbolic and Algebraic Computation (ISSAC)*, Montreal, July 1995, 187-194.

36. Kapur, D., and Saxena, T., Sparsity considerations in the Dixon resultant formulation, Proc. *ACM Symposium on Theory of Computing (STOC)*, Philadelphia, May 1996, 184-191

37. Kapur, D., and Saxena, T., Extraneous factors in the Dixon resultant formulations. Proc. *Intl. Symp. on Symbolic and Algebraic Computation (ISSAC)*, Hawaii, July 1997, 141-148.

38. Kapur D., Saxena T. and Yang L., Algebraic and geometric reasoning using Dixon resultants. Proc. *Intl. Symp. on Symbolic and Algebraic Computation (ISSAC)*, Oxford, July 1994, 99-107.

39. Kapur, D., and Wan, H., Refutational proofs of geometry theorems via characteristic set computation. Proc. *Intl. Symp. on Symbolic and Algebraic Computation (ISSAC)*, Tokyo, August 1990, 277-284.

40. Knuth, D.E., *Seminumerical Algorithms: The Art of Computer Programming*. Vol. 2, Second Edition, Addison Wesley, 1980, 407-408.

41. Ko, H.P., and Hussain, M.A., A study of Wu's method—A method to prove certain theorems in elementary geometry. Proc. *1985 Congressus Numerantium*, 1985,

42. Kutzler, B., and Stifter, S., Automated geometry theorem proving using Buchberger's algorithm. Proc. *1986 Symposium on Symbolic and Algebraic Computation (SYMSAC 86)*, Waterloo, 1986, 209-214.

43. Lazard D., Generalized Stewart platform: How to compute with rigid motions? *Proc. IMACS-SC*, Lille, June 1993, 85-88.

44. Li, H., and Cheng, M., Proving theorems in elementary geometry with Clifford algebraic method. Preprint, MMRC, Academia Sinica, China, 1995.

45. Mourrain B., The 40 "generic" positions of a parallel robot. *Proc. ACM Intl. Symp. on Symbolic and Algebraic Computation (ISSAC)*, Kiev, July 1993.

46. Mundy, J., L., Zisserman, A., Towards a new framework for vision. *Geometric Invariance in Computer Vision* (eds. Mundy and Zisserman), MIT Press, 1992, 1-39.

47. Recio, T., and Velez-Melon, M.P., Automatic discovery of theorems in elementary geometry. Presented at the *Workshop on Automated Deduction in Geometry*, Toulouse, Sept 1996.

48. Rege, A., A complete and practical algorithm for geometric theorem proving. Proc *11th Annual Symp. on Computational Geometry*, Vancouver, June 1995, 277-286.

49. Richter-Gebert, J., Mechanical theorem proving in projective geometry. *Algebraic Approaches to Geometric Reasoning* (eds. Hong, Wang, and Winkler), A special issue of Annals of Mathematics and Artificial Intelligence, 1995.

50. Ritt, J.F., *Differential Algebra*. AMS Colloquium Publications, 1950.

51. Ronga F., and Vust T., Stewart platforms without computer. *Preprint*, Department of Mathematics, University of Geneva, 1992.

52. Saxena, T., *Efficient Variable Elimination using Resultants*. Ph.D. Thesis, Dept. of Computer Science, State University of New York, Albany, Nov. 1996.

53. Stifter, S., Geometry theorem proving in vector spaces by means of Gröbner bases. Proc. *Intl. Symp. on Symbolic and Algebraic Computation (ISSAC)*, Kiev, July 1993, 301-310.

54. Sykes, M., *A Source Book of Problems for Geometry, based upon Industrial Design and Architectural Ornament*. Boston, Allyn and Bacon, 1912.

55. Tarski, A., A Decision Method for Elementary Algebra and Geometry. U. of Calif. Press, 1948; 2nd edition, 1951.

56. Wan, H., *Refutational Proofs of Geometry Theorems using the Characteristic Set Method*. MS project, RPI, Troy, NY, 1987.

57. Wang, D., Proving-by-examples method and inclusion of varieties. *Kexue Tongbao*, 33, 1988, 1121-1123.

58. Wang, D., A new theorem discovered by computer prover. *J. of Geometry*, 36, 1989, 173-182.

59. Wang, D., Algebraic factoring and geometry theorem proving. Proc. *12th Intl. Conf. on Automated Deduction (CADE-12)*, Nancy, LNAI 814, 1994, 386-400.

60. Wang, D., Geometry machines: From AI to SMC. Proc. *Intl. Conf. on Artificial Intelligence and Symbolic Mathematical Computation (AISMC-3)*, Steyr, Austria, 1996, LNCS 1138, 213-239.

61. Wang, D., and Gao, X.-S., Geometry theorems proved mechanically using Wu's method—Part on Euclidean geometry. *MM Research Preprints*, 2, 1987, 75-106.

62. Wu, W., On the decision problem and the mechanization of theorem proving in elementary geometry. *Scientia Sinica* 21 (1978) 150-172. Also in *Automated Theorem Proving: After 25 Years, Contemporary Mathematics* 29 (eds. Bledsoe and Loveland), 1984, 213-234.

63. Wu, W., Some recent advances in mechanical theorem proving of geometries. *Automated Theorem Proving: After 25 Years, Contemporary Mathematics* 29 (eds. Bledsoe and Loveland), 1984, 235-241.

64. Wu, W., Basic principles of mechanical theorem proving in geometries. *J. of System Sciences and Mathematical Sciences* 4 (3), 1984, 207-235. Also in *J. of Automated Reasoning* 2, 1986, 221-252.

65. Wu, W., A mechanization method of geometry and its applications. *J. of System Sciences and Mathematical Sciences* 6 (3), 1986, 204-216.

66. Wu, W., On reducibility problem in mechanical theorem proving of elementary geometries. *Chinese Quart. J. Mathematics*, 2, 1987, 1-20.

67. Wu, W., On zeros of algebraic equations—An application of Ritt's principle. *Kexue Tongbao*, 31 (1), 1986, 1-5.

68. Wu, W., *Mechanical Theorem Proving in Geometries: Basic Principles.* Translated from Chinese by Jin and Wang, Springer Verlag, 1994.

69. Yang, L., Zhang, J.-Z., and Hou, X.-R., An efficient decomposition algorithm for geometry theorem proving without factorization. Proc. *ASCM'95*, Beijing, August 1995, 33-41.

70. Yang, L., Zhang, J.-Z., and Li, C.-Z., A prover for parallel numerical verification of a class of constructive geometry theorems. Proc. *Intl. Workshop on Mechanization of Mathematics* (ed. Wu and Cheng), Beijing, China, July 1992, 244-250.

71. Zhang, J.-Z., Chou, S.-C., and Gao, X.-S., Automated production of traditional proofs for theorems in Euclidean geometry I. *Algebraic Approaches to Geometric Reasoning* (eds. Hong, Wang, and Winkler), A special issue of Annals of Mathematics and Artificial Intelligence, 1995, 109-137.

72. Zhang, J.-Z., Yang, L., and Deng, M.-K., The parallel numerical method of mechanical theorem proving. *Theoretical Computer Science,* 74, 1990, 253-271.

Extended Dixon's Resultant and Its Applications

Quoc-Nam Tran *

Research Institute for Symbolic Computation (RISC–Linz)
Johannes Kepler University
A–4040 Linz, Austria
E-mail: tqnam@risc.uni-linz.ac.at

Abstract. Dixon's resultant method is an efficient way of simultaneously eliminating several variables from a system of nonlinear polynomial equations at a time. However, the method only works for systems of $n + 1$ generic n-degree polynomials in n variables and does not work for most algebraic and geometric problems. In this paper, by using techniques from pseudoinverse theory and linear transformations, the author extends Dixon's resultant method to an arbitrary system of $n + 1$ nontrivial polynomials in n variables where the Dixon matrix can be singular or even nonsquare. The extended method does not require any precondition – this is the main contribution of the paper. The extended method can be used efficiently as an elimination method in geometric reasoning, computer aided geometric design (CAGD) and solid modeling. Several examples show that the new method works well also in situations where other methods (of the same subject) may fail to give a correct answer.

1 Introduction

Solving problems from geometric reasoning, CAGD, solid modeling, etc., one often has to eliminate several variables from a system of equations (see [Cho88], [Far88], [Hof89], [HWW95] and [PW95]). The Gröbner bases method [Buc85], Ritt-Wu's characteristic sets method [Rit50, Wu94] and successive Sylvester resultant computations [vdW40] can be used to achieve the demands. However, they are not efficient enough and require a lot of memory space for some large problems.

Meanwhile, Dixon's resultant method is an efficient way to simultaneously eliminate several variables from a system of nonlinear polynomial equations at a time. In contrast with successive resultant computation techniques which eliminate variables one by one and compute numerous intermediate resultants before finally computing the resultant of the whole set, Dixon's resultant method creates a much smaller matrix and avoids such intermediate results, and hence is much faster and also saves memory space. By Dixon's resultant we mean a nonzero multiple of the standard resultant because it provides a solvability criterion

* Supported by the Austrian Science Foundation (FWF) project HySaX, Proj. Nr. P11160-TEC

for the existence of solutions other than certain known solutions. (A standard resultant furnishes a solvability criterion for the existence of any solutions.)

Unfortunately, Dixon's resultant method does not work for most algebraic and geometric problems. Originally, Dixon's resultant method was for obtaining the resultant of a set of three generic bidegree polynomials (see [Dix08b]). This is a generalization of Cayley's formulation [Cay65] of Bezout's efficient method for computing the resultant of two univariate polynomials. Dixon had generalized his method to any $n + 1$ generic n-degree polynomials in n variables – a condition which cannot be satisfied by most problems arising in practice. For most practical problems, the matrix set up in Dixon's method – the Dixon matrix – becomes singular or even nonsquare. As a consequence, Dixon's resultant vanishes identically, without providing any information about the common solutions of equations. There are some efforts to obtain nonzero conditions by perturbing the polynomial system (see [Chi90]). However, these efforts suffer from the facts that the method is not automatic, requires human expertise and enlarges the size of Dixon matrix with larger entries. Another approach of Kapur et al. (see [KSY94], [KS95]) is based on the idea of checking if there exists a submatrix of the Dixon matrix with one column less and strictly smaller rank than the Dixon matrix when the Dixon matrix is square and singular. However, this approach requires a precondition on the system of equations.

In this paper, by using techniques from pseudoinverse theory and linear transformations, the author overcomes the above obstacles and extends Dixon's resultant method to an arbitrary system of $n+1$ nonlinear polynomials in n variables where the Dixon matrix can be singular or even nonsquare. Our method does not require any precondition – this is the main contribution of the paper. Several examples show that the new method works well also in situations where other methods may fail to give a correct answer. Note that this approach can also be used to extend other multivariate resultant methods such as Macaulay resultant or sparse resultant. However, extracting the extraneous factors from Dixon's resultants is a difficult task and still needs further investigations.

The structure of the paper is as follows. In the next section, we summarize some well-known results from Dixon's resultant method. In Section 3, we review some basic definitions and fundamental properties from the theory of pseudo-inverses. In Section 4, we present our main theoretical results together with our algorithms to compute extended Dixon's resultants. In Section 5, we give a comparison between the new method and other known methods of the same subject. Several examples show that the new method works well while other methods may fail to give a correct answer. In Section 6, we present our experiences with some practical problems in geometric reasoning, CAGD and solid modeling.

2   Preliminaries about Dixon's Resultant

We first review the fundamental results of Dixon's resultant method – an efficient way to simultaneously eliminate several variables from a system of nonlinear polynomial equations at a time. Readers can find all of the details in [Dix08b, Dix08a, Dix09b, Dix09a], [Chi90] and [CG95].

2.1 Three Generic Bidegree Polynomials

Dixon formulated several resultant expressions for systems of polynomial equations satisfying various restrictions ([Dix08b, Dix08a, Dix09b, Dix09a]). One such restriction requires that there be a system of three generic bidegree polynomial equations

$$
\mathcal{F} = \begin{cases} p_1 \equiv \sum_{ij} a_{ij} x_1^i x_2^j = 0, \\ p_2 \equiv \sum_{ij} b_{ij} x_1^i x_2^j = 0, \\ p_3 \equiv \sum_{ij} c_{ij} x_1^i x_2^j = 0, \end{cases} \tag{1}
$$

where a_{ij}, b_{ij}, c_{ij} are distinct indeterminates, $m \geq i \geq 0$, $n \geq j \geq 0$. With this restriction, Dixon neatly extended Sylvester's dialytic method and Cayley's statement for Bezout's method ([Sal64]) from two equations to three equations.

By embedding the affine plane (x_1, x_2) in the projective plane (w, x_1, x_2), we can consider the system $\mathcal{F}$, after homogenization by a variable w, $\mathcal{F}' = \{P_1 \equiv \sum_{ij} a_{ij} w^{m+n-i-j} x_1^i x_2^j, P_2 \equiv \sum_{ij} b_{ij} w^{m+n-i-j} x_1^i x_2^j, P_3 \equiv \sum_{ij} c_{ij} w^{m+n-i-j} x_1^i x_2^j\}$ as a system of three curves in the projective plane. Common solutions of these equations are given by the common solutions of two general curves $aP_1 + bP_2 + cP_3 = 0$ and $a'P_1 + b'P_2 + c'P_3 = 0$. It is easy to verify that $(0, 1, 0)$ is an n-fold point and $(0, 0, 1)$ is an m-fold point of each of the three curves. Hence, two general curves have at least n^2 intersections at $(0, 1, 0)$ and at least m^2 intersections at $(0, 0, 1)$. We shall refer to these $m^2 + n^2$ intersections at $(0, 1, 0)$ and $(0, 0, 1)$ as the canonical solutions of the three bidegree equations. Since these canonical solutions exist, the standard resultant of three bidegree equations vanishes. However, the Dixon bidegree resultant allows these canonical solutions and thus provides a solvability criterion for solutions other than these known solutions. By Bezout's intersection theorem, two degree $m+n$ curves should have $(m+n)^2$ intersections. Thus there may be as many as $2mn$ noncanonical solutions. Since a solution of any two of the three bidegree equations induces a linear relation in the coefficients of the other bidegree equation, $2mn$ noncanonical solutions induce a degree $2mn$ relation in the coefficients of the other bidegree equation.

By multiplying each of the bidegree polynomials in $\mathcal{F}$ by each of the $2mn$ monomials $x_1^i x_2^j$, $0 \leq i \leq 2m - 1$, $0 \leq j \leq n - 1$, we obtain $6mn$ polynomials in the $6mn$ monomials $x_1^i x_2^j$, $0 \leq i \leq 3m - 1$, $0 \leq j \leq 2n - 1$. This is the main idea of Dixon's dialytic method. Thus, the coefficient matrix of the spawned equations is a square matrix of order $6mn$. The determinant of this coefficient matrix has the term $a_{mn}^{2mn} b_{mn}^{mn} b_{00}^{mn} c_{00}^{2mn}$ and is of degree $2mn$ in the coefficients of each of the bidegree polynomials in $\mathcal{F}$. Thus by Cayley's proof ([Sal64]), this order $6mn$ determinant is a bidegree resultant for the three bidegree equations $\mathcal{F}$.

Alternately, Dixon considered the determinant

$$\Delta(x_1, x_2, \alpha_1, \alpha_2) = \begin{vmatrix} p_1(x_1, x_2) & p_2(x_1, x_2) & p_3(x_1, x_2) \\ p_1(\alpha_1, x_2) & p_2(\alpha_1, x_2) & p_3(\alpha_1, x_2) \\ p_1(\alpha_1, \alpha_2) & p_2(\alpha_1, \alpha_2) & p_3(\alpha_1, \alpha_2) \end{vmatrix}$$

where α_1, α_2 are new variables and $p_i(\alpha_1, \alpha_2)$ stands for uniformly replacing x_j by α_j for $1 \leq j \leq 2$ in p_i. The product $(x_1 - \alpha)(x_2 - \beta)$ divides $\Delta(x_1, x_2, \alpha_1, \alpha_2)$ because $\Delta(x_1, x_2, \alpha_1, \alpha_2) = 0$ when $x_1 = \alpha$ or $x_2 = \beta$. Hence, in general,

$$\delta(x_1, x_2, \alpha_1, \alpha_2) = \frac{\Delta(x_1, x_2, \alpha_1, \alpha_2)}{(x_1 - \alpha)(x_2 - \beta)}$$

is actually a polynomial of degree $m - 1$ in x_1, $2n - 1$ in x_2, $2m - 1$ in α and $n - 1$ in β. In terms of matrices we have

$$\delta(x_1, x_2, \alpha_1, \alpha_2) = \left(\alpha^{2m-1}\beta^{n-1} \ldots \alpha^{2m-1} \ldots \beta^{n-1} \ldots 1\right) \cdot D \cdot \begin{pmatrix} x_1^{m-1} x_2^{2n-1} \\ \vdots \\ x_1^{m-1} \\ \vdots \\ x_2^{2n-1} \\ \vdots \\ 1 \end{pmatrix}$$

where D is the order $2mn$ coefficient matrix.

If $(\bar{x}_1, \bar{x}_2)$ is a solution of $\mathcal{F}$, we have $\Delta(\bar{x}_1, \bar{x}_2, \alpha_1, \alpha_2) = 0$ and consequently $\delta(\bar{x}_1, \bar{x}_2, \alpha_1, \alpha_2)$ vanishes. This means the coefficients of the monomials $\alpha^i \beta^j$, $0 \leq i \leq 2m - 1$, $0 \leq j \leq n - 1$ in $\delta(\bar{x}_1, \bar{x}_2, \alpha_1, \alpha_2)$ are all zero. That is

$$D \cdot (\bar{x}_1^{m-1} \bar{x}_2^{2n-1} \ldots \bar{x}_1^{m-1} \ldots \bar{x}_2^{2n-1} \ldots 1)^T = 0.$$

By Cayley's proof, we see that $|D|$ is a Dixon's resultant for the system $\mathcal{F}$. Notice that this construction of Dixon's resultant is very similar to the Cayley construction of the Bézout univariate resultant.

2.2 An Insufficient Solvability Criterion

In general, without the generic restriction, the general curves $aP_1 + bP_2 + cP_3 = 0$ and $a'P_1 + b'P_2 + c'P_3 = 0$ may intersect more then n^2 times at $(0, 1, 0)$ or more then m^2 times at $(0, 0, 1)$, then the bidegree equations may have excessive canonical solutions. This happens if and only if the curves P_1, P_2 and P_3 have common tangents at the singular points $(0 : 1 : 0)$ or $(0, 0, 1)$. If there are excessive canonical solutions, the bidegree Dixon resultant vanishes because some of the $2mn$ solutions it accounts for have already occurred.

2.3 Adaptability of Dixon's Method

Without the bidegree restriction, the Dixon 3×3 determinant method for constructing an order $2mn$ determinant in general produces a square matrix of order $n(3n-1)/2$ for three degree n polynomials. However, the determinant of this matrix vanishes identically when $n > 2$. Thus in general, we cannot use the method to create a nonzero multiple of the resultant when $n > 2$. But for constructing resultants for three equal-degree equations, Dixon offer another dialytic method.

Let the coefficient of $\alpha^i \beta^j$ in $\delta(x_1, x_2, \alpha, \beta)$ be $D_{i,j}$. We know that $D_{i,j}$ is linear in the coefficient of each p_1, p_2 and p_3 and is a polynomial in x_1, x_2 of degree at most $2n - 2$. Since for any solution $(\bar{x}_1, \bar{x}_2)$ of $p_1 = 0$, $p_2 = 0$ and $p_3 = 0$, $\delta(\bar{x}_1, \bar{x}_2, \alpha, \beta)$ vanishes identically, $D_{i,j}$ must vanish as well. Dixon's method will try to spawn two distinct sets of equations from the three given equations of common degree n such that a solution of the given equations is also a solution of each spawned equation. One set of equations is generated dialytically by multiplying the given equations with the $\binom{n}{2}$ monomials of degree $n - 2$ to obtain $3\binom{n}{2}$ equations of degree $2n - 2$. The equations $D_{i,j} = 0$, $i + j < n$, also of degree $2n - 2$ with coefficients linear in the coefficients of each original equation, serve as another set of $\binom{2n}{2} - 3\binom{n}{2}$ nondialytic equations. Thus Dixon's resultant for three equal-degree equations is the determinant $|D_n|$ where D_n is the coefficient matrix of

$$D_n[1^{2n-2} \cdots x_1^{2n-2} \cdots x_2^{2n-2}]^T = \begin{bmatrix} 1^{2n-2} \cdot p_1 & \cdots x_1^{2n-2} \cdot p_2 \cdots & x_2^{2n-2} \cdot p_3 \\ D_{0,0} D_{1,0} D_{0,1} & \cdots D_{n-1,0} \cdots & D_{0,n-1} \end{bmatrix}^T .$$

Dixon had also generalized his method to any $n+1$ generic n-degree polynomials in n variables as follows. Let $\mathcal{F} = \{p_1(x_1, \ldots, x_n), \ldots, p_{n+1}(x_1, \ldots, x_n)\}$ be a set of $n+1$ generic n-degree polynomials in n variables, i.e. there exist nonnegative integers $m_1, m_2, \ldots, m_n$ such that $p_j = \sum_{i_1=0}^{m_1} \cdots \sum_{i_n=0}^{m_n} a_{j,i_1,\ldots,i_n} x_1^{i_1} \cdots x_n^{i_n}$, for $1 \leq j \leq n + 1$ where $a_{j,i_1,\ldots,i_n}$ are distinct indeterminates. We form an $(n + 1) \times (n + 1)$ determinant $\Delta(x_1, \ldots, x_n, \alpha_1, \ldots, \alpha_n)$ as

$$\begin{vmatrix} p_1(x_1, x_2, \ldots, x_n) & \cdots & p_{n+1}(x_1, x_2, \ldots, x_n) \\ p_1(\alpha_1, x_2, \ldots, x_n) & \cdots & p_{n+1}(\alpha_1, x_2, \ldots, x_n) \\ p_1(\alpha_1, \alpha_2, \ldots, x_n) & \cdots & p_{n+1}(\alpha_1, \alpha_2, \ldots, x_n) \\ \cdots & \cdots & \cdots \\ p_1(\alpha_1, \alpha_2, \ldots, \alpha_n) & \cdots & p_{n+1}(\alpha_1, \alpha_2, \ldots, \alpha_n) \end{vmatrix}$$

where $\alpha_1, \alpha_2, \ldots, \alpha_n$ are n new variables and $p_i(\alpha_1, \alpha_2, \ldots, \alpha_k, x_{k+1}, \ldots, x_n)$ stands for uniformly replacing x_j by α_j for $1 \leq j \leq k$ in p_i. Each of $x_i = \alpha_i$, for all $1 \leq i \leq n$, is a zero of Δ, so they can be removed by dividing Δ by $\prod_{i=1}^{n}(x_i - \alpha_i)$. Let

$$\delta(x_1, \ldots, x_n, \alpha_1, \ldots, \alpha_n) = \frac{\Delta(x_1, \ldots, x_n, \alpha_1, \ldots, \alpha_n)}{(x_1 - \alpha_1) \cdots (x_n - \alpha_n)} .$$

The polynomial δ, called the Dixon polynomial of $\mathcal{F}$, is of degree $((n + 1 - i) \cdot m_i) - 1$ in α_i and $(i \cdot m_i) - 1$ in x_i, for $1 \leq i \leq n$.

Any common zero of $\mathcal{F}$, say $(x_1 = c_1, \ldots, x_n = c_n)$ makes the Dixon polynomial vanish, no matter what the values of $\alpha_1, \ldots, \alpha_n$ are; hence, all the coefficients of the various power products of $\alpha_1, \ldots, \alpha_n$ in the Dixon polynomial vanish. Let $\mathcal{E}'$ be the set of all polynomials in $x_1, \ldots, x_n$ which are coefficients of the power products of $\alpha_1, \ldots, \alpha_n$ in δ. This set then has exactly

$$\prod_{i=1}^{n}((n+1-i) \cdot m_i) = n! \cdot \prod_{i=1}^{n} m_i = s$$

equations (one for each power product of $\alpha_1, \ldots, \alpha_n$), each of which is of degree $(i \cdot m_i) - 1$ in x_i. Also, there are

$$\prod_{i=1}^{n}(i \cdot m_i) = n! \cdot \prod_{i=1}^{n} m_i = s$$

power products in $x_1, \ldots, x_n$ in the equations of $\mathcal{E}'$. Let D be the $s \times s$ coefficient matrix of $\mathcal{E}'$. Then

$$\mathcal{E}' \equiv DX = 0, \tag{2}$$

where X is the vector of the power products of $x_1, \ldots, x_n$. If each power product of $x_1, \ldots, x_n$ (including $x_1^0 \cdots x_n^0 = 1$) is viewed as a new variable, say v_i, then we get a set $\mathcal{E}$ of s homogeneous linear equations in s variables

$$\mathcal{E} \equiv Dv = 0. \tag{3}$$

If there exists a common affine zero of $\mathcal{F}$, say $(x_1 = c_1, \ldots, x_n = c_n)$, then this also is a solution to $\mathcal{E}'$. This results in a nontrivial solution to $\mathcal{E}$, i.e. $(v_1 = 1, v_2 = c_1, \ldots)$. Hence, as before, if $\mathcal{F}$ has a common affine zero, then $\mathcal{E}$ has a nontrivial solution, implying that the determinant of D is zero. This gives a necessary condition on the coefficients of $p_1, \ldots, p_{n+1}$ for them to have a common zero. We call D the Dixon matrix and its determinant Dixon's resultant.

However, for most algebraic and geometric problems the method does not work as the polynomials are not generic and the Dixon matrix becomes singular or even nonsquare. As a consequence, Dixon's resultant vanishes identically, without providing any information about the common solutions of equations.

3 Pseudoinverse Matrices

It seems that the concept of generalized inverses was first mentioned in print in 1903 by Fredholm [Fre03]. The existence of pseudoinverses of matrices was first noted by E. H. Moore (see [Moo20]), who defined a unique inverse (called by him the general reciprocal) for every finite matrix (square or nonsquare). In 1950, Bjerhammar rediscovered Moore's pseudoinverse and also noted the relationship of generalized inverses to solutions of linear systems (see [Bje51b], [Bje51a], [Bje68]). In 1955, Penrose [Pen55] sharpened and extended Bjerhammar's results on linear systems and showed that Moore's pseudoinverse, for a given matrix A, is the unique matrix A^+ satisfying the four equations (4)–(7) of the next section. The latter discovery has been so important and fruitful that this unique inverse is now commonly called the Moore–Penrose inverse or pseudoinverse.

The theory of pseudoinverses (more precisely the pseudoinverse of the Jacobian matrices) has been successfully used in [Tra95, Tra97b] for finding the roots of an arbitrary system of equations. In this paper, we will use the theory of pseudoinverses to solve a quite different problem, namely computing Dixon's resultant.

3.1 Definitions and Fundamental Properties

We use the following definitions and properties from the theory of pseudoinverses. More details can be found in [Pen55], [BO71] or [BIG74]. All matrices in this section are defined over the complex number field. Clearly, analogous results can be obtained by assuming the matrices are defined over the real number field.

The following matrix equations are used to define generalized inverses of matrices where $()^*$ denotes the conjugate transpose of the matrix:

$$AXA = A, \tag{4}$$
$$XAX = X, \tag{5}$$
$$(AX)^* = AX, \tag{6}$$
$$(XA)^* = XA. \tag{7}$$

Definition 3.1 *A pseudoinverse of a matrix $A \in \mathbb{C}^{m \times n}$ is a matrix $X \in \mathbb{C}^{n \times m}$, denoted in particular by A^+, satisfying (4), (5), (6) and (7).*

Given a matrix M of rank r, one can transform M to the form $\begin{pmatrix} I_r & O \\ O & O \end{pmatrix}$ by elementary row and column operations. So there exist nonsingular matrices P and Q such that

$$PMQ = \begin{pmatrix} I_r & O \\ O & O \end{pmatrix}.$$

Letting

$$M^{(1)} = Q \begin{pmatrix} I_r & O \\ O & W \end{pmatrix} P,$$

where W is arbitrary, it is easy to verify that $M^{(1)}$ satisfies (4). Now with a given matrix $A \in \mathbb{C}^{m \times n}$, let $X = A^{(1,2,4)}.A.A^{(1,2,3)}$, where $A^{(1,2,4)} = A^*(AA^*)^{(1)}$ and $A^{(1,2,3)} = (A^*A)^{(1)}A^*$. The identity $(BAA^* - CAA^*).(B - C)^* = (BA - CA).(BA - CA)^*$ along with the fact that $AA^* = 0$ implies $A = 0$ gives us the following result:

$$\text{if } BAA^* = CAA^* \text{ then } BA = CA. \tag{8}$$

Similarly, we have

$$\text{if } BA^*A = CA^*A \text{ then } BA^* = CA^*, \tag{9}$$
$$\text{if } A^*AB = A^*AC \text{ then } AB = AC, \tag{10}$$
$$\text{if } AA^*B = AA^*C \text{ then } A^*B = A^*C. \tag{11}$$

By using (8), (9), (10) and (11), it is easy to verify that $A^{(1,2,4)}$ satisfies (4), (5) and (7); $A^{(1,2,3)}$ satisfies (4), (5) and (6); and X satisfies (4), (5), (6) and (7). Moreover, if X and Y are two matrices satisfying (4), (5), (6) and (7). Then

$$
\begin{aligned}
X &= X(AX)^* = XX^*A^* = X(AX)^*(AY)^* \\
&= XAY = (XA)^*(YA)^*Y = A^*Y^*Y = (YA)^*Y \\
&= Y.
\end{aligned}
$$

Therefore, we have the following theorem on the existence and uniqueness of the pseudoinverse of a matrix.

Theorem 3.1 *For any matrix A there exists a pseudoinverse matrix A^+. Moreover, the pseudoinverse is unique.*

In the following lemma, we give certain properties of the pseudoinverse of a matrix A. The proof of the lemma is straightforward and can be found in [BO71].

Lemma 3.1 *Let $A \in \mathbb{C}^{m \times n}$, $\lambda \in \mathbb{C}$. Then*

1. $(A^+)^+ = A$,
2. $(A^*)^+ = (A^+)^*$,
3. $(AA^*)^+ = (A^+)^*A^+$ and $(A^*A)^+ = A^+(A^+)^*$,
4. $\operatorname{rank} A^+ = \operatorname{rank} A = \operatorname{rank} A^+A$,
5. $(\lambda A)^+ = \lambda^+ A^+$, where λ^+ is λ^{-1} if $\lambda \neq 0$ and 0 otherwise,
6. $A^+ = (A^*A)^+A^* = A^*(AA^*)^+$,
7. AA^+, A^+A, $I - AA^+$ and $I - A^+A$ are Hermitian and idempotent,
8. $A^+A = I_n$ if and only if A is of full column rank,
9. $AA^+ = I_m$ if and only if A is of full row rank.

Note that, in general we may not have $(AB)^+ = B^+A^+$.

4 Extending Dixon's Resultant

Remember that without the generic restriction, the matrix set up in Dixon's resultant method – the Dixon matrix – becomes singular or even nonsquare. As a consequence, Dixon's resultant vanishes identically, without providing any information about the common solutions of equations. Another problem is that in Equation (2), vector X of the power products of $x_1, \ldots, x_n$ may not include $x_1^0 \cdots x_n^0 = 1$. Therefore, the existence of a common affine zero of $\mathcal{F}$ may result in a trivial solution of $\mathcal{E}$ in Equation (3) and hence the determinant $|D| = 0$ does not even give a necessary condition on the coefficients of $p_1, \ldots, p_{n+1}$ for them to have a common zero.

To deal with singular or nonsquare Dixon matrix, we use the results from the theory of pseudoinverses as follows. Let

$$
Dx = b
$$

be a system of simultaneous linear equations, where b is a given vector and x is an unknown vector. If D is nonsingular then there exists a unique solution for x, given by

$$x = D^{-1}b.$$

In the general case, when D may be singular or nonsquare, there may sometimes be no solutions or a multiplicity of solutions. If the system has many solutions, we may desire to have not just one solution but a characterization of all solutions. It is easy to verify that if $D^{(1)}$ is any matrix satisfying $DD^{(1)}D = D$, then $Dx = b$ has a solution if and only if

$$DD^{(1)}b = b,$$

and in this case the most general solution is

$$x = D^{(1)}b + (I - D^{(1)}D)y,$$

where y is arbitrary. We will call such a matrix $D^{(1)}$ a generalized inverse of D. Unlike the case of the nonsingular matrix, which has a single unique inverse for all purposes, there are different generalized inverses for different purposes. However, in this paper, we are only interested in a generalized inverse whose existence is unique for every matrix (square or nonsquare), namely the Moore–Penrose generalized inverse or pseudoinverse as in Section 2.

   To deal with trivial solutions of $\mathcal{E}$ in Equation (3), we will use the means of linear transformations. Actually, the author has found a linear transformation which does not change the resultant of the transformed system while it secures that $\mathcal{E}$ of the transformed system will not have any trivial solutions.

Definition 4.1 *Let $\mathcal{F} \subset K[x_1, \ldots, x_n]$ be a system of $n + 1$ equations in n variables, where K is an integral domain. We call $\mathcal{F}$ a **regular** system if and only if every equation of $\mathcal{F}$ has a nonzero term of the power product $x_1^0 \cdots x_n^0$. In other words, the system is regular with respect to the homogenous variable.*

We have the following fundamental results.

Lemma 4.1 *Every system of equations in $K[x_1, \ldots, x_n]$, where K is an integral domain, can be transformed into a regular system by using a linear transformation.*

PROOF. Given a system $\mathcal{F} = \{p_1, \ldots, p_{n+1}\}$. We construct a linear transformation as follows.

$$\begin{aligned}
w &= w, \\
x_1 &= x_1' + c_1 w, \\
x_2 &= x_2' + c_2 w, \\
&\cdots \\
x_n &= x_n' + c_n w,
\end{aligned}$$

where w is the homogeneous variable, $c_i \in K$, $1 \le i \le n$ and $p_j(c_1, \ldots, c_n) \ne 0$, $1 \le j \le n + 1$. It is easy to verify that the transformed system is regular. $\square$

Lemma 4.2 *The standard resultant of the transformed system $\mathcal{F}'$ (by applying the above linear transformation to the system of equations $\mathcal{F}$) is equal to the standard resultant of the system $\mathcal{F}$. Moreover, $\mathcal{E}$ in Equation (3) of the transformed system $\mathcal{F}'$ will not have trivial solutions.*

PROOF. Given a system $\mathcal{F} = \{p_1, \dots, p_{n+1}\}$. The standard resultant of the transformed system $\mathcal{F}'$ is

$$\text{Res}(\mathcal{F}') = \text{Res}(\mathcal{F}) \cdot \det(\Theta)^{d_1 \cdots d_{n+1}}$$

where d_j is the degree of p_j, $1 \leq j \leq n+1$, and $\det(\Theta)$ stands for the determinant of the matrix of the coefficients of the linear transformation. Since $\det(\Theta) = 1$, $\text{Res}(\mathcal{F}') = \text{Res}(\mathcal{F})$. It is easy to verify that $\mathcal{E}$ in Equation (3) of the transformed system $\mathcal{F}'$ will not have any trivial solutions.
□

From now on, without loss of generality, we assume that the system of equations is regular; hence, when D is nonsingular the determinant $|D| = 0$ always gives a necessary condition on the coefficients of p_1, ..., p_{n+1} for them to have a common zero.

We now use the above techniques to extend Dixon's resultant method for an arbitrary system $\mathcal{F} \subset K[x_1, \dots, x_n]$ of $n + 1$ equations in n variables, where K is a field of any characteristic.

4.1 Polynomials with Coefficients in a Field

We construct Δ as in Section 2, dividing its determinant by $(x_1 - \alpha_1) \cdots (x_n - \alpha_n)$ to get the Dixon polynomial δ. This polynomial may now have a degree less than or equal to $((n+1-i) \cdot d_{\max_i}) - 1$ in each α_i and less than or equal to $(i \cdot d_{\max_i}) - 1$ in each x_i, where $d_{\max_i} = \max(\text{degree}(p_1, x_i), \dots, \text{degree}(p_{n+1}, x_i))$, for all $1 \leq i \leq n$. We construct the set of linear equation $\mathcal{E}$ from the Dixon polynomial as before, except that now $\mathcal{E}$ may have less than or equal to s equations in less than or equal to s variables. We continue to call the coefficient matrix of $\mathcal{E}$ the Dixon matrix D of size $s_1 \times s_2$ and $v = (v_1, \dots, v_{s_2})$ the s_2-tuple of the power products of $x_1, \dots, x_n$ which are viewed as new variables. Note that the Dixon matrix now may be singular or even nonsquare.

Theorem 4.1 *Given a system $\mathcal{F} \subset K[x_1, \dots, x_n]$ of $n + 1$ equations in n variables, $Dx = 0$ has a nonzero root if and only if*

$$\det(D^+ D) = 0,$$

where D^+ is the pseudoinverse of D of size $s_2 \times s_1$.

PROOF. Assume that $D\bar{x} = 0$ then $D^+ D\bar{x} = 0$ where $\bar{x}$ is a nonzero root of D. Hence $\det(D^+ D) = 0$. Conversely, if $\det(D^+ D) = 0$ then there exists a nonzero root $\bar{x}$ of $D^+ D$. Therefore $D\bar{x} = DD^+ D\bar{x} = 0$. □

Since every matrix in $K^{s_1 \times s_2}$ has its unique pseudoinverse (see Theorem 3.1) and we can even obtain the pseudoinverse in an algorithmic manner (see [Tra96]), the theorem means that we have a necessary and sufficient condition for the existence of a nonzero root of the system $\mathcal{E}$.

Corollary 4.1 *Let $\mathcal{F} \subset K[x_1, \ldots, x_n]$ be a system of $n + 1$ equations in n variables. Then*

$$\det(D^+ D) = 0.$$

is a necessary condition for the existence of a common root of $\mathcal{F}$.

PROOF. Straightforward from Theorem 4.1. $\square$

We now give an algorithm to compute a necessary condition for the existence of a common root of an arbitrary system of $n + 1$ polynomials in n variables.

Algorithm 4.1

Inputs: A system $\mathcal{F}$ of $n + 1$ nonzero polynomials in n variables.
Output: A necessary condition for the solvability of the system $\mathcal{F}$.

1. **If $\mathcal{F}$ is irregular then**
   Regularize the system by using a linear transformation as showed in Lemma 4.1.
   Fi;
2. *Setup the Dixon matrix D of $\mathcal{F}$ and let D^+ be the pseudoinverse of D;*
3. **Return** $(\det(D^+ D))$.

Note that $\det(D^+ D) = 0$ is a necessary condition for the existence of a common root of the system $\mathcal{F}$.

4.2 Polynomials with Coefficients in a Polynomial Ring

We now extend the coefficient field of the polynomials into a polynomial ring $K[y_1, \ldots, y_m]$ where K is a field of any characteristic and the new indeterminates $y_1, \ldots, y_m$ are adjoined to the field K. Any ring homomorphism $\varphi : K[y_1, \ldots, y_m] \to K$ such that $y_i \mapsto a_i$ for some $a_i \in K$, $1 \le i \le m$, is called a specialization mapping. We also call $\varphi(D) = (\varphi(d_{ij}))$ a specialization of matrix $D = (d_{ij})$. Note that, from now on, any affine zero of $\mathcal{F}$, if it exists, is not a solution to $\mathcal{E}$ but one of its specialization.

Lemma 4.3 *Let D, F, G be matrices of appropriate sizes with entries being in $K[y_1, \ldots, y_m]$; a, b, c be in K; and φ be a specialization mapping. Then we have:*

1. $\varphi(a \cdot F \cdot G) = a \cdot \varphi(F) \cdot \varphi(G)$,
2. $\varphi(b \cdot F + c \cdot G) = b \cdot \varphi(F) + c \cdot \varphi(G)$,
3. $\det(\varphi(D)) = \varphi(\det(D))$, *and* $\operatorname{rank}\varphi(D) \le \operatorname{rank}D$,
4. $(\varphi(F))^+ = \varphi(F^+)$.

PROOF. (1), (2) and (3) are obvious. For proving (4), it is easy to check that $\varphi(F^+)$ satisfies Equation (4)–(7). $\square$

Let $\mathcal{F} \subset K[y_1, \ldots, y_m][x_1, \ldots, x_n]$ be a nonempty set of $n+1$ equations in n variables $x_1, \ldots, x_n$. For any m-tuple $(a_1, \ldots, a_m) \in K^m$, we denote the corresponding specialization mapping defined by the tuple φ_a.

Theorem 4.2 *Let $(a_1, \ldots, a_m)$ be an m-tuple in K^m, $\mathcal{F} \subset K[y_1, \ldots, y_m][x_1, \ldots, x_n]$ be a nonempty set of $n+1$ equations in n variables $x_1, \ldots, x_n$. $\varphi_a(D)$ has a nonzero root if and only if*

$$\varphi_a(\det(D^+ D)) = 0,$$

PROOF. Let w be a nonzero root of $\varphi_a(D)$. Then

$$(\varphi_a(D))^+ \varphi_a(D)w = \varphi_a(D^+ D)w = 0$$

and hence

$$\varphi_a(\det(D^+ D)) = \det(\varphi_a(D^+ D)) = 0,$$

where $\det(D^+ D)$ is a polynomial in $K[y_1, \ldots, y_m]$. Conversely, if $\varphi_a(\det(D^+ D)) = 0$ then there exists a nonzero affine zero w of $(\varphi_a(D))^+ \varphi_a(D)$. Therefore,

$$\varphi_a(D)(\varphi_a(D))^+ \varphi_a(D)w = \varphi_a(D)w = 0.$$

$\square$

As a consequence of this theorem, for getting a necessary condition for the existence of a common root of $\mathcal{F}$, Corollary 4.1 and Algorithm 4.1 remain valid with K is replaced by $K[y_1, \ldots, y_m]$.

4.3 A More Efficient Algorithm

In this section, we investigate some further properties of the Dixon matrices. Based upon these properties, we are able to elaborate a more efficient algorithm to compute a necessary condition for the existence of a common root of an arbitrary system of equations. Remember that with the use of the linear transformation in Lemma 4.1 we can always assume that the system of equations is regular.

Corollary 4.2 *Given a system $\mathcal{F} \subset K[y_1, \ldots, y_m][x_1, \ldots, x_n]$ of $n+1$ equations in n variables. If $D \in K(y_1, \ldots, y_m)_{s_2}^{s_1 \times s_2}$ is of full column rank then there exists a nontrivial necessary condition for the existence of a common root of the system $\mathcal{F}$. Moreover, for any square submatrix D' of the same rank with D, $\det(D') = 0$ is such a necessary condition.*

PROOF. From the previous lemma, $\det(D^+ D) = 0$ is such a necessary condition. Let (a, b) be a nontrivial common root of $\mathcal{F}$, D' be any nonsingular submatrix of D. Then $\operatorname{rank}\varphi_a(D) = \operatorname{rank}\varphi_a(D') < \operatorname{rank}D$ and hence $\varphi_a(D')$ must be singular. Therefore $\varphi_a(\det(D')) = 0$ where $\det(D')$ is not identical to zero. $\square$

In case the Dixon matrix is not of full column rank, we need to use the full-rank factorization techniques as in [Tra96]. In that, for any matrix D, there exist matrices F and G such that

- $D = FG$,
- F is of full-column rank,
- G is of full-row rank.

In fact, we can choose F to be any matrix whose columns are a basis for the range space of D. Then the matrix G is uniquely determined. Alternately, we can choose G be any matrix whose rows are a basis for the row space of D. The matrix G can be obtained by, for example, performing elementary row operations on the matrix D. Then the matrix F is uniquely determined. Actually, the full-rank factorization is very useful for improving our method as we will see in the next lemmas.

Lemma 4.4 *If* $\text{rank}\varphi_a(D) < \text{rank}D$ *for a common root* (a, b) *of* $\mathcal{F}$ *then there exists a nontrivially necessary condition. Moreover, for any square submatrix* D' *of the same rank with* D, $\det(D') = 0$ *is such a necessary condition.*

PROOF. We can extend D' to G, a matrix whose rows are a basis for the row space of D. Let $D = FG$ be the corresponding full-rank factorization of D for an appropriate matrix F. Then $\text{rank}\varphi_a(G) < \text{rank}G$ and hence $\varphi_a(\det(D')) = 0$ $\square$

Lemma 4.5 *Let* $D = FG$ *be a full-rank factorization of* D. *If* $\text{rank}\varphi(D) = \text{rank}D$ *for a common root of* $\mathcal{F}$ *then* $\varphi(D) = \varphi(F)\varphi(G)$ *is a full-rank factorization of* $\varphi(D)$.

PROOF. Otherwise, $\text{rank}\varphi(D)$ would less than $\text{rank}D$. $\square$

We have the following similar corollary for matrices of full row rank.

Corollary 4.3 *Let* $\mathcal{F}$ *be a system of* $n + 1$ *equations in* n *variables. If* D *is of full row rank then there exists a nontrivial necessary condition for the existence of a common root of the system* $\mathcal{F}$. *Moreover, for any square submatrix* D' *of the same rank as* D, $\det(D') = 0$ *is such a necessary condition.*

We now give a more efficient algorithm to compute a necessary condition for the existence of a common root of an arbitrary system of $n + 1$ polynomials in n variables.

Algorithm 4.2

Inputs: A system $\mathcal{F}$ *of* $n + 1$ *nonlinear polynomials in* n *variables.*
Output: A necessary condition for the existence of a common root of the system $\mathcal{F}$.

1. **If** $\mathcal{F}$ *is irregular* **then**

   Regularize the system by using a linear transformation as showed in Lemma 4.1.

   Fi;
2. *Setup the Dixon matrix D of $\mathcal{F}$;*
3. **If** D *is of full rank* **then**

   (a) *Compute a square submatrix D' of the same rank with D by using, for example modular and multivariate sparse interpolation methods.*

   (b) **Return**$(\det(D'))$.

   Else *Compute a full-rank factorization $D = FG$.*

   (a) *Compute a square submatrix G' of the same rank with G by using, for example modular and multivariate sparse interpolation methods.*

   (b) **Return**$(\det(G'))$.

   Fi.

Lemma 4.6 (Correctness) *The Algorithm 4.2 computes a necessary condition for the existence of a common root of an arbitrary system of $n + 1$ nonzero polynomials in n variables.*

PROOF. In [Tra96], the author has proved that for any matrix D, there exist matrices F and G such that

- $D = FG$,
- F is of full-column rank,
- G is of full-row rank.

Roughly speaking, we can choose F to be any matrix whose columns are a basis for the range space of D. Then the matrix G is uniquely determined. Alternately, we can choose G be any matrix whose rows are a basis for the row space of D. The matrix G can be obtained by, for example, performing elementary row operations on the matrix D. Then the matrix F is uniquely determined. Moreover, the null space of D is also the null space of G. $\square$

REMARK: The author would like to emphasize that the growth of the number of terms, as well as the size of the coefficients of the entries of the Dixon matrix will crucially influence the efficiency of the algorithm. To overcome this difficulty, recently in [Tra97a] the author presents an algorithm based on modular and multivariate sparse interpolation methods which works on the actual terms of the determinant polynomial.

5   Comparison with Other Methods

Several examples have been found to show that the new method in this paper works well also in situations where other methods of the same subject (e.g. [KSY94], [KS95]) may fail to give a correct answer. Here we psesent two of the examples. These examples fall into two categories.

In the first category, the system of equations has a trivial common root while the Dixon matrix is of full rank or even nonsingular. Therefore, by simply computing the determinant of a rank submatrix of the Dixon matrix we will obtain a wrong answer (see Example 5.1).

In the second category, the system of equations has a nontrivial common root while the Dixon matrix is of positive rank. Therefore, by just computing the determinant of a rank submatrix of the Dixon matrix it may lead to a wrong answer (see Example 5.2).

Example 5.1. Consider the following system of equations

$$0 = -83y^4x - 74x^2 + 59y^2 + 43y^4 - 95xy^2 + 97y,$$
$$0 = 77x^4y - 81xy - 2y^4 + 47xy^2 - 68y - 61x,$$
$$0 = -65x^2y + 59y^3 + 75x^2y^3 + 72y^4 - 39y - 91x.$$

The system clearly has a common root at $(0,0)$. The Dixon matrix is a matrix of size 22×22 and of full rank. That means, if we simply compute the determinant of a rank submatrix of the Dixon matrix by using other methods (such as [KSY94], [KS95]) we will receive an answer that the Dixon's resultant is

18411626220399729306220095784777325940330701744234894130749815473850
17499453403271089231410402782515623175537619428255392812933300000,

which is a nonzero constant.

Meanwhile, by using the new method in this paper, we first regularize the system getting a regular system with the same resultant. Using the linear transformation

$$\{x = x - 1, y = y + 1\},$$

where $p_j(-1, 1) \neq 0$, $1 \leq j \leq 3$, we obtain a regular system

$$0 = -522xy - 332xy^3 - 30x + 909y + 126y^4 - 593xy^2 + 910y^2 - 74x^2$$
$$+ 504y^3 - 83xy^4 + 303,$$
$$0 = 102 - 295xy + 462x^2y - 308x^3y - 403x - 12y - 2y^4 + 47xy^2 - 59y^2$$
$$+ 462x^2 - 8y^3 - 308x^3 + 77x^4 + 77x^4y,$$
$$0 = -320xy + 75x^2y^3 + 160x^2y + 225x^2y^2 - 150xy^3 - 111x + 586y$$
$$+ 72y^4 - 450xy^2 + 834y^2 + 10x^2 + 422y^3 + 193.$$

The Dixon matrix of this regular system is a matrix of size 23×25 and of rank 22. The determinant $\det(D^+D)$ is 0.

Example 5.2. Consider the following system of equations

$$0 = 68x^3y^2 - 14080xy + 7218x^2 - 5286x^2y + 3840xy^2 - 17x^4y - 98x^2y^3$$
$$+ 1254x^2y^2 - 544x^3y - 352xy^3 + 1020x^3 + 3230y^2 + 17184x$$
$$- 11228y + 51x^4 - 312y^3 + 13038,$$
$$0 = 62xy^3 - 558xy^2 + 1674xy - 2325x + 185y^3 - 1665y^2 + 5035y - 5512$$
$$- 55x^3 - 330x^2,$$
$$0 = 2118xy + 45x^2 - 105x^2y - 1251xy^2 - 5x^2y^3 - 28xy^4 + 45x^2y^2$$
$$+ 316xy^3 - 3663y^2 - 1359x + 5979y + 4y^5 - 116y^4 + 1001y^3 - 3573.$$

The system has common roots at $(\alpha, 3)$ where α is a root of $55x^3 + 330x^2 + 651x + 397 = 0$. The Dixon matrix of this regular system is a matrix of size 28×29 and of rank 22.

If we simply compute the determinant of a rank submatrix of the Dixon matrix, we will receive an answer that the Dixon's resultant is

$$-162932754006623509451797022712582192590783304854806114448834129357977729628086574855858261483673600000000000.$$

Meanwhile, the determinant $\det(D^+D)$ is 0.

The following corollary shows that the result in [KSY94] is a special case of our results.

Corollary 5.1 *If there exists a submatrix $\bar{D}$ of size $s_1 \times (s_2 - 1)$ by deleting the ith column of D such that $\mathrm{rank}\bar{D} < \mathrm{rank}D$ then there exists a nontrivial necessary condition for the fact that the system $\mathcal{F}$ has a nontrivial common root. (Don't forget the assumption that the system is regular.) Moreover, for any square submatrix D' of the same rank with D, $\det(D') = 0$ is such a necessary condition.*

PROOF. We have $\mathrm{rank}\varphi_a(\bar{D}) = \mathrm{rank}\varphi_a(D) < \mathrm{rank}D$. The assertion comes directly from Lemma 4.4. $\square$

In the next section, we discuss some practical examples in geometric reasoning, CAGD and solid modeling.

6 Experiments

We have implemented the algorithms using CASA, a system for computational algebra and constructive algebraic geometry developed at RISC-Linz and based upon the kernel of the Maple language (i.e. we have our own implementation of algorithms for fundamental computations such as Gröbner bases, Puiseux expansions, resultants, etc. – see [TW97b] and [TW97a] for more details). All the experiments were carried out on a Sun Ultra-1 workstation under SunOS 5.5 with 128MB RAM.

Example 6.1. Given an algebraic surface in parametric form

$$x_1 = \frac{y_1^2 - y_2^2}{y_2}, \quad x_2 = \frac{y_1^2 - y_2^2}{y_1}, \quad x_3 = \frac{1}{y_1 \cdot y_2}.$$

The objective is to find an implicit form for the surface. In order to avoid the base points, we first transform the system of equations into

$$\begin{aligned}
q_1 &= x_1 \cdot y_2 - (y_1^2 - y_2^2), \\
q_2 &= x_2 \cdot y_1 - (y_1^2 - y_2^2), \\
q_3 &= x_3 \cdot y_2 - 1, \\
q_4 &= t \cdot y_1 \cdot y_2 - 1.
\end{aligned}$$

The objective now is to eliminate t, y_2 and y_1 without any constraints on them.

We first compute the Dixon matrix which turns out to be a matrix of size 4×4 and of full column rank. Therefore, by the Lemma 4.2 for any square submatrix D' of the same rank with D, $\det(D') = 0$ is a necessary condition. We perform a Gaussian elimination process, compute the determinant and remove the extra factors getting an implicit form for the surface

$$x_2^4 - x_3 \cdot x_2^3 \cdot x_1^3 - 2 \cdot x_2^2 \cdot x_1^2 + x_1^4.$$

The total time taken was 0.080 second. The total memory space requirements were 298KB.

Example 6.2. Let ABC be a triangle, a, b and c be the length of the sides BC, AC and AB, a_i and a_e be the length of internal (AD) and external (AE) bisectors of angle A, and b_e be the length of the external angle bisector (BF) of angle B.

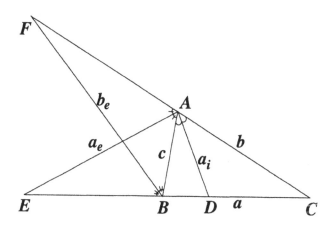

The objective is to determine that given general value of the three angle bisectors, whether one can draw the triangle using only a compass and a ruler. It has been proved in [Her75] that this is possible if and only if we can express a in terms of a_i, a_e and b_e, and the expression is of degree 2^m in a for some integer m. The problem can be transformed into an algebraic form as follows.

$$q_1 = a_i^2 \cdot (b + c)^2 - c \cdot b \cdot (c + b - a) \cdot (c + b + a),$$
$$q_2 = a_e^2 \cdot (c - b)^2 - c \cdot b \cdot (a + b - c) \cdot (c - b + a),$$
$$q_3 = b_e^2 \cdot (c - a)^2 - a \cdot c \cdot (a + b - c) \cdot (c + b - a).$$

The objective now is to eliminate b and c without any constraints on them. This example has been used in [GW95] and [KSY94]. In that the authors report solving the problem took about 19 hours on their implementation in Lisp on a

SUN 4 workstation and 501 seconds on their implementation in Maple on a SUN sparc 10 respectively.

We first compute the Dixon matrix which turns out to be a matrix of size 13×14 and of full row rank. Since we implicitly assume that b and c are not zero, by the Corollary 4.3 for any square submatrix D' of the same rank with D, $\det(D') = 0$ is a necessary condition. We perform a Gaussian elimination process, compute the determinant and remove the extra factors getting an expression in a, a_i, a_e and b_e of 330 terms and of degree 20 in a. That means given a_i, a_e and b_e, the triangle cannot be constructed using only a compass and a ruler. The total time taken was 57.80 second. The total memory space requirements were 318,337KB.

We present some more examples. In Table 1 we report the matrix sizes, the norm bounds, the term bounds, the computing times (in second) and the memory consumptions (in KB) of the extended Dixon's resultant method. Note that, the method is far more efficient than other elimination methods such as the Gröbner Bases method and the successive resultant method. In fact, most of the examples cannot be carried out by the Gröbner Bases method or the successive resultant method.

Example 6.3 ([Kal91]). Find an implicit form for the surface given in parametric form

$$x_1 = \frac{y_1^2 - y_2^2}{y_2}, \quad x_2 = \frac{y_1^2 - y_2^2}{y_1}, \quad x_3 = \frac{1}{y_1 - y_2}.$$

Example 6.4 ([Kal91]). Find an implicit form for the surface given in parametric form

$$x_1 = \frac{y_2 + 2y_1^4 - 1}{y_2 - y_1 - 2}, \quad x_2 = \frac{y_1^2 y_2 + 1}{y_1}, \quad x_3 = \frac{1}{y_1 \cdot y_2}.$$

Example 6.5. Determine if the following system has a common root.

$$q_1 = x_2^2 + x_1^2 - 1,$$
$$q_2 = x_1 \cdot x_2 - 1,$$
$$q_3 = x_2^2 - x_1.$$

Fig. 1. Some Experiences

Example	matrix size	norm bound	term bound	time (second)	mem (KB)
E.g. 6.1	4	10^4	10^3	0.08	298
E.g. 6.2	14	10^{13}	10^{13}	57.80	318300
E.g. 6.3	5	10^3	10^2	0.09	422
E.g. 6.4	13	10^6	10^4	0.32	1136
E.g. 6.5	5	10^5	10^1	0.03	161

7 Conclusion

By using techniques from pseudoinverse theory and linear transformations, we extended Dixon's resultant method to an arbitrary system of $n + 1$ nonlinear polynomials in n variables where the Dixon matrix can be singular or even nonsquare. Our method does not require any precondition. The method has been successfully used to solve the problem of implicitization and other nontrivial problems in geometric reasoning, CAGD and solid modeling. Several examples show that the new method works well also in situations where other methods (of the same subject) may fail to give a correct answer. However, extracting the extraneous factors from Dixon's resultants is a difficult task and still needs further investigations.

References

[BIG74]   A. Ben-Israel and T. N. E. Greville. *Generalized Inverses: Theory and Applications*. John Wiley & sons, New York, 1974.

[Bje51a] A. Bjerhammar. Applications of calculus of matrices to method of least squares with special reference to geodetic calculations. Technical Report 49, Kungl. Tekn. Högsk. Handl., 1951.

[Bje51b] A. Bjerhammar. Rectangular reciprocal matrices with special reference to geodetic calculations. *Bull. Géodésique*, pages 188–220, 1951.

[Bje68]   A. Bjerhammar. A generalized matrix algebra. Technical Report 124, Kungl. Tekn. Högsk. Handl., 1968.

[BO71]   T. L. Boullion and P. L. Odell. *Generalized Inverse Matrices*. Wiley–Interscience, New York, 1971.

[Buc85]   B. Buchberger. Gröbner Bases: An algorithmic method in polynomial ideal theory. In N. K. Bose, editor, *Multidimensional Systems Theory*, chapter 6, pages 184–232. Reidel Publishing Company, Dodrecht, 1985.

[Cay65]   A. Cayley. On the theory of elimination. *Cambridge and Dublin Math. J.*, III:210–270, 1865.

[CG95]   E.-W. Chionh and R. N. Goldman. Elimination and resultants – multivariate resultants. *IEEE Computer Graphics and Applications*, 15(2):60–69, 1995.

[Chi90]   E.-W Chionh. *Base points, resultants, and the implicit representation of rational Surfaces*. PhD thesis, Dept. Comp. Sci., Uni. of Waterloo, 1990.

[Cho88]   S.-C. Chou. *Mechanical Geometry Theorem Proving*. Reidel Publishing Company, Dordrecht, 1988.

[Dix08a] A.-L. Dixon. The eliminant of three quantics in two independent variables. *Proc. London Math. Soc.*, 7:49–69, 473–492, November 1908.

[Dix08b] A.-L. Dixon. On a form of the elimination of two quantics. *Proc. London Math. Soc.*, 6:468–478, June 1908.

[Dix09a] A.-L. Dixon. The eliminant of the equations of four quadric surfaces. *Proc. London Math. Soc.*, 8:340–352, December 1909.

[Dix09b] A.-L. Dixon. Symbolic expressions for the eliminant of two binary quantics. *Proc. London Math. Soc.*, 8:265–276, June 1909.

[Far88]   G. Farin. *Curves and Surfaces for Computer Aided Geometric Design - A Practical Guide*. Academic Press, New York, 1988.

[Fre03]   I. Fredholm. Sur une classe d'équations fonctionnelles. *Acta Math.*, 27:365–390, 1903.

[GW95]   X. S. Gao and D. K. Wang. On the automatic derivation of a set of geometric formulae. *J. Geometry*, 53:79–88, 1995.

[Her75]   I. N. Herstein. *Topics in Algebra*. John Wiley & Sons, USA, second edition, 1975.

[Hof89]   C. M. Hoffmann. *Geometric and Solid Modeling – An Introduction*. Morgan Kaufmann Publishers, Inc., San Mateo, California, 1989.

[HWW95] H. Hong, D. Wang, and F. Winkler, editors. *Algebraic Approaches to Geometric Reasoning*. Baltzer Science Publisher, Amsterdam, 1995. Special issue of Ann. of Math. and Artif. Intell., 13(1,2).

[Kal91]   M. Kalkbrener. *Three Contributions to Elimination Theory*. PhD thesis, Research Institute for Symbolic Computation, Univ. Linz, Linz, Austria, 1991.

[KS95]   D. Kapur and T. Saxena. Comparison of various multivariate resultant formulations. In *The proceedings of ISSAC'95*, pages 187–194, 1995.

[KSY94] D. Kapur, T. Saxena, and L. Yang. Algebraic and geometric reasoning using Dixon resultants. In *The proceedings of ISSAC'94*, pages 99–107, 1994.

[Moo20] E. H. Moore. On the reciprocal of the general algebraic matrix (abstract). *Bull. Amer. Math. Soc.*, 26:394–395, 1920.

[Pen55]   R. Penrose. A generalized inverse for matrices. *Proc. Cambridge Philos. Soc.*, 51:406–413, 1955.

[PW95]   J. Pfalzgraf and D. Wang, editors. *Automated Practical Reasoning*. Texts and Monographs in Symbolic Computation. Springer-Verlag, Wien, 1995.

[Rit50]   J. F. Ritt. *Differential Algebra*, volume 33. AMS Colloquium Publications, New York, 1950.

[Sal64]   George Salmon. *Lessons Introductory to the Modern Higher Algebra*. Chelsea Publishing Company, Bronx, New York, fifth edition, 1964.

[Tra95]   Quoc-Nam Tran. A hybrid symbolic-numerical method for tracing surface-to-surface intersections. In *The proceedings of ISSAC'95*, pages 51–58, 1995.

[Tra96]   Quoc-Nam Tran. *A Hybrid Symbolic-Numerical Approach in Computer Aided Geometric Design (CAGD) and Visualization*. PhD thesis, Research Institute for Symbolic Computation (RISC–Linz), University of Linz, Austria, 1996.

[Tra97a] Quoc-Nam Tran. Extending Dixon's resultant using pseudoinverse matrices and its apllications to geometric reasonning. Technical Report 97–11, RISC-Linz, The University of Linz, Austria, 1997. Talk at "Automated Deduction in Geometry", September 1996, Toulouse, France

[Tra97b] Quoc-Nam Tran. Extending Newton's method for finding the roots of an arbitrary system of equations and its applications. *International Journal of Modeling and Simulation*, 17(4), 1997. To appear.

[TW97a] Quoc-Nam Tran and Franz Winkler. Casa reference manual (version 2.3). Technical report, RISC-Linz, The University of Linz, Austria, 1997.

[TW97b] Quoc-Nam Tran and Franz Winkler. An overview of CASA – a system for computational algebra and constructive algebraic geometry. In T. Racio T. V. Effelterre and F. Winkler, editors, *The Symbolic and Algebraic Computation (SAC) Newsletter*, volume 2. Stichting CAN, Computer Algebra Nederland, Universiteit van Amsterdam, the Netherlands, 1997.

[vdW40] B. L. van der Waerden. *Moderne Algebra*, volume 2. Springer Verlag, Berlin, second edition, 1940.

[Wu94]   W.-T. Wu. *Mechanical Theorem Proving in Geometries*. Texts and Monographs in Symbolic Computation. Springer-Verlag, Wien, 1994. Translated from Chinese by X. Jin and D. Wang.

Computational Geometry Problems in REDLOG

Thomas Sturm and Volker Weispfenning

Fakultät für Mathematik und Informatik
Universität Passau, D-94030 Passau, Germany
sturm@fmi.uni-passau.de, weispfen@alice.fmi.uni-passau.de

Abstract. We solve algorithmic geometrical problems in real 3-space or the real plane arising from applications in the area of CAD, computer vision, and motion planning. The problems include parallel and central projection problems, shade and cast shadow problems, reconstruction of objects from images, offsets of objects, Voronoi diagrams of a finite family of objects, and collision of moving objects. Our tools are real elimination algorithms implemented in the REDUCE package REDLOG. In many cases the problems can be solved uniformly in unspecified parameters. The power of the method is illustrated by examples many of which have been outside the scope of real elimination methods so far.

1 Introduction

This note is concerned with a number of computational problems in real 3-space or the real plane that arise from applications mainly in the area of CAD, computer vision, and motion planning. The problems are of the following types:

Parallel projection Given an object A and a parametric light ray r, compute the parallel projection of A in direction r onto a plane perpendicular to r. Compute the shaded and lighted parts of A. Compute the shadow cast by A onto some other object B. Reconstruct object A, or determine features of A from its image under parallel projection.

Central projection We treat the same problems as with parallel projection, but wrt. central projection arising from a punctual light source.

Offsets Given a closed object A and a positive real r. Compute the r-offset of A: This is the set of points having distance exactly r from A. Here distance may be Euclidean distance or distance in some other norm.

Voronoi diagrams Given finitely many pairwise disjoint closed objects A_1, ..., A_n, compute their Voronoi diagram: This is the set of all points that have equal minimal distance to at least two different objects A_i and A_j. Again distances may be measured according to various norms.

Collision problems Given two moving objects A and B, a starting position and velocity vector for A, and a parametric starting position and velocity vector for B. Determine the conditions on the parameters under which the objects A and B will eventually collide, and under these conditions the time and place, where the collision will happen.

Among these problems, projections have obvious applications in the area of CAD and computer vision. Offsets are important special cases of Minkowski sums required for robot motion planning, cf. [BKOS97]. There has also been considerable research on using offset computations for rounding and blending of solids in CSG-based CAD systems, cf. [Ros85, RR86, SW97]. Such problems fall in the field of mathematical morphology, cf. [Ser82], which is originally motivated by texture analysis for geological applications. Voronoi diagrams have numerous applications in various fields such as social geography, physics, astronomy, and again robotics, cf. [BKOS97]. The relevance of collision problems, e.g. for robot motion planning, is again obvious.

Throughout the paper, we will assume that all inputs to our computations are objects in the real plane or in real 3-space given by a "low-degree" semi-algebraic description, i.e. a Boolean combination of polynomial equations and inequalities in two or three variables, respectively. These equations and inequalities will be quadratic in most cases. Frequently we will allow unspecified real parameters in such a description; then we obtain a solution for a whole parametric family of problems. Our methods do not impose any restrictions on the degree of parameters. Similarly, the computed output objects will also be given in such a description, where, however, the polynomial degrees may and will exceed 2 as a rule.

We will show that all the problem types listed above can be modeled as *extended quantifier elimination* problems in real algebra. Roughly speaking, extended quantifier elimination differs from ordinary quantifier elimination by the fact that the procedure provides—besides a quantifier-free equivalent to the input formula—sample parametric answers for values of the variables in the outmost existential quantifier block. In the most frequent case of a purely existential input formula extended quantifier elimination can be construed as parametric constraint solving. Extended quantifier elimination is implemented in the REDLOG [DS96, DS97b] package of REDUCE for input formulas with low-degree quantified variables in several variants.

The variant that we will use almost exclusively is *generic quantifier elimination*. Here the system will automatically exclude a measure zero set of degenerate real parameter values wherever this supports the elimination process. Since REDLOG does, however, explicitly output its non-degeneracy assumptions, the user can easily rerun the quantifier elimination for specific degenerate parameter values that may be of interest to him.

Many of the examples computed below involve a significant number of parameters besides the variables to be eliminated. Examples of this kind have been inaccessible to other implemented real elimination methods.

The plan of the paper is as follows: In Section 2 we give a general survey on the REDLOG package, and roughly sketch the REDLOG algorithms for quantifier elimination, extended quantifier elimination, generic quantifier elimination, and Gröbner simplification. The following sections are devoted to the different types of geometrical problems mentioned above. For each problem class we first describe its modeling by first-order formulas. Then we discuss specific instances in this problem class, and their automatic solution using the REDLOG package. All computations have been performed on a SUN SPARC-4 workstation.

2 The REDUCE package REDLOG

REDLOG [DS96, DS97b] stands for REDUCE logic system. It extends the computer algebra system REDUCE by symbolic algorithms on first-order formulas wrt. temporarily fixed first-order languages and theories. For the purpose of this paper we are interested in the theory of real closed fields over the language of ordered rings. In contrast to constraint logic programming systems, [Col90], the algebraic component is not only used for supporting the logical engine but the largest part of the logical algorithms is defined and implemented in terms of algebraic algorithms.

The algorithms implemented in REDLOG include the following:

- Several techniques for the *simplification* of quantifier-free formulas. The simplifiers do not only operate on the Boolean structure of the formulas but also discover algebraic relationships (see Section 2.4). For the notion of simplification and a detailed description of the implemented techniques cf. [DS97c].
- *Quantifier elimination*
    • For formulas obeying certain degree restrictions for the quantified variables, we use a technique based on elimination set ideas [Wei88, LW93, Wei97b] (see below).
    • In addition, there is an interface to Hoon Hong's QEPCAD [HCJE93] package implementing a complete quantifier elimination.
- *Generic quantifier elimination* (see Section 2.3).
- *Extended* variants of both classical and generic quantifier elimination (see Section 2.2).
- Linear *optimization* using quantifier elimination techniques [Wei94].
- CNF/DNF computation including both Boolean and algebraic simplification [DS97c].
- Several other *normal form* computations, e.g., prenex normal form computation minimizing the number of quantifier changes.
- A lot of useful tools for constructing, decomposing, and analyzing formulas.

REDLOG has been applied successfully for solving non-academic problems, mainly for the simulation and error diagnosis of physical networks [Wei97c].

Applications inside the scientific community include the following:

- Control theory [ADY+96].
- Stability analysis for PDE's [HLS97].
- Geometric reasoning [DSW96].
- Parametric scheduling.
- Non-convex parametric linear and quadratic optimization [Wei94], transportation problems [LW93].
- Real implicitization of algebraic surfaces.

– Computation of comprehensive Gröbner bases.
– Implementation of guarded expressions for coping with degenerate cases in the evaluation of algebraic expressions [CJ92, DS97a].

For non-commercial use the REDLOG source code is freely available on the www.[1] The rest of this section is devoted to a survey of the REDLOG procedures used throughout this paper.

We consider polynomial equations and inequalities $f \; R \; 0$, where f is a multivariate polynomial with rational coefficients and $R \in \{=, \geq, \leq, >, <, \neq\}$. A *quantifier-free formula* ψ is a Boolean combination of such equations and inequalities obtained by applying negation "¬," conjunction "∧," and disjunction "∨." We call ψ of degree d in a variable x if all polynomials occurring in ψ have an x-degree of at most d.

2.1 Quantifier elimination

Suppose that a quantifier-free formula ψ is quadratic, i.e. of degree 2, in some variable x, and denote $\exists x \big(\psi(x, u_1, \ldots, u_n)\big)$ by $\varphi(u_1, \ldots, u_n)$. Then the algorithm given in [Wei97b] computes from φ a quantifier-free formula $\varphi'(u_1, \ldots, u_n)$ not containing x such that over the ordered field of the reals we have the equivalence

$$\varphi(u_1, \ldots, u_n) \longleftrightarrow \varphi'(u_1, \ldots, u_n).$$

In other words, for arbitrary values $a_1, \ldots, a_n \in \mathbb{R}$ of the u_i, the assertion $\varphi'(a_1, \ldots, a_n)$ holds in $\mathbb{R}$ iff there exists $b \in \mathbb{R}$ such that $\psi(b, a_1, \ldots, a_n)$ holds in $\mathbb{R}$. The elimination of a universal quantifier can be reduced to that of an existential quantifier using the equivalence

$$\forall x \psi \longleftrightarrow \neg \exists x \neg \psi.$$

Several quantifiers can be eliminated one by one starting with the innermost one provided that the elimination result for some inner quantifier still obeys the degree restrictions. The process sketched above is referred to as *quantifier elimination*. Note that the implementation in REDLOG includes various heuristics for coping with formulas violating the degree restrictions, e.g. polynomial factorization, cf. [DSW96] for details.

2.2 Extended quantifier elimination

For eliminating a quantifier $\exists x$ from $\exists x \psi$, our elimination algorithm proceeds as follows: All equations and inequalities $f(x, u_1, \ldots, u_n) \; R \; 0$ contained in ψ are renormalized wrt. x obtaining, e.g. in the case of a quadratic constraint:

$$a(u_1, \ldots, u_n)x^2 + b(u_1, \ldots, u_n)x + c(u_1, \ldots, u_n) \; R \; 0.$$

---

[1] http://www.fmi.uni-passau.de/~redlog/

From these renormalized constraints we compute, using the well-known solution formulas for quadratic and linear equations, a finite set T of *test terms* not containing x such that

$$\exists x \psi \longleftrightarrow \bigvee_{t \in T} \psi[x /\!/ t].$$

Here $[x /\!/ t]$ denotes a modified substitution of t for x with the following features (cf. [Wei97b] for details):

- We can, semantically correct, substitute terms t involving square roots such that the substitution result is a well-formed formula not containing any square root.
- We can, semantically correct, substitute infinite elements $t = \pm \infty$ and terms $t = s \pm \varepsilon$ involving infinitesimal elements without these non-standard elements occurring in the substitution result.
- Besides substituting, we also add conditions and case distinctions wrt. the validity of substitution terms. For instance, with terms arising from the quadratic solution formula, we add the condition $a \neq 0 \wedge b^2 - 4ac \geq 0$. That is, the constraint is actually quadratic and the discriminant is non-negative.

By keeping track of the terms t substituted during the elimination process, we obtain—instead of the quantifier-free equivalent $\psi' = \bigvee_{i=1}^{k} \psi[x /\!/ t_i]$—a scheme

$$\begin{bmatrix} \psi[x /\!/ t_1] & x = t_1 \\ \vdots & \vdots \\ \psi[x /\!/ t_k] & x = t_k \end{bmatrix}$$

including satisfying sample points. This process of *extended quantifier elimination* can also be repeated for several existential quantifiers. The result then is a set of conditions each associated with an answer for each eliminated variable obtained by resubstitution.

Note that the sample points t_i can include the symbols ∞ and ε. The former has to be read as "a real number x_0 such all $x \geq x_0$ satisfy ψ." In an analogous way, ε stands for a "small enough" positive real. Unfortunately, nothing can be said about the order between several non-standard symbols in the output. For identifying equal non-standard symbols after resubstitution, all such symbols are indexed.

2.3 Generic quantifier elimination

Very much of the complexity of our quantifier elimination procedure arises from the case distinctions wrt. parametric coefficient expressions being zero or not. *Generic quantifier elimination* assumes for purely parametric expressions to be non-zero wherever this supports the elimination process. The assumptions are collected and, besides the elimination result, returned to the user. It has turned out that in most cases these assumptions are *non-degeneracy conditions* (ND-conditions), cf. [DSW96]. Generic quantifier elimination can, of course, be combined with extended quantifier elimination.

Besides the automatic generation of assumptions, all REDLOG procedures used throughout this paper allow to pass a set of polynomial equations and inequalities as an optional *background theory* argument.

2.4 Gröbner simplification

During the quantifier elimination procedures, REDLOG applies a simplification procedure to the obtained (sub)results. Besides this fast, though still sophisticated, *standard simplifier*, REDLOG provides several advanced simplifiers that can be explicitly applied to final elimination results. Of particular importance for our purposes is a simplifier using *Gröbner basis* [Buc65] methods.

We illustrate the technique by means of a very simple example: Consider as input formula $xy + 1 \neq 0 \lor yz + 1 \neq 0 \lor x - z = 0$. It can be rewritten as

$$xy + 1 = 0 \land yz + 1 = 0 \longrightarrow x - z = 0.$$

Reducing the conclusion modulo the Gröbner basis $\{x - z, yz + 1\}$ of the premises, we obtain the equivalent formula $xy + 1 = 0 \land yz + 1 = 0 \longrightarrow 0 = 0$, which can in turn be easily simplified to "true." For details on the algorithm actually used cf. [DS97c].

3 Parallel Projection

We consider semi-algebraic objects A, B, ... in real 3-space $S = \mathbb{R}^3$ given by corresponding defining quantifier-free formulas $\alpha(u, v, w)$, $\beta(u, v, w)$, ...

The *boundary* b(A) of A can be described by a quantifier-free formula α'_b equivalent to

$$\alpha_b(x, y, z) \equiv \alpha(x, y, z) \land \forall \varepsilon \big[\varepsilon > 0 \longrightarrow \exists u \exists v \exists w \big(\neg \alpha(u, v, w) \land$$
$$-\varepsilon < x - u < \varepsilon \land -\varepsilon < y - v < \varepsilon \land -\varepsilon < z - w < \varepsilon\big)\big].$$

In a similar way, β'_b denotes a quantifier-free formula describing the boundary b(B) of object B, and so on.

Let now $r = (k, l, m)$ be a non-zero vector in S. Interpreting r as the direction of light coming in parallel rays from an infinitely far light source, we can model various concepts of image and shadow as follows: A point $(x, y, z) \in S$ is the image of a point $(u, v, w) \in S$ under parallel projection along r iff

$$\exists t(t > 0 \land x = u + kt \land y = v + lt \land z = w + mt).$$

Hence the region SR(A) in space S that is shaded by object A can be described by the formula

$$(x, y, z) \in \mathrm{SR}(A) \iff \alpha_{\mathrm{SR}}(x, y, z) \equiv \exists u \exists v \exists w \exists t \big(\alpha(u, v, w) \land t > 0 \land$$
$$x = u + kt \land y = v + lt \land z = w + mt).$$

So by determining a quantifier-free equivalent $\alpha'_{SR}(x, y, z)$ to the formula α_{SR} above, we get a semi-algebraic description of the region $SR(A)$.

Similarly, we can describe the shaded part $Sh(A)$ and the lighted part $Li(A)$ of the boundary of object A:

$$(x, y, z) \in Sh(A) \iff \alpha_b(x, y, z) \wedge \exists u \exists v \exists w \exists t \big(\alpha(u, v, w) \wedge t > 0 \wedge$$
$$x = u + kt \wedge y = v + lt \wedge z = w + mt\big),$$

$$(x, y, z) \in Li(A) \iff \alpha_b(x, y, z) \wedge \neg \exists u \exists v \exists w \exists t \big(\alpha(u, v, w) \wedge t > 0 \wedge$$
$$x = u + kt \wedge y = v + lt \wedge z = w + mt\big).$$

For strictly convex objects C, we can also define the boundary $SL(C)$ between $Sh(C)$ and $Li(C)$. This definition works even for convex, but not strictly convex objects, with a non-degenerate light direction r:

$$(x, y, z) \in SL(C) \iff \gamma_b(x, y, z) \wedge \neg \exists u \exists v \exists w \exists t \big(\gamma(u, v, w) \wedge t \neq 0 \wedge$$
$$x = u + kt \wedge y = v + lt \wedge z = w + mt\big).$$

Suppose now that B is a second object in S. Then the shadow $CS(A, B)$ cast by object A onto the boundary of B can be described by the following formula:

$$(x, y, z) \in CS(A, B) \iff \beta_b(x, y, z) \wedge \exists u \exists v \exists w \exists t \big(\alpha(u, v, w) \wedge t \geq 0 \wedge$$
$$x = u + kt \wedge y = v + lt \wedge z = w + mt\big).$$

This applies in particular to the case when B is a plane perpendicular to the light ray r given by the equation $kx + ly + mz = p$. In this case the shadow cast onto B by A is identical to the image of A viewed in direction r. This image is called the *aspect* of A. For simplicity, we may take B as a plane through the origin and drop the condition $t \geq 0$. Then we get the following formula for the aspect of A in direction $r = (k, l, m)$:

$$(x, y, z) \in Asp(r, A) \iff kx + ly + mz = 0 \wedge \exists u \exists v \exists w \exists t \big(\alpha(u, v, w) \wedge$$
$$x = u + kt \wedge y = v + lt \wedge z = w + mt\big).$$

Finally we may use extended quantifier elimination to reconstruct features of the object A and its orientation relative to the projection plane from its aspect. To this end we assume that our object belongs to a parametric class of objects, e.g. cuboids, balls, or ellipsoids, described by a quantifier-free formula $\alpha(u, v, w, a_1, \ldots, a_n)$. The real parameters $a_1, \ldots, a_n$ describe the dimensions and orientation of the object.

In general, we can put our coordinate system into such a position that the light ray projects the object along the Z-axis onto the X-Y-plane. Since the image of the object is invariant under translation of the object along the light ray, we may assume in addition that some reference point of the object is placed in the origin of the coordinate system. Suppose now that we have a quantifier-free formula $\alpha_0(x, y)$ which we assume to describe $Asp((0, 0, 1), A)$. Then an application of extended quantifier elimination to the formula

$$\exists w \exists a_1 \ldots \exists a_n \forall x \forall y \big(\alpha(x, y, w, a_1, \ldots, a_n) \longleftrightarrow \alpha_0(x, y)\big)$$

will yield the result "false" if α_0 does not describe the aspect of an object in the given class; otherwise it will yield "true" together with a list of possible real values for $a_1, \ldots, a_n$.

Among the problems described in this section, feature reconstruction is by far the hardest one. In order to obtain suitable results within a tolerable amount of time, one has to put more intelligence into the coding of the problem than with the general technique described above.

For recovering, e.g., the dimensions of a *rectangular solid* A and its orientation in space from its aspect, we proceed as follows: Consider A to be generated by pairwise perpendicular vectors

$$e_1 = (e_{11}, e_{12}, e_{13}), \quad e_2 = (e_{21}, e_{22}, e_{23}), \quad e_3 = (e_{31}, e_{32}, e_{33})$$

with one corner located at the origin. Its corners $x = (x_1, x_2, x_3)$ are described by the formula

$$e_1 e_2 = e_1 e_3 = e_2 e_3 = 0 \wedge \exists \lambda_1 \exists \lambda_2 \exists \lambda_3 \left(\gamma(\underline{\lambda}, \underline{e}) \right),$$

where $\gamma(\underline{\lambda}, \underline{e}) \equiv \lambda_1, \lambda_2, \lambda_3 \in \{0, 1\} \wedge x = \lambda_1 e_1 + \lambda_2 e_2 + \lambda_3 e_3$. The direction of the light is along the x_3-axis. The projection plane P is the x_1-x_2-plane.

Provided that A is in generic position, i.e., not parallel to any of the axes, its aspect is the convex hull of 6 points with two possible choices for points which cannot be observed:

1. All e_{i3} coordinates have the same sign. Then the images of the origin and of its opposite are in the interior of the aspect.
2. Up to a permutation of the e_i, the sign of e_{13} differs from that of e_{23} and e_{33}. Then the image of e_1 and that of its opposite $e_2 + e_3$ are in the interior of the aspect.

The concrete coordinate system is chosen by the observer of the aspect, who will exclude the first of the two cases above by placing one aspect corner into the origin.

We generate by quantifier elimination (255 ms) a quantifier-free generic aspect description ι' from the following formula:

$$\iota(e_{11}, e_{12}, \ldots, e_{33}, i_1, i_2) \equiv e_1 e_2 = e_1 e_3 = e_2 e_3 = 0 \wedge$$
$$\exists x_1 \exists x_2 \exists x_3 \exists \lambda_1 \exists \lambda_2 \exists \lambda_3 \left(\gamma(\underline{\lambda}, \underline{e}) \wedge \right.$$
$$\left. \neg(\lambda_1 \neq \lambda_2 = \lambda_3) \wedge i_1 = x_1 \wedge i_2 = x_2 \right).$$

For a given semi-algebraic description α_0 of aspect corners we can now reconstruct the original cuboid by applying extended quantifier elimination to the following *reconstruction* formula:

$$\varrho(\alpha_0) \equiv \exists e_{11} \exists e_{12} \exists e_{13} \exists e_{21} \exists e_{22} \exists e_{23} \exists e_{31} \exists e_{32} \exists e_{33} \forall i_1 \forall i_2 (\iota' \longleftrightarrow \alpha_0).$$

For correct generic aspect descriptions α_0 this elimination will yield "true" together with suitable vectors e_1, e_2, e_3 as answer. If α_0 does not describe the aspect of a cuboid in generic position the elimination will yield "false."

Examples computed with REDLOG

For our first Examples 1–5, we consider *a parametric quadric* in space of the form $\{(u, v, w) \in S \mid au^2 + bv^2 + cw^2 \leq 1\}$ and a light ray (k, l, m). Depending on the signs of the parameters a, b, c, this quadric is an ellipsoid, a cylinder, or a hyperboloid. Similar results as below have been obtained with other parametric quadrics.

Example 1 (Shadowed region). The following formula describes the region shadowed by the ellipsoid:

$$\alpha_{\mathrm{SR}}(x, y, z, k, l, m) \equiv \exists u \exists v \exists w \exists t (t > 0 \wedge au^2 + bv^2 + cw^2 \leq 1 \wedge$$
$$u + tk = x \wedge v + tl = y \wedge w + tm = z).$$

We obtain after 442 ms a quantifier-free equivalent containing 6 atomic formulas, and the ND-condition $ak^2 + bl^2 + cm^2 \neq 0$. Fig. 1 illustrates how such a computation looks like in REDLOG.

Example 2 (Shaded part of the surface). The shaded part of the surface is described by the following formula:

$$\sigma(x, y, z, a, b, c, k, l, m) \equiv ax^2 + by^2 + cz^2 = 1 \wedge$$
$$\exists u \exists v \exists w \exists t (t > 0 \wedge au^2 + bv^2 + cw^2 \leq 1 \wedge$$
$$u + tk = x \wedge v + tl = y \wedge w + tm = z).$$

We obtain after 629 ms a quantifier-free description containing 9 atomic formulas, valid under the ND-condition $ak^2 + bl^2 + cm^2 \neq 0$.

Example 3 (Lighted part of the surface). We use the following formula to describe the lighted part of the surface of our quadric:

$$\lambda(x, y, z, a, b, c, k, l, m) \equiv ax^2 + by^2 + cz^2 = 1 \wedge$$
$$\neg \exists u \exists v \exists w \exists t (t > 0 \wedge au^2 + bv^2 + cw^2 \leq 1 \wedge$$
$$u + tk = x \wedge v + tl = y \wedge w + tm = z).$$

We obtain after 561 ms a quantifier-free description in 5 atomic formulas valid under the ND-condition $ak^2 + bl^2 + cm^2 \neq 0$.

Example 4 (Boundary between shaded and lighted part). Describing the boundary between the lighted and the shaded part of the quadric by

$$\beta(x, y, z, a, b, c, k, l, m) \equiv ax^2 + by^2 + cz^2 = 1 \wedge$$
$$\neg \exists u \exists v \exists w \exists t (t \neq 0 \wedge au^2 + bv^2 + cw^2 \leq 1 \wedge$$
$$u + tk = x \wedge v + tl = y \wedge w + tm = z)$$

we obtain within 476 ms the quantifier-free equivalent

$$ak^2 + bl^2 + cm^2 \geq 0 \wedge akx + bly + cmz = 0 \wedge ax^2 + by^2 + cz^2 - 1 = 0$$

valid under the ND-condition $ak^2 + bl^2 + cm^2 \neq 0$.

```
9: ells1:= ex({u,v,w,t},t>0 and a*u**2+b*v**2+c*w**2 <= 1 and
9: u + t*k = x and v + t*l = y and w + t*m = z)))$

10: rlgqe(ells1);
---- (ex u v w t) [BFS: depth 4]
-- left: 4
[11g]
-- left: 3
[11g]
-- left: 2
[11g]
-- left: 1
[1e#q!] [DEL:0/4]
{{a*k**2 + b*l**2 + c*m**2 <> 0},
a*k**2 + b*l**2 + c*m**2 <= 0 or
a*x**2 + b*y**2 + c*z**2 - 1 < 0 or
a**2*b*k**4*y**2 - 2*a**2*b*k**3*l*x*y +
a**2*b*k**2*l**2*x**2 + a**2*c*k**4*z**2 -
2*a**2*c*k**3*m*x*z + a**2*c*k**2*m**2*x**2 -
a**2*k**4 + a*b**2*k**2*l**2*y**2 -
2*a*b**2*k*l**3*x*y + a*b**2*l**4*x**2 +
2*a*b*c*k**2*l**2*z**2 - 2*a*b*c*k**2*l*m*y*z +
2*a*b*c*k**2*m**2*y**2 - 2*a*b*c*k*l**2*m*x*z -
2*a*b*c*k*l*m**2*x*y + 2*a*b*c*l**2*m**2*x**2 -
2*a*b*k**2*l**2 + a*c**2*k**2*m**2*z**2 -
2*a*c**2*k*m**3*x*z + a*c**2*m**4*x**2 -
2*a*c*k**2*m**2 + b**2*c*l**4*z**2 -
2*b**2*c*l**3*m*y*z + b**2*c*l**2*m**2*y**2 -
b**2*l**4 + b*c**2*l**2*m**2*z**2 -
2*b*c**2*l*m**3*y*z + b*c**2*m**4*y**2 -
2*b*c*l**2*m**2 - c**2*m**4 <= 0 and
a**2*k**3*x + a*b*k**2*l*y + a*b*k*l**2*x +
a*c*k**2*m*z + a*c*k*m**2*x + b**2*l**3*y +
b*c*l**2*m*z + b*c*l*m**2*y + c**2*m**3*z > 0 or
a*k*x + b*l*y + c*m*z > 0 and
a*x**2 + b*y**2 + c*z**2 - 1 = 0}$

Time: 442 ms plus GC time: 68 ms

11: rlatnum(second(ws));

6$
```

Fig. 1. A generic quantifier elimination in REDLOG, showing the successive elimination of quantified variables and the elimination technique used in each case. The first entry of the output list contains the ND-conditions assumed during the elimination; the second entry is the quantifier-free formula equivalent to the input formula under the ND-conditions.

Example 5 (Aspect). The aspect of our quadric can be described by the following formula:

$$\varphi(x,y,z,a,b,c,k,l,m) \equiv \exists u \exists v \exists w \exists t (au^2 + bv^2 + cw^2 \leq 1 \wedge u + tk = x \wedge$$
$$v + tl = y \wedge w + tm = z \wedge kx + ly + mz = 0).$$

We obtain after 442 ms a quantifier-free equivalent with 6 atomic formulas valid under the ND-condition $ak^2 + bl^2 + cm^2 \neq 0$.

Example 6 (Intersection aspect). We compute the aspect of the intersection line of two parametric tubes crossing perpendicularly but not necessarily centrally:

$$\tau(x,y,z,a,b,c,k,l,m) \equiv \exists u \exists v \exists w \exists t \big((u-a)^2 + (v-b)^2 = d_1^2 \,\wedge$$
$$u^2 + w^2 = d_2^2 \wedge x = u + kt \wedge y = v + lt \,\wedge$$
$$z = w + mt \wedge kx + ly + zw = 0\big).$$

The result obtained after 357 ms is the following quantifier-free description valid under the ND-conditions $m \neq 0$ and $z \neq 0$:

$$a^2 m^2 z^2 + 2ak^2 mxz + 2aklmyz + 2akmz^3 - 2am^2 xz^2 + b^2 m^2 z^2 +$$
$$2bklmxz + 2bl^2 myz + 2blmz^3 - 2bm^2 yz^2 - d_1^2 m^2 z^2 + k^4 x^2 +$$
$$2k^3 lxy + 2k^3 xz^2 + k^2 l^2 x^2 + k^2 l^2 y^2 + 2k^2 lyz^2 - 2k^2 mx^2 z +$$
$$k^2 z^4 + 2kl^3 xy + 2kl^2 xz^2 - 4klmxyz - 2kmxz^3 + l^4 y^2 + 2l^3 yz^2 -$$
$$2l^2 my^2 z + l^2 z^4 - 2lmyz^3 + m^2 x^2 z^2 + m^2 y^2 z^2 = 0 \,\wedge$$
$$d_2^2 m^2 z^2 - k^4 x^2 - 2k^3 lxy - 2k^3 xz^2 - k^2 l^2 y^2 - 2k^2 lyz^2 -$$
$$k^2 m^2 x^2 + 2k^2 mx^2 z - k^2 z^4 - 2klm^2 xy + 2klmxyz + 2kmxz^3 -$$
$$l^2 m^2 y^2 - m^2 x^2 z^2 = 0.$$

Example 7 (Solid reconstruction). As an example for recovering a rectangular solid, consider the aspect given in Fig. 2. Its corners have the following semi-algebraic description:

$$\alpha_0 \equiv (13i_1 - 1124 = 0 \wedge 13i_2 - 1960 = 0) \vee$$
$$(13i_1 + 176 = 0 \wedge 13i_2 - 1050 = 0) \vee$$
$$(13i_1 + 436 = 0 \wedge 13i_2 - 530 = 0) \vee (i_1 - 120 = 0 \wedge i_2 - 110 = 0) \vee$$
$$(i_1 - 100 = 0 \wedge i_2 - 70 = 0) \vee (i_1 = 0 \wedge i_2 = 0).$$

Extended quantifier elimination applied to $\varrho(\alpha_0)$ yields after 15 674 ms "true" together with the correct reconstruction pictured in Fig. 3:

$$e_1 = (20, 40, -96), \quad e_2 = \left(-\frac{436}{13}, \frac{530}{13}, 10\right), \quad e_3 = (100, 70, 50).$$

4   Central Projection

In this section we consider the same problems as in the previous one but for central projection: We again consider, objects A, B, ... described by quantifier-free formulas α, β, ...; their boundaries $b(A)$, $b(B)$, ... are described by formulas α_b, β_b, ...

Let $(q, r, s) \in S$ denote the position of the punctual light source in space. Then a point $(x, y, z) \in S$ is the image of a point $(u, v, w) \in S$ under central projection from point (q, r, s) iff

$$\exists t\big(t > 0 \wedge x = t(u - q) \wedge y = t(v - r) \wedge z = t(w - s)\big).$$

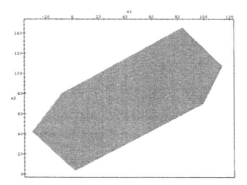

Fig. 2. A sample aspect of a rectangular solid in generic position.

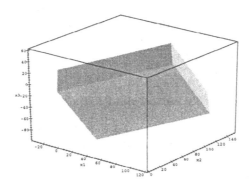

Fig. 3. Rectangular solid reconstructed from the top aspect given in Fig. 2.

Hence the region SR(A) in space that is shaded by object A under central projection from (q, r, s) is described by the formula

$$(x, y, z) \in \text{SR}(A) \iff \alpha_{\text{SR}}(x, y, z) \equiv \exists u \exists v \exists w \exists t \big(\alpha(u, v, w) \wedge t > 0 \wedge$$
$$x = t(u - q) \wedge y = t(v - r) \wedge$$
$$z = t(w - s)\big).$$

By determining a quantifier-free equivalent α'_{SR} to the formula α_{SR} above, we get a semi-algebraic description of the region SR(A).

In a similar way our descriptions given in the previous section of the shaded part, the lighted part, and of the boundary between these parts can be adapted to central projection. For the latter we again have to exclude degenerate situations, where the surface of the object contains more than one point on some light ray:

$$(x, y, z) \in \text{Sh}(A) \iff \alpha_b(x, y, z) \wedge \exists u \exists v \exists w \exists t \big(\alpha(u, v, w) \wedge t > 0 \wedge$$
$$x = t(u - q) \wedge \ y = t(v - r) \wedge z = t(w - s)\big),$$

$$(x, y, z) \in \text{Li}(A) \iff \alpha_b(x, y, z) \wedge \neg \exists u \exists v \exists w \exists t \big(\alpha(u, v, w) \wedge t > 0 \wedge$$
$$x = t(u - q) \wedge y = t(v - r) \wedge z = t(w - s)\big),$$

$$(x, y, z) \in \text{SL}(A) \iff \alpha_b(x, y, z) \wedge \neg \exists u \exists v \exists w \exists t \big(\alpha(u, v, w) \wedge t \neq 0 \wedge$$
$$x = t(u - q) \wedge y = t(v - r) \wedge z = t(w - s)\big).$$

Let B be another object in S. Then the shadow $\text{CS}(A, B)$ cast by object A onto the boundary of B can be described by the following formula:

$$(x, y, z) \in \text{CS}(A, B) \iff \beta_b(x, y, z) \wedge \exists u \exists v \exists w \exists t \big(\alpha(u, v, w) \wedge t \geq 0 \wedge$$
$$x = t(u - q) \wedge y = t(v - r) \wedge z = t(w - s)\big).$$

This applies in particular to the case when B is a plane given by the equation $kx + ly + mz = p$. In this case the shadow cast onto B by A is identical to the *image* $\text{Im}(A, B)$ of A on B under central projection from (q, r, s). The following formula describes this image:

$$(x, y, z) \in \text{Im}(A, B) \iff kx + ly + mz = p \wedge \exists u \exists v \exists w \exists t \big(\alpha(u, v, w) \wedge t \geq 0 \wedge$$
$$x = t(u - q) \wedge y = t(v - r) \wedge z = t(w - s)\big).$$

In the previous section, we have recovered a rectangular solid from its aspect. We now turn to a more difficult problem, where we wish to recover a cuboid from a central projection. To make the problem easier, we consider a wire frame instead of a solid avoiding visibility considerations. Still the cuboid will not be uniquely determined by one wire frame image.

We modify our aspect case model as follows: The cuboid generated by the vectors e_1, e_2, e_3 is shifted from the origin by some translation vector $v = (v_1, v_2, v_3)$. The projection plane is the x_1-x_2 plane. The focal point is, e.g., $(0, 0, 5)$. Our idea is that the focal point lies between the image and the projection plane modeling photography: We take photos along the x_3-axis. The 3-space origin is the center of our photo.

A generic image description ι' is derived by applying quantifier elimination (765 ms) to the following formula:

$$\iota \equiv e_1 e_2 = e_1 e_3 = e_2 e_3 = 0 \wedge \exists x_1 \exists x_2 \exists x_3 \exists \lambda_1 \exists \lambda_2 \exists \lambda_3 \exists k (\lambda_1, \lambda_2, \lambda_3 \in \{0, 1\} \wedge$$
$$x = \lambda_1 e_1 + \lambda_2 e_2 + \lambda_3 e_3 + v \wedge i_1 = kx_1 \wedge i_2 = kx_2 \wedge 0 = 5 + k(5 - x_3)).$$

For a given semi-algebraic image description π_0, we can recover information about the original wire frame by applying extended quantifier elimination to the following reconstruction formula:

$$\exists e_{11} \exists e_{12} \exists e_{13} \exists e_{21} \exists e_{22} \exists e_{23} \exists e_{31} \exists e_{32} \exists e_{33} \exists v_1 \exists v_2 \exists v_3 \forall i_1 \forall i_2 (\iota' \iff \pi_0).$$

In practice, the elimination can be supported by fixing the image of the shifted origin v. Compare Example 10 for details.

With two images taken simultaneously from different locations, a cuboid can be reconstructed uniquely provided that the choices for the v, e_1, e_2, e_3 are consistent. Unfortunately, in a corresponding reconstruction formula for a stereo camera one cannot play the trick of fixing the images of v indicated above, because one cannot automatically determine which points correspond in two different projection images.

Examples computed with REDLOG

Example 8. This example is inspired by an example discussed in [KM88]. We consider the central projection of the line

$$a_1 u + b_1 v + c_1 w = d_1 \wedge a_2 u + b_2 v + c_2 w = d_2$$

from a punctual light source located at $(0, 0, f)$ onto the projection plane $w = 0$:

$$\varphi_1(x, y, a_1, \ldots, d_2, f) \equiv \exists u \exists v \exists w \big(x(f - w) = fu \wedge y(f - w) = fv \wedge$$
$$a_1 u + b_1 v + c_1 w = d_1 \wedge a_2 u + b_2 v + c_2 w = d_2 \big).$$

Generic quantifier elimination with the theory $\{f > 0\}$ yields after 221 ms

$$a_1 c_2 fx - a_1 d_2 x - a_2 c_1 fx + a_2 d_1 x +$$
$$b_1 c_2 fy - b_1 d_2 y - b_2 c_1 fy + b_2 d_1 y + c_1 d_2 f - c_2 d_1 f = 0$$

subject to the input theory plus the ND-condition $a_1 x + b_1 y - c_1 f \neq 0$. Given another line

$$a_3 u + b_3 v + c_3 w = d_3 \wedge a_4 u + b_4 v + c_4 w = d_4$$

in space, we can read off from the elimination result above the necessary and sufficient condition for the images of the two lines to be parallel under the corresponding ND-conditions, viz.:

$$(a_1 c_2 f - a_1 d_2 - a_2 c_1 f + a_2 d_1)(b_3 c_4 f - b_3 d_4 - b_4 c_3 f + b_4 d_3) =$$
$$(a_3 c_4 f - a_3 d_4 - a_4 c_3 f + a_4 d_3)(b_1 c_2 f - b_1 d_2 - b_2 c_1 f + b_2 d_1)$$

provided that all bracketed items are non-zero.

Example 9. We compute the central projection of the unit ball in arbitrary position. Again, the light source is located at $(0, 0, f)$, and the projection plane is given by $w = 0$:

$$\varphi_2(x, y, a, b, c, f) \equiv \exists u \exists v \exists w \big(x(f - w) = fu \wedge y(f - w) = fv \wedge$$
$$(u - a)^2 + (v - b)^2 + (w - c)^2 = 1 \big).$$

Generic quantifier elimination with the theory $\{f > 0\}$ yields after 187 ms:

$$a^2 f^2 + a^2 y^2 - 2abxy + 2acfx - 2af^2 x + b^2 f^2 + b^2 x^2 + 2bcfy -$$
$$2bf^2 y + c^2 x^2 + c^2 y^2 - 2cfx^2 - 2cfy^2 + f^2 x^2 + f^2 y^2 - f^2 - x^2 - y^2 \leq 0$$

without any additional ND-condition.

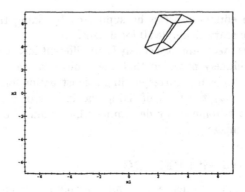

Fig. 4. A photo of the wire frame of the cuboid in Fig. 3.

Example 10 (Recovering a cuboid). Fig. 4 shows a photo taken of the wire frame of the cuboid in Fig. 3 shifted from the origin by $v = (100, 200, 300)$. The semi-algebraic image description is as follows:

$$
\begin{aligned}
\pi_0 \equiv\ &(3367i_1 - 12120 = 0 \wedge 3367i_2 - 22800 = 0)\ \vee \\
&(923i_1 - 2164 = 0 \wedge 923i_2 - 4040 = 0)\ \vee \\
&(2717i_1 - 5620 = 0 \wedge 2717i_2 - 18250 = 0)\ \vee \\
&(793i_1 - 864 = 0 \wedge 793i_2 - 3130 = 0)\ \vee \\
&(249i_1 - 1100 = 0 \wedge 249i_2 - 1550 = 0)\ \vee \\
&(69i_1 - 200 = 0 \wedge 23i_2 - 90 = 0)\ \vee \\
&(199i_1 - 600 = 0 \wedge 199i_2 - 1200 = 0)\ \vee \\
&(59i_1 - 100 = 0 \wedge 59i_2 - 200 = 0).
\end{aligned}
$$

In the reconstruction formula we support the elimination procedure by fixing the image point $\left(\frac{100}{59}, \frac{200}{59}, 0\right)$ to be the image the shifted origin:

$$
\exists e_{11} \exists e_{12} \exists e_{13} \exists e_{21} \exists e_{22} \exists e_{23} \exists e_{31} \exists e_{32} \exists e_{33} \exists v_1 \exists v_2 \exists v_3 \forall i_1 \forall i_2
$$
$$
\left[(\iota' \longleftrightarrow \pi_0) \wedge \exists k \left(\frac{100}{59} = kv_1 \wedge \frac{200}{59} = kv_2 \wedge 0 = 5 + k(5 - v_3) \right) \right].
$$

We obtain after $1\,952\,671$ ms (33 minutes) the quantifier-free equivalent "true" together with the following answer:

$$
v = \left(\frac{5\infty_1}{2}, 5\infty_1, \frac{59\infty_1 + 40}{8} \right), \quad
e_1 = \left(\frac{5\infty_1}{2}, \frac{7\infty_1}{4}, \frac{5\infty_1}{4} \right),
$$
$$
e_2 = \left(\frac{\infty_1}{2}, \infty_1, \frac{-12\infty_1}{5} \right), \quad
e_3 = \left(\frac{-109\infty_1}{130}, \frac{53\infty_1}{52}, \frac{\infty_1}{4} \right).
$$

For $\infty_1 = 40$ this is our original cuboid.

Without fixing the image of v, i.e., using

$$\exists e_{11}\exists e_{12}\exists e_{13}\exists e_{21}\exists e_{22}\exists e_{23}\exists e_{31}\exists e_{32}\exists e_{33}\exists v_1\exists v_2\exists v_3\forall i_1\forall i_2(\iota' \longleftrightarrow \pi_0)$$

as reconstruction formula, the elimination takes $42\,998\,933$ ms (approximately $12\,h$) yielding a correct reconstruction with different choices for the vectors.

5   Offsets

Let $A \subseteq S$ be an object that constitutes a non-empty closed set. Fix a metric $d : S \to \mathbb{R}$ that induces the usual topology on S. Then the distance of a point $(x, y, z) \in S$ from A is defined as

$$d\big((x, y, z), A\big) := \min\{\, d\big((x, y, z), (u, v, w)\big) \mid (u, v, w) \in A \,\}.$$

The existence of the minimum is guaranteed by the assumption that A is closed in S and that d is continuous. Let now r be a non-negative real number. Then the r-offset of A is defined as

$$A_r = \{\, (x, y, z) \in S \mid d\big((x, y, z), A\big) = r \,\}.$$

We generalize the notion of an r-offset: The $\leq$-r-offset $A_{\leq, r}$ of A and the $\geq$-r-offset $A_{\geq, r}$ are defined by $d\big((x, y, z), A\big) \leq r$ and $d\big((x, y, z), A\big) \geq r$, respectively. We then obviously have $A_r = A_{\leq, r} \cap A_{\geq, r}$.

For a fixed metric d on S, suppose we have formulas

$$\delta_{\leq}(x, y, z, u, v, w, r) \quad \text{and} \quad \delta_{\geq}(x, y, z, u, v, w, r)$$

describing the facts that $d\big((u, v, w), (x, y, z)\big) \leq r$ and $d\big((u, v, w), (x, y, z)\big) \geq r$, respectively. Then the following formulas describe various offsets of A:

$$\alpha_{\leq, r}(x, y, z) \equiv \exists u \exists v \exists w \big(\alpha(u, v, w) \wedge \delta_{\leq}(x, y, z, u, v, w, r)\big),$$
$$\alpha_{\geq, r}(x, y, z) \equiv \forall u \forall v \forall w \big(\alpha(u, v, w) \longrightarrow \delta_{\geq}(x, y, z, u, v, w, r)\big),$$
$$\alpha_r(x, y, z) \equiv \alpha_{\leq, r}(x, y, z) \wedge \alpha_{\geq, r}(x, y, z).$$

We give the relevant formulas $\delta_{\geq}(x, y, z, u, v, w, r)$ for some common metrics on S. The formulas $\delta_{\leq}(x, y, z, u, v, w, r)$ are formed similarly:

– for the Euclidean (L_2) metric:

$$(x - u)^2 + (y - v)^2 + (z - w)^2 \geq r^2;$$

– for the "Manhattan" (L_1) metric:

$$\exists k \exists l \exists m \Big((k \geq 0 \wedge (k = x - u \vee k = u - x)) \wedge$$
$$(l \geq 0 \wedge (l = y - v \vee l = v - y)) \wedge$$
$$(m \geq 0 \wedge (m = z - w \vee m = w - z)) \wedge k + l + m \geq r\Big);$$

— for the maximum (L_∞) metric:

$$r \leq x - u \vee r \leq u - x \vee r \leq y - v \vee r \leq v - y \vee r \leq z - w \vee r \leq w - z.$$

Notice that the formulas α_r for the r-offset involve 3 existential and 3 universal quantifiers and hence pose a non-trivial problem for quantifier elimination.

For certain smooth algebraic surfaces $A : f(u, v, w) = 0$ without singularities in S we have much simpler formulas for A_r: Suppose the minimal radius of the curvature of A is less than or equal to r. Then the point $(u, v, w) \in A$ of minimal distance to a given point $(x, y, z) \in S$ is determined by the fact that the vector $(x-u, y-v, z-w)$ is perpendicular to the tangent plane of A at (u, v, w). In other words this vector is a real multiple of the gradient vector ∇f. As a consequence we may in this case define α_r^* instead of α_r:

$$\alpha_r^*(x, y, z) \equiv \exists u \exists v \exists w \exists t \Big(f(u, v, w) = 0 \wedge \delta(x, y, z, u, v, w, r) \wedge$$

$$t\frac{\partial f(u, v, w)}{\partial u} = x - u \wedge$$

$$t\frac{\partial f(u, v, w)}{\partial v} = y - v \wedge$$

$$t\frac{\partial f(u, v, w)}{\partial w} = z - w\Big).$$

Here $\delta \equiv \delta_\leq \wedge \delta_\geq$ describes the fact that $d\big((u, v, w), (x, y, z)\big) = r$. For a well-behaved algebraic surface with singularities, α_r^* can be combined with α_r restricted to the singularities.

The idea of using the gradient vectors can be extended to smooth algebraic curves $A : f(u, v, w) = 0 \wedge g(u, v, w) = 0$ without singularities in S, where we assume again that the radius of the curvature of A is less than or equal to r. We then have the following description α_r^{**} of the r-offset:

$$\alpha_r^{**}(x, y, z) \equiv \exists u \exists v \exists w \exists s \exists t \Big(f(u, v, w) = 0 \wedge \delta(x, y, z, u, v, w, r) \wedge$$

$$s\frac{\partial f(u, v, w)}{\partial u} + t\frac{\partial g(u, v, w)}{\partial u} = x - u \wedge$$

$$s\frac{\partial f(u, v, w)}{\partial v} + t\frac{\partial g(u, v, w)}{\partial v} = y - v \wedge$$

$$s\frac{\partial f(u, v, w)}{\partial w} + t\frac{\partial g(u, v, w)}{\partial w} = z - w\Big).$$

For certain special objects in 3-space there are even simpler descriptions available: Consider, e.g., the r_2-offset of a circle $x^2 + y^2 = r_1^2$ wrt. the Euclidean metric, which is a torus T. It can be defined by the following formula τ with only one quantifier:

$$\tau(x, y, z) \equiv \exists u \big(u^2 + z^2 = r_2^2 \wedge (r_1 + u)^2 = x^2 + y^2 \big).$$

The idea behind this description is to regard T as a union of circles in planes parallel to the X-Y-plane.

Example 14 below suggests to use iterated offset computations for the problem of rounding corners and edges, cf. [Hof96]. For more details cf. [SW97].

The concept of an r-offset in the plane instead of 3-space is defined analogously. It should be obvious, how to modify the formulas given above for the planar case. Compare [PP97] and [Wan93] for a discussion of offsets and examples, using other methods.

Examples computed with REDLOG

Example 11. We compute a quantifier-free description of the torus given by

$$\tau(x, y, z, r_1, r_2) \equiv \exists u \left(u^2 + z^2 = r_2^2 \wedge (r_1 + u)^2 = x^2 + y^2 \right).$$

Quantifier elimination with subsequent Gröbner basis simplification yields within 1241 ms the following quantifier-free description:

$$r_1^4 - 2r_1^2 r_2^2 - 2r_1^2 x^2 - 2r_1^2 y^2 + 2r_1^2 z^2 + r_2^4 - 2r_2^2 x^2 - \\ 2r_2^2 y^2 - 2r_2^2 z^2 + x^4 + 2x^2 y^2 + 2x^2 z^2 + y^4 + 2y^2 z^2 + z^4 = 0 \wedge r_2^2 - z^2 \geq 0.$$

Example 12. We compute r-offsets of the cross consisting of the X-axis and the Y-axis in S. Its offset wrt. the Euclidean metric and the L_1 metric is described by the following formulas φ_2 and φ_1, respectively:

$$\varphi_2(x, y, z, r) \equiv \exists u \exists v \big((u = 0 \vee v = 0) \wedge (x - u)^2 + (y - v)^2 + z^2 = r^2 \big) \wedge \\ \forall u \forall v \neg \big((u = 0 \vee v = 0) \wedge (x - u)^2 + (y - v)^2 + z^2 < r^2 \big),$$

$$\varphi_1(x, y, z, r) \equiv \exists u \exists v \big((u = 0 \vee v = 0) \wedge \exists k \exists l \exists m (\\ k \geq 0 \wedge (k = x - u \vee k = u - x) \wedge \\ l \geq 0 \wedge (l = y - v \vee l = v - y) \wedge \\ m \geq 0 \wedge (m = z - w \vee m = w - z) \wedge k + l + m = r)) \wedge \\ \forall u \forall v \neg \big((u = 0 \vee v = 0) \wedge \exists k \exists l \exists m (\\ k \geq 0 \wedge (k = x - u \vee k = u - x) \wedge \\ l \geq 0 \wedge (l = y - v \vee l = v - y) \wedge \\ m \geq 0 \wedge (m = z - w \vee m = w - z) \wedge k + l + m < r)).$$

Quantifier elimination with subsequent Gröbner basis simplification applied to φ_2 yields after 1292 ms:

$$(r^2 - x^2 - z^2 = 0 \wedge x^2 - y^2 < 0 \wedge x \neq 0 \wedge y \neq 0) \vee \\ (r^2 - x^2 - z^2 = 0 \wedge x^2 - y^2 < 0 \wedge x = 0) \vee \\ (r^2 - y^2 - z^2 = 0 \wedge x^2 - y^2 \geq 0 \wedge x \neq 0 \wedge y \neq 0) \vee \\ (r^2 - y^2 - z^2 = 0 \wedge x^2 - y^2 \geq 0 \wedge y = 0).$$

The offset thus consists of segments of the tubes $r^2 - x^2 - z^2 = 0$ and $r^2 - y^2 - z^2 = 0$ separated by the planes $X = Y$ and $X = -Y$ at which they exactly fit together. Fig. 5 pictures this for $r = 4$.

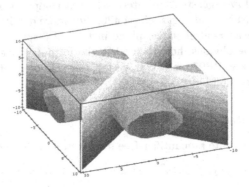

Fig. 5. 4-Offset of the X-axis and the Y-axis together with the planes $X = Y$ and $X = -Y$ separating the relevant segments of the tubes.

For φ_1 we obtain without Gröbner simplification after 5.3 s a quantifier-free formula with 632 atomic subformulas. The Gröbner simplifier does not yield a simplified equivalent for this result. It fails in performing an initial Boolean normal form, i.e. CNF or DNF, computation step, which is necessary with the current version of REDLOG, cf. [DS97c].

Example 13. We compute, wrt. the Euclidean metric, the r-offset of the parabola in S given by the equations $v = u^2$ and $w = 0$:

$$\varphi(x, y, z, r) \equiv \exists u \exists v \exists s (v = u^2 \wedge$$
$$(x - u)^2 + (y - v)^2 + z^2 = r^2 \wedge 2su = x - u \wedge -s = y - v).$$

The generic elimination can eliminate only v and leaves a formula in which u occurs in a conjunction of a quartic equation $f_1(u) = 0$ and a cubic equation $f_2(u) = 0$. We conjunctively add $h(u) = 0$, where $h(u)$ is the quadratic remainder of $f_1(u)/f_2(u)$. The generic elimination of u can then be completed automatically, yielding a formula with 7 atomic subformulas under the ND-condition $2y - 1 \neq 0$. The total time is 5.8 s.

Example 14. Consider the rectangle A of side lengths 1 and 2 described by the formula

$$\alpha(u, v) \equiv 0 \leq u \leq 1 \wedge 0 \leq v \leq 2.$$

We want to round the corners of A by circular segments of radius $0 < r < \frac{1}{2}$.

For this purpose, let B be the closure of the complement of A. It is described by $\beta(u, v) \equiv \neg(0 < u < 1 \wedge 0 < v < 2)$. The $\geq$-r-offset $B_{\geq,r}$ of B is described by the formula

$$\beta_{\geq,r}(u, v, r) \equiv \forall u \forall v (\beta \longrightarrow (x - u)^2 + (y - v)^2 \geq r^2)).$$

Quantifier elimination using the theory $\{0 < r, r < \frac{1}{2}\}$ with subsequent Gröbner simplification yields after 153 ms the following quantifier-free description:

$$\beta'_{\geq,r}(x,y,r) \equiv r^2 - x^2 + 2x - 1 \leq 0 \wedge r^2 - x^2 - y^2 + 4y - 4 \leq 0 \wedge$$
$$r^2 - x^2 - y^2 \leq 0 \wedge r^2 - x^2 \leq 0 \wedge r^2 - y^2 + 4y - 4 \leq 0 \wedge$$
$$r^2 - y^2 \leq 0 \wedge x - 1 < 0 \wedge x > 0 \wedge y - 2 < 0 \wedge y > 0.$$

Under our theory $\{0 < r, r < \frac{1}{2}\}$ we can transform this by some easy hand computations including completion of squares into the following simpler form:

$$\beta''_{\geq,r}(x,y,r) \equiv x \geq r \wedge 1 - r \geq x \wedge y \geq r \wedge 2 - r \geq y.$$

The correctness of our simplification can be proved automatically in less than 17 ms by applying quantifier elimination to

$$\forall x \forall y \forall r \left(0 < r < \frac{1}{2} \longrightarrow (\beta'_{\geq,r} \longleftrightarrow \beta''_{\geq,r})\right).$$

The final result of rounding is obtained as the $\leq$-r-offset

$$(B_{\geq,r})_{\leq,r}$$

of $B_{\geq,r}$. The corresponding quantifier elimination, again with the theory $\{0 < r, r < \frac{1}{2}\}$, delivers after 323 ms the following quantifier-free description of the rounded rectangle:

$$(r^2 + 2rx + 2ry - 6r + x^2 - 2x + y^2 - 4y + 5 \leq 0) \vee$$
$$(r^2 + 2rx - 2ry - 2r + x^2 - 2x + y^2 + 1 \leq 0) \vee$$
$$(r^2 - 2rx + 2ry - 4r + x^2 + y^2 - 4y + 4 \leq 0) \vee$$
$$(r^2 - 2rx - 2ry + x^2 + y^2 \leq 0) \vee$$
$$(r + x - 1 \leq 0 \wedge r - x \leq 0 \wedge r + y - 2 \leq 0 \wedge r - y \leq 0) \vee$$
$$(2rx - 2r + x^2 - 2x + 1 \leq 0 \wedge r + y - 2 \leq 0 \wedge r - y \leq 0) \vee$$
$$(2rx - x^2 \geq 0 \wedge r + y - 2 \leq 0 \wedge r - y \leq 0) \vee$$
$$(2ry - 4r + y^2 - 4y + 4 \leq 0 \wedge r + x - 1 \leq 0 \wedge r - x \leq 0) \vee$$
$$(2ry - y^2 \geq 0 \wedge r + x - 1 \leq 0 \wedge r - x \leq 0).$$

It contains 20 atomic formulas. The shortest description we are able to produce by hand contains 12 atomic formulas. For fixed values of r the equivalence between the automatic result and our hand formulation can be shown by quantifier elimination in less than one minute.

6   Voronoi Diagrams and Equi-Distance Surfaces

Let $F = \{A_1, \ldots, A_n\}$ be a finite family of objects that constitute non-empty closed sets in space. Fix a metric $d : S \to \mathbb{R}$ that induces the usual topology on S. Then the *Voronoi diagram* $\mathrm{Vor}(F)$ wrt. d of F is defined as the set of all $(x, y, z) \in S$ for which there are $1 \le i < j \le n$ such that

$$d\big((x, y, z), A_i\big) = d\big((x, y, z), A_j\big) = \min\big\{\, d\big((x, y, z), A_k\big) \mid 1 \le k \le n \,\big\}.$$

In the special case $F = \{A_1, A_2\}$, the Voronoi diagram is the *equi-distance surface* of the objects A_1 and A_2.

A similar definition applies to Voronoi diagrams in the plane. Compare [PS85] for a discussion of Voronoi diagrams in the plane.

Suppose that we have quantifier-free formulas $\alpha_i(u, v, w)$ describing the objects A_i for $1 \le i \le n$. As sketched in the previous section, there are formulas $\delta(x, y, z, u, v, w, r)$ and $\delta_\ge(x, y, z, u, v, w, r)$ describing, under the given metric d, the fact that $d\big((x, y, z), (u, v, w)\big) = r$ and $d\big((x, y, z), (u, v, w)\big) \ge r$, respectively. From these we can construct formulas

$$\delta_i(x, y, z, r) \quad \text{and} \quad \delta_{\ge, i}(x, y, z, r)$$

describing the fact that $d\big((x, y, z), A_i\big) = r$ and $d\big((x, y, z), A_i\big) \ge r$, respectively. Based on these intermediate formulas, the Voronoi diagram $\mathrm{Vor}(F)$ can be described by the following formula:

$$\bigvee_{\substack{i,j=1 \\ i \ne j}}^{n} \exists r \Big(\delta_i(x, y, z, r) \wedge \delta_j(x, y, z, r) \wedge \bigwedge_{\substack{k=1 \\ k \ne i,j}}^{n} \delta_{\ge, k}(x, y, z, r) \Big).$$

Examples computed with REDLOG

Example 15. In the real plane, the Voronoi diagram of the three points $(-1, 0)$, $(1, 0)$, $(1, 1)$ is described by the formula

$$
\begin{aligned}
\varphi(x, y) \equiv \exists r \big(& (r^2 = (x-1)^2 + y^2 = (x+1)^2 + y^2 \wedge r^2 \le (x-1)^2 + (y-1)^2) \vee \\
& (r^2 = (x-1)^2 + y^2 = (x-1)^2 + (y-1)^2 \wedge r^2 \le (x+1)^2 + y^2) \vee \\
& (r^2 = (x+1)^2 + y^2 = (x-1)^2 + (y-1)^2 \wedge r^2 \le (x-1)^2 + y^2) \big).
\end{aligned}
$$

Quantifier elimination with subsequent Gröbner simplification yields the following quantifier-free description (170 ms):

$$
\begin{aligned}
& (4x + 2y - 1 = 0 \wedge 4y^2 - 4y + 5 \ge 0 \wedge 2y - 1 \ge 0) \vee \\
& (4x^2 - 8x + 5 \ge 0 \wedge x \ge 0 \wedge 2y - 1 = 0) \vee (x = 0 \wedge 2y - 1 \le 0).
\end{aligned}
$$

Example 16. The Voronoi diagram of the unit circle and the lines $X = -2$ and $X = 2$ in the plane is described by the formula

$$\varphi(x,y) \equiv \exists r \exists u \exists v \, [u^2 + v^2 = 1 \wedge vx = uy \wedge$$
$$((r^2 = (x+2)^2 = (x-2)^2 \wedge r^2 \leq (x-u)^2 + (y-v)^2) \vee$$
$$(r^2 = (x+2)^2 = (x-u)^2 + (y-v)^2 \wedge r^2 \leq (x-2)^2) \vee$$
$$(r^2 = (x-2)^2 = (x-u)^2 + (y-v)^2 \wedge r^2 \leq (x+2)^2))].$$

Generic quantifier elimination yields after 1122 ms a formula with 16 atomic subformulas plus the ND-condition $x \neq 0$. Subsequent Gröbner simplification produces the following result (153 ms):

$$(6x - y^2 + 9 = 0 \wedge x < 0) \vee (2x - y^2 + 1 = 0 \wedge x < 0) \vee$$
$$(6x + y^2 - 9 = 0 \wedge x > 0) \vee (2x + y^2 - 1 = 0 \wedge x > 0).$$

The following two examples concern the computation of equi-distance surfaces in space, cf. [Hof96].

Example 17. Find a description of the surface defined by equal distance to the unit ball and a parametric plane $X = a$:

$$\varphi_1(x,y,z,a) \equiv \exists u \exists v \exists w \, [(x-a)^2 = (x-u)^2 + (y-v)^2 + (z-w)^2 \wedge$$
$$u^2 + v^2 + w^2 = 1 \wedge uy = vx \wedge uz = wx \wedge vz = wy \wedge$$
$$(xu > 0 \vee (x = 0 \wedge u = 0)) \wedge (yv > 0 \vee (y = 0 \wedge v = 0)) \wedge$$
$$(zw > 0 \vee (z = 0 \wedge w = 0))].$$

Generic quantifier elimination produces after 1037 ms a formula φ_1' with 8 atomic subformulas under the ND-condition $x \neq 0$. Fixing $a = 5$ and applying the Gröbner simplifier, we obtain the following φ_1'':

$$(x^2 - 6x \geq 0 \wedge 12x + y^2 + z^2 - 36 = 0 \wedge x < 0) \vee$$
$$(x^2 - 4x \leq 0 \wedge 8x + y^2 + z^2 - 16 = 0 \wedge x < 0) \vee$$
$$(x^2 - 6x \leq 0 \wedge 12x + y^2 + z^2 - 36 = 0 \wedge x - 6 \neq 0 \wedge x > 0) \vee$$
$$(x^2 - 4x \geq 0 \wedge 8x + y^2 + z^2 - 16 = 0 \wedge x - 4 \neq 0 \wedge x > 0).$$

We suspect that one of the two contained equations describes the Voronoi diagram: Applying quantifier elimination to

$$\forall x \forall y \forall z \, (x \neq 0 \longrightarrow (\varphi_1'' \longleftrightarrow 12x + y^2 + z^2 - 36 = 0))$$

we in fact obtain "true" after 170 ms. Fig. 6 pictures our special case $a = 5$.

Example 18. We compute the equi-distance surface for the X-axis and the line $X = Y$, $Z = 1$:

$$\varphi_2(x,y,z) \equiv \exists u \, (y^2 + z^2 = (x-u)^2 + (y-u)^2 + (z-1)^2 \wedge x - u = -(y-u)).$$

Elimination yields $x^2 - 2xy - y^2 - 4z + 2 = 0$ after 17 ms. This surface is pictured in Fig. 7.

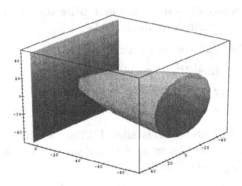

Fig. 6. The plane $X = 5$ and its equi-distance surface with the unit ball.

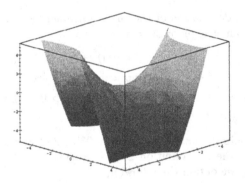

Fig. 7. The equi-distance surface of X-axis and the line $X = Y$, $Z = 1$.

7 Collision Problems

Let A, B be objects moving in space from their initial position with time. The question is, whether A and B will collide, and—if yes—when and where is a collision going to happen?

In the easiest case, A and B move along straight lines with constant velocities. Suppose $\alpha(x, y, z)$ and $\beta(x, y, z)$ describe the objects A and B at time $t = 0$, and (k, l, m), (p, q, r) are the velocity vectors for the motion of A and B, respectively. Then a collision situation of A and B is characterized by both objects having a point in common. This is described by the formula $\psi \equiv \exists t \big(t > 0 \wedge \varphi(t) \big)$, where

$$\varphi(t) \equiv \exists x \exists y \exists z \big(\alpha(x - kt, y - lt, z - mt) \wedge \beta(x - pt, y - qt, z - rt) \big).$$

If we eliminate from ψ the four quantifiers by extended quantifier elimination, we get either "false," i.e., A and B will never collide, or "true" together with a common point and the corresponding time. In general, this does not describe the first collision of A and B. In order to obtain time and location of the first contact of the two objects, we have to apply quantifier elimination with answer to the formula $\psi^* \equiv \exists t \left[t > 0 \wedge \varphi(t) \wedge \forall t' \left(0 < t' < t \longrightarrow \neg \varphi(t') \right) \right]$.

Examples computed with REDLOG

Example 19. Consider the unit disk moving with velocity 1 along the X-axis, and a square with center at $(0, -8)$ moving with velocity vector (a, b) from its starting position. A collision situation is described by

$$\varphi(t, a, b) \equiv \exists x \exists y (-1 \leq x - at \leq 1 \wedge -9 \leq y - bt \leq -7 \wedge (x - t)^2 + y^2 \leq 1).$$

For obtaining necessary and sufficient conditions on the parametric velocity vector (a, b) for a collision to take place, we apply generic quantifier elimination to the formula

$$\psi(a, b) \equiv \exists t \left(t > 0 \wedge \varphi(t, a, b) \right).$$

This yields after 1292 ms a quantifier-free formula $\psi'(a, b)$ with 69 atomic subformulas under the ND-condition $a^2 - 2a + b^2 + 1 \neq 0$, which obviously excludes only the degenerate velocity vector $(a, b) = (0, 0)$.

This example is generalized from a decision problem in [CH91], where the parameters a, b were both fixed to $\frac{17}{16}$. We plug into φ these original values for a and b yielding

$$\varphi_0(t) \equiv \varphi \left(t, \frac{17}{16}, \frac{17}{16} \right).$$

The result of quantifier elimination with answer applied to the corresponding

$$\psi_0 \equiv \exists t (t > 0 \wedge \varphi_0)$$

is "true" together with the answer $t = \frac{144}{17}$ and $(x, y) = (\frac{144}{17}, 0)$. This extended quantifier elimination takes 85 ms.

Next we check whether this describes the first collision by eliminating quantifiers with answer from

$$\psi_0^* \equiv \exists t \left[t > 0 \wedge \varphi_0(t) \wedge \forall t' \left(0 < t' < t \longrightarrow \neg \varphi_0(t') \right) \right].$$

The result obtained after 629 ms is "true" together with the answer $t = \frac{96}{17}$ and $(x, y) = (\frac{96}{17}, -1)$ showing the true first collision.

The next example is a very strong 3-dimensional generalization of the previous one.

Example 20. Starting with its center located at the origin, a solid parametric quadric $a_1 x^2 + b_1 y^2 + c_1 z^2 \leq 1$ with main axes parallel to the coordinate axes moves with velocity 1 along the X-axis. A cube with side length 2 moves starting with its center at (a, b, c) with constant parametric velocity (k, l, m) through space: The input formula is $\psi \equiv \exists t(t > 0 \wedge \varphi)$, where

$$\varphi(t, a_1, b_1, c_1, a, b, c, k, l, m) \equiv \exists x \exists y \exists z (a_1 (x - t)^2 + b_1 y^2 + c_1 z^2 \leq 1 \wedge$$
$$a - 1 \leq x - kt \leq a + 1 \wedge$$
$$b - 1 \leq y - lt \leq b + 1 \wedge$$
$$c - 1 \leq z - mt \leq c + 1).$$

Note that $\psi(a_1, b_1, c_1, a, b, c, k, l, m)$ has 9 parameters. Generic quantifier elimination yields a quantifier-free formula with 2792 atomic subformulas under 7 ND-conditions in 43.4 s.

We fix the dimensions of the quadric and the starting position for the cube yielding

$$\varphi_0(t, k, l, m) \equiv \varphi(t, 3, 2, 1, 1, -10, 15, k, l, m)$$

and $\psi_0(k, l, m) \equiv \exists t(t > 0 \wedge \varphi_0(t, k, l, m))$ as corresponding input formula. This reduces the timing for generic quantifier elimination to 9.4 s yielding only 336 atomic formulas under 5 ND-conditions. For this specialized situation, we ask for a velocity vector such that a collision takes place not later than at the time $t = 10$:

$$\psi_1 \equiv \exists k \exists l \exists m \exists t (0 < t \leq 10 \wedge \varphi_0(t, k, l, m)).$$

Applying extended quantifier elimination, we obtain after 272 ms "true" together with the answer $(k, l, m) = (1, \frac{11}{10}, -\frac{7}{5})$, $t = 10$, and $(x, y, z) = (10, 0, 0)$.

Example 21. We consider two planes circling in descending spirals in a cylinder above a circle with radius r. Each plane has a given position and constant horizontal and vertical velocity. The question is, whether the two planes are going to crash within a given number of laps.

All distances are given in meters, velocities are given in meters per second. For velocities and positions we take coordinates (C_1, C_2), where C_1 is the position on the circumference, i.e. the distance on the circumference of the circle from a fixed reference point, and C_2 is the height. We consider two models of the situation:

Model 1 In the simpler model planes are modeled as points and a collision occurs if the two points coincide. Table 1 summarizes the data involved in this model (and also in the other one). A collision is then described by the following formula:

$$\varphi_1 \equiv p_2 + tv_2 = q_2 + tw_2 \wedge \bigvee_{k,l=0}^{n} (2\pi rk \leq tv_1 < 2\pi r(k + 1) \wedge$$
$$2\pi rl \leq tw_1 < 2\pi r(l + 1) \wedge p_1 + tv_1 - 2\pi rk = q_1 + tw_1 - 2\pi rl).$$

Model 2 In the second, more realistic, model we prescribe, in addition, horizontal and vertical safety distances, cf. Table 2. A "collision" occurs, if the

Table 1. Data in both models for plane collision; k, l, and n are not first-order variables but concrete integers.

t	time
r	radius of the circle
(p_1, p_2)	starting position of plane 1
(q_1, q_2)	starting position of plane 2
(v_1, v_2)	velocity of plane 1
(w_1, w_2)	velocity of plane 2
k	number of complete laps of plane 1
l	number of complete laps of plane 2
n	a bound on the number of complete laps of both planes

Table 2. Additional data for Model 2.

w	horizontal safety distance
h	vertical safety distance

points representing the planes violate both safety distances. The description of a collision then reads as follows:

$$\varphi_2 \equiv -h \leq p_2 + tv_2 - (q_2 + tw_2) \leq h \wedge$$
$$\bigvee_{k,l=0}^{n} \left(2\pi rk \leq tv_1 < 2\pi r(k+1) \wedge 2\pi rl \leq tw_1 < 2\pi r(l+1) \wedge \right.$$
$$\left. -w \leq p_1 + tv_1 - 2\pi rk - (q_1 + tw_1 - 2\pi rl) \leq w \right).$$

One may be tempted to replace the disjunctions by an unbounded existential quantification $\exists k \exists l$. This is, however, not possible in out context, because these quantifiers would range over integers. In principle mixed real/integer elimination is possible, cf. [Wei90, Wei97a].

For obtaining a well-formed input formula we have to fix n to a natural number. Setting $n = 3$, φ_1 has 81 atomic subformulas in 10 parameters plus the symbol π. Our algorithms do not know anything about π. It plays the role of just another parameter.

Eliminating the time by applying generic quantifier elimination to $\psi_1 \equiv \exists t \varphi_1$, we obtain after 1870 ms necessary and sufficient conditions

$$\psi_1'(r, p_1, p_2, q_1, q_2, v_1, v_2, w_1, w_2, \pi)$$

in the other parameters for a collision to take place. This ψ_1' is valid under the obtained ND-condition $v_2 - w_2 \neq 0$. It contains 80 atomic formulas.

Proceeding the same way for the second model, we obtain after 48790 ms a quantifier-free formula

$$\psi_2'(r, p_1, p_2, q_1, q_2, v_1, v_2, w_1, w_2, h, w, \pi)$$

without any ND-condition. This ψ_2 contains 4597 atomic formulas.

Next we fix in φ_1 and φ_2 the following values for the radius of the circle, the starting positions, both velocities of plane 1, the horizontal velocity of plane 2, and the horizontal and vertical safety distances yielding $\varphi_1^*(t, w_2, \pi)$ and $\varphi_2^*(t, w_2, \pi)$, respectively:

$$r = 10000,$$
$$(p_1, p_2) = (0, 9000),$$
$$(q_1, q_2) = (2000, 10000),$$
$$(v_1, v_2) = (100, -3),$$
$$w_1 = 50,$$
$$h = 50,$$
$$w = 500.$$

Applying extended quantifier elimination to $\exists t \varphi_1^*$ we ask for vertical velocities w_2 of plane 2 that will lead to a collision within at most 3 laps of each plane. We obtain after 391 ms two possible conditions on w_2 together with corresponding answers for the collision times:

$$\begin{bmatrix} 10\pi w_2 + 30\pi + w_2 + 28 = 0 \wedge 5\pi - 1 > 0 & t = 400\pi + 40 \\ 5\pi - 1 > 0 \wedge w_2 + 28 = 0 & t = 40 \end{bmatrix}.$$

This means in rounded numerical values that a collision occurs within the first 3 laps if either $w_2 \approx -3.77$ or $w_2 = -28$. In the first case the collision time is $t \approx 1296.64$, in the second case it is $t = 40$.

The corresponding call applied to $\exists t \varphi_2^*$ takes 1683 ms yielding 6 conditions involving w_2 and π together with suitable answers for t. At least two of these cases have conditions γ which cannot hold for the true π. This is automatically found out by deciding $\exists w_2 \exists \pi (3.1 < \pi < 3.2 \wedge \gamma)$ via quantifier elimination, which takes 8.7 s for all 6 cases together.

Finally, we check the danger of a collision for concrete vertical speed ranges: Deciding the formula

$$\exists w_2 \exists \pi \exists t (3.1 < \pi < 3.2 \wedge w_2 \geq -3.8 \wedge \varphi_2^*)$$

by quantifier elimination yields "true" after 799 ms, while for

$$\exists w_2 \exists \pi \exists t (3.1 < \pi < 3.2 \wedge w_2 \geq -3.7 \wedge \varphi_2^*)$$

we obtain "false" after 835 ms. Taking these results together, some vertical speed $-3.8 \leq w_2 < -3.7$ will—modulo the approximate π—lead to a collision during the first 3 laps. The prescription $w_2 \geq -3.7$, in contrast, is collision-safe for this period.

References

[ADY+96] Chaouki T. Abdallah, Peter Dorato, Wei Yang, Richard Liska, and Stanly Steinberg. Applications of quantifier elimination theory to control system design. In *Proceedings of the 4th IEEE Mediterranean Symposium on Control and Automation*, pages 340–345. IEEE, 1996.

[BKOS97] Mark de Berg, Marc van Kreveld, Mark Overmars, and Otfried Schwarzkopf. *Computational Geometry, Algorithms and Applications*. Springer, Berlin, Heidelberg, New York, 1997.

[Buc65] Bruno Buchberger. *Ein Algorithmus zum Auffinden der Basiselemente des Restklassenringes nach einem nulldimensionalen Polynomideal*. Doctoral dissertation, Mathematisches Institut, Universität Innsbruck, Innsbruck, Austria, 1965.

[CH91] George E. Collins and Hoon Hong. Partial cylindrical algebraic decomposition for quantifier elimination. *Journal of Symbolic Computation*, 12(3):299–328, September 1991.

[CJ92] Robert M. Corless and David J. Jeffrey. Well ... it isn't quite that simple. *ACM SIGSAM Bulletin*, 26(3):2–6, August 1992.

[Col90] Alain Colmerauer. Prolog III. *Communications of the ACM*, 33(7):70–90, July 1990.

[DS96] Andreas Dolzmann and Thomas Sturm. Redlog user manual. Technical Report MIP-9616, FMI, Universität Passau, D-94030 Passau, Germany, October 1996. Edition 1.0 for Version 1.0.

[DS97a] Andreas Dolzmann and Thomas Sturm. Guarded expressions in practice. In Wolfgang W. Küchlin, editor, *Proceedings of the 1997 International Symposium on Symbolic and Algebraic Computation (ISSAC 97)*, pages 376–383, New York, July 1997. ACM, ACM Press.

[DS97b] Andreas Dolzmann and Thomas Sturm. Redlog: Computer algebra meets computer logic. *ACM SIGSAM Bulletin*, 31(2):2–9, June 1997.

[DS97c] Andreas Dolzmann and Thomas Sturm. Simplification of quantifier-free formulae over ordered fields. *Journal of Symbolic Computation*, 24(2):209–231, August 1997.

[DSW96] Andreas Dolzmann, Thomas Sturm, and Volker Weispfenning. A new approach for automatic theorem proving in real geometry. Technical Report MIP-9611, FMI, Universität Passau, D-94030 Passau, Germany, May 1996. To appear in the Journal of Automated Reasoning.

[HCJE93] Hoon Hong, George E. Collins, Jeremy R. Johnson, and Mark J. Encarnacion. QEPCAD interactive version 12. Kindly provided to us by Hoon Hong, September 1993.

[HLS97] Hoon Hong, Richard Liska, and Stanly Steinberg. Testing stability by quantifier elimination. *Journal of Symbolic Computation*, 24(2):161–187, August 1997. Special Issue on applications of quantifier elimination.

[Hof96] Christoph M. Hoffmann. How solid is solid modeling. In Ming C. Lin and Dinesh Manocha, editors, *Applied Computational Geometry*, volume 1148 of *Lecture Notes in Computer Science*, pages 1–8, Berlin, Heidelberg, 1996. Springer.

[KM88] Deepak Kapur and Joseph L. Mundy. Wu's method and its application to perspective viewing. *Artificial Intelligence*, 37(1-3):15–36, December 1988. Special volume on geometric reasoning.

[LW93]   Rüdiger Loos and Volker Weispfenning. Applying linear quantifier elimination. *The Computer Journal*, 36(5):450–462, 1993. Special issue on computational quantifier elimination.

[PP97]   Martin Peternell and Helmut Pottmann. Computing rational parametrizations of canal surfaces. *Journal of Symbolic Computation*, 23(2–3):255–266, February–March 1997.

[PS85]   Franco P. Preparata and Michael I. Shamos. *Computational Geometry—An Introduction*. Texts and monographs in computer science. Springer, New York, 1985.

[Ros85]   Jaroslaw R. Rossignac. *Blending and Offsetting Solid Models*. Ph.D. thesis, Department of Electrical Engineering, College of Engineering and Applied Science, University of Rochester, Rochester, New York 14627, July 1985.

[RR86]   Jaroslaw R. Rossignac and Aristides A. G. Requicha. Offsetting operations in solid modelling. *Computer Aided Geometric Design*, 3(2):129–148, August 1986.

[Ser82]   Jean Serra. *Image Analysis and Mathematical Morphology*. Academic Press, New York, 1982.

[SW97]   Thomas Sturm and Volker Weispfenning. Rounding and blending of solids by a real elimination method. In Achim Sydow, editor, *Proceedings of the 15th IMACS World Congress on Scientific Computation, Modelling, and Applied Mathematics (IMACS 97)*, volume 2, pages 727–732, Berlin, August 1997. IMACS, Wissenschaft & Technik Verlag.

[Wan93]   Dongming Wang. An elimination method based on Seidenberg's theory and its applications. In F. Eyssette and A. Galligo, editors, *Computational Algebraic Geometry (MEGA '92)*, pages 301–328. Birkhäuser Verlag, 1993.

[Wei88]   Volker Weispfenning. The complexity of linear problems in fields. *Journal of Symbolic Computation*, 5(1):3–27, February 1988.

[Wei90]   Volker Weispfenning. The complexity of almost linear diophantine problems. *Journal of Symbolic Computation*, 10(5):395–403, November 1990.

[Wei94]   Volker Weispfenning. Parametric linear and quadratic optimization by elimination. Technical Report MIP-9404, FMI, Universität Passau, D-94030 Passau, Germany, April 1994. To appear in the Journal of Symbolic Computation.

[Wei97a] Volker Weispfenning. Complexity and uniformity of elimination in presburger arithmetic. In Wolfgang W. Küchlin, editor, *Proceedings of the 1997 International Symposium on Symbolic and Algebraic Computation (ISSAC 97)*, pages 48–53, New York, July 1997. ACM, ACM Press.

[Wei97b] Volker Weispfenning. Quantifier elimination for real algebra—the quadratic case and beyond. *Applicable Algebra in Engineering Communication and Computing*, 8(2):85–101, February 1997.

[Wei97c] Volker Weispfenning. Simulation and optimization by quantifier elimination. *Journal of Symbolic Computation*, 24(2):189–208, August 1997. Special issue on applications of quantifier elimination.

Probabilistic Verification of Elementary Geometry Statements

Giuseppa Carrá Ferro[1], Giovanni Gallo[1] and Rosario Gennaro[2]

[1] Dep. of Mathematics, University of Catania, Viale Doria 6, 95125 Catania, Italy
[2] IBM T.J. Watson Research Center, Yorktown Heights, NY 10598, U.S.A.

Abstract. In this paper a probabilistic approach to automated theorem proving in elementary geometry is shown. Bounds on the effective Hilbert Nullstellensatz and on the degree of a Ritt characteristic set are used together with Schwartz's probabilistic results on polynomial identities.

1 Introduction

One of the very first principles that we all learn in high school while approaching Euclid's geometry is: "Examples teach us nothing". In reality for most geometry theorems examples provide the bases to conjecture a statement and encourage the learner to look for a rigorous proof of it. In this presentation we address the question: "How much could we trust an example?" A way to answer to such a question quantitatively is to use the language of probability. The question could be hence reformulated in "What is the probability that a property verified for N examples is a theorem?"

In 1980 J.T. Schwartz [19] in a now classical paper answered an analogous question about polynomial identities: "If two polynomials agree on a set of N random points what is the probability that they are identical?" In the same paper applications in theorem proving were discussed and sketched.

Wu's procedure [4, 24, 25, 26, 27] to prove geometric statements provides a quick way to transfer Schwartz's results to the above question. Indeed the truth in all cases and the generic truth in almost all cases of a statement in Wu's formulation turn to be equivalent with the vanishing of a polynomial. If an example verifies a given statement, it corresponds to a zero of a certain polynomial. In presence of a large enough number of such positive examples (or, equivalently, of a large enough number of zeroes of the polynomial) it is possible to estimate the probability that the theorem is true and generically true.

Other approaches, based on different constructive methods in computer algebra, have also been proposed for Elementary Geometry Theorem Proving: for instance, B. Kutzler and S. Stifter [15] and D. Kapur [12] have proposed methods based on Gröbner bases; G. Carrá and G. Gallo [3] have devised a method using the dimension of the underlying algebraic variety; L. Yang and J.Z. Zhang [30], D. Wang [21, 22], D. Kapur, T. Saxena and L. Yang [13] have used elimination techniques and the resultant theory; J.W. Hong [11] has introduced a seminumerical algorithm using an interesting gap theorem and "proof-by-example" techniques, that were partially clarified in [20]. All the methods mentioned use the same basic technique: translate the hypotheses of a geometric statement into a set of multivariate polynomials and study the relationships between the algebraic variety described by them and the hypersurface described by the polynomial that algebraically expresses the thesis.

Closer to the approach followed here in J.Z. Zhang, L. Yang, M.K. Deng [6, 28, 29, 31] it has been proposed to study the algebraic varieties corresponding to hypotheses and thesis constructing a certain number of points on the hypotheses variety (i.e., a certain number N of examples calculated from the degrees of the hypothesis polynomials) and checking if they are zeroes of the thesis by using parallel numerical verification. No quantitative bounds were provided and we do not know of any further improvement to this paper. Our approach, hence, is quite well known, by now, to the community of people interested to algebraic methods for theorem proving. We do not know, however, of any quantitative analysis similar to the one reported here. We use the bounds of Brownawell and Kollar on the exponent of a polynomial in the radical of an ideal [1], [14] and the bounds of Gallo and Mishra [9, 10] on the degree of Wu-Ritt's characteristic sets to give an accurate estimate of the probability that a given statement is a theorem after N randomly chosen positive examples have been observed.

Our results could be summarized as follows. Let $B = 2P(3)^{C+1}$ and let $D = c2^{C^3}3^{C^3}C^{C^2}$ where c is a constant and P, C denote respectively the number of points and the number of circle or straight line constructions in a statement of plane Euclidean Geometry. Let h be the number of independent parameters needed to completely specify an example. Let J be a set of $2B$ (respectively $2D$) elements in an algebraically closed field k. If N examples, obtained taking from J, at random, h elements, verify the statement, then, with probability greater than $1 - 2^{-N}$, the statement is true (respectively generically true).

It is important to observe that in this paper we make use of the effective Hilbert's Nullstellensatz and of Wu's procedure to get a probability estimate but our probabilistic estimate applies once the statement under consideration has been formalized in a set of polynomial equations and the verification of a theorem for each example can be done numerically, provided that exact arithmetic is used.

In order to make this paper self-contained the next three sections present, with no proofs, the bounds in [1], [14], [10] and [19] that are needed to get the main results of this paper. After analyzing with some detail, in a successive section, the translation into an algebraic language of a geometric statement, our main results are stated and proved.

2 Nullstellensatz and Effective Bounds

The following theorem is well known in algebraic geometry and it relates polynomials having the same zeroes.

Theorem 1 (Hilbert's Nullstellensatz). *Let k be an algebraically closed field, let $I = (f_1, \ldots, f_s)$ be an ideal in $k[x_1, \ldots, x_n]$ and let $f \in k[x_1, \ldots, x_n]$. Every* n-tuple $(a_1, \ldots, a_n) \in k^n$ solution of the system $\{f_1 = \ldots = f_s = 0\}$ is a solution of $f = 0$ if and only if $f \in \operatorname{rad}(I)$, which is equivalent to the existence of a positive integer q, such that $f^q \in I$.

Effective upper bounds for the integer q have been studied. The following upper bound is originally due to [1].

Theorem 2. Let $I = (f_1, \ldots, f_s)$ be an ideal in $k[x_1, \ldots, x_n]$, where k is an algebraically closed field, and $\deg(f_i) \leq d$, $1 \leq i \leq s$. If $f \in k[x_1, \ldots, x_n]$ has degree less than d then $f \in \operatorname{rad}(I)$ if and only if there exists $q \in \mathbb{N}$ such that $f^q = \sum_{i=1}^{r} b_i f_i$, where $q \leq (\min\{s, n\} + 1)(n + 2)(d + 1)^{\min\{s,n\}+1}$ and $\deg(b_i f_i) \leq (\min\{s, n\} + 1)(n + 2)(d + 1)^{\min\{s,n\}+2}$ for all i.

The following sharp bound is due to [14]; other sharp bounds in any characteristic are due to [2] and to [7]; let

$$N'(n, d_1, \ldots, d_s) = \begin{array}{ll} d_1 \cdots d_s & if \quad s \leq n \\ d_1 \cdots d_{n-1} d_s & if \ s > n - 1 \\ d_1 + d_s - 1 & if \ s > n = 1. \end{array}$$

Theorem 3. Given $f_1, \ldots, f_s$ and f in $k[x_1, \ldots, x_n]$, assume that f vanishes on all common zeroes of $f_1, \ldots, f_s$ in the algebraic closure of k. Let $d_i = \deg(f_i)$ and assume that none of the d_i is 2. Then one can find $b_1, \ldots, b_s$ in $k[x_1, \ldots, x_n]$ and $q \in \mathbb{N}$ satisfying $f^q = \sum_{i=1}^{r} b_i f_i$, such that $q \leq N'(n, d_1, \ldots, d_s)$ and $\deg(b_i f_i) \leq (1 + \deg(f))N'(n, d_1, \ldots, d_s)$.

By Theorems 2 and 3, if $s \leq n$ we have the following bounds.

Theorem 4. Let $I = (f_1, \ldots, f_s)$ be an ideal in $k[x_1, \ldots, x_n]$, where k is an algebraically closed field and $s \leq n$. Let $d' = \max\{\deg(f_i): i = 1, \ldots, s, 3\}$. If $f \in k[x_1, \ldots, x_n]$ has degree δ then $f \in \operatorname{rad}(I)$ if and only if there exists $q \in \mathbb{N}$ such that $f^q = \sum_{i=1}^{r} b_i f_i$, where $q \leq (d')^s$ and $\deg(b_i f_i) \leq (1 + \delta)(d')^s$ for all i.

3 Wu-Ritt Characteristic Sets of Polynomials and Bounds on the Pseudo-remainder

A full account of Wu-Ritt method and of its theoretical bases is given in [24, 25, 26]. There are many implementations of Wu's algorithm, among them an implementation that through empirical evidence shows the power and reliability of Wu's techniques is described in [4] and in [5]. Our probabilistic version of Wu's prover makes also use of the degree bounds proved in [9] and in [10].

The heart of Wu's procedure is the computation of the so called Wu-Ritt characteristic sets: the concept of a characteristic set was introduced in the late forties by J.F. Ritt [17, 18] in the setting of differential algebra. Wu has showed its importance and usefulness in the contest of commutative algebra.

Let $k[x_1, \ldots, x_n]$ be the ring of polynomials in n variables, with coefficients in a field k. Consider a fixed ordering on the set of variables;

$$x_1 \prec x_2 \prec \cdots \prec x_n.$$

Let $f \in k[x_1, \ldots, x_n]$ be a multivariate polynomial with coefficients in k. For $1 \leq j \leq n$ $\deg_{x_j}(f)$, is defined to be the maximum degree of the indeterminate x_j in f. The *class* of f with respect to a given ordering is defined as follows: if $f \in k$ its class is 0, otherwise it is the minimum integer i such that $f \in k[x_1, \ldots, x_i]$. The *Tdegree of* f is defined to be $\mathrm{Tdeg}(f) = \sum_{j=1}^{n} \deg_{x_j}(f)$.

Given a polynomial f of class j its *initial polynomial*, $\mathrm{In}(f)$, is defined to be the polynomial $I_d(x_1, \ldots, x_{j-1})$ that is the coefficient of the highest power of x_j in f.

Consider two polynomials f and $g \in k[x_1, \ldots, x_n]$, with $\mathrm{class}(f) = j$. Then there exist two polynomials q and r, and an integer α such that the following equation holds:

$$\mathrm{In}(f)^\alpha g = qf + r, \qquad r = \mathrm{prem}(g, f), \tag{1}$$

where $\deg_{x_j}(r) < \deg_{x_j}(f)$ and $\alpha \leq \deg_{x_j}(g) - \deg_{x_j}(f) + 1$.

If α is assumed to be the smallest possible power satisfying this equation then q and r are uniquely determined. r is called the pseudo-remainder of g with respect to f and denoted with $r = \mathrm{prem}(g, f)$. A polynomial g is *reduced* with respect to f if $g = \mathrm{prem}(g, f)$.

A sequence of polynomials $F = \langle f_1, f_2, \ldots, f_r \rangle \subseteq k[x_1, \ldots, x_n]$ is said to be an *ascending set* (or *chain*), if one of the following two conditions holds:

1. $r = 1$ and f_1 is not identically zero;
2. $r > 1$, and $0 < \mathrm{class}(f_1) < \mathrm{class}(f_2) < \cdots < \mathrm{class}(f_r) \leq n$, and each f_i is reduced with respect to the preceding f_j's $(1 \leq j < i)$.

Every ascending set is finite and has at most n elements.

The pseudo-remainder concept can be generalized as follows: consider an ascending set $F = \langle f_1, f_2, \ldots, f_r \rangle \subseteq k[x_1, \ldots, x_n]$, and a polynomial $g \in k[x_1, \ldots, x_n]$. Then there exists a sequence of polynomials (called a pseudo-remainder sequence) $g_0, g_1, \ldots, g_r = g$, such that for each $1 \leq i \leq r$, the following equation holds,

$$(\exists!\, q_i')\, (\exists!\, \alpha_i)\, \left[\mathrm{In}(f_i)^{\alpha_i} g_i = q_i' f_i + g_{i-1} \right]$$

where g_{i-1} is reduced with respect to f_i and α_i assumes the smallest possible power, achievable. Thus, the pseudo-remainder sequence is uniquely determined.

$$\mathrm{In}(f_r)^{\alpha_r} \mathrm{In}(f_{r-1})^{\alpha_{r-1}} \cdots \mathrm{In}(f_1)^{\alpha_1} g = \sum_{i=1}^{r} q_i f_i + g_0. \tag{2}$$

The polynomial $g_0 \in k[x_1, \ldots, x_n]$ is said to be the generalized pseudo-remainder of g with respect to the ascending set F.

We say a polynomial g is reduced with respect to an ascending set F if it is equal to its pseudo-remainder with respect to F.

Given two polynomials f_1 and $f_2 \in k[x_1, \ldots, x_n]$, we say f_1 is of lower rank than f_2, $f_1 \prec f_2$, if either (i) $\mathrm{class}(f_1) < \mathrm{class}(f_2)$, or (ii) $\mathrm{class}(f_1) = \mathrm{class}(f_2) = i$ and $\deg_{x_i}(f_1) < \deg_{x_i}(f_2)$.

Note that there are distinct polynomials f_1 and f_2 that are not comparable under the preceding order. In this case, f_1 and f_2 are said to be of the same rank, $f_1 \sim f_2$.

Given two ascending sets $F = \langle f_1, \ldots, f_r \rangle$ and $G = \langle g_1, \ldots, g_s \rangle$, we say F is of lower rank than G, $F \prec G$, if one of the following two conditions is satisfied

1. There exists an index $i \leq \min\{r, s\}$ such that $\forall 1 \leq j < i$ $f_j \sim g_j$ and $f_i \prec g_i$;
2. $r > s$ and $\forall 1 \leq j \leq s$ $f_j \sim g_j$.

Note that there are distinct ascending sets F and G that are not comparable under the preceding order. In this case $r = s$, and $(\forall 1 \leq j \leq s)$ $f_j \sim g_j$, and F and G are said to be of the same rank, $F \sim G$.

Definition 5 (Characteristic Set). Let I be an ideal in $k[x_1, \ldots, x_n]$. Consider the family of all ascending sets, each of whose components is in I,

$$\mathbf{S}_I = \left\{ F = \langle f_1, \ldots, f_r \rangle \colon F \text{ is an ascending set and } f_i \in I, 1 \leq i \leq r \right\}.$$

A minimal element in $\mathbf{S}_I$ (with respect to the $\prec$ order on ascending sets) is said to be a *characteristic set* of the ideal I.

Definition 6 (Independent Variables and Dimension). A set of variables $x_{i(1)}, \ldots, x_{i(l)}$ are said to be *independent* with respect to an ideal I in $k[x_1, \ldots, x_n]$ if $I \cap k[x_{i(1)}, \ldots, x_{i(l)}] = (0)$ and $I \cap k[x_{i(1)}, \ldots, x_{i(l)}, x_j] \neq (0)$ for each $j \neq i(1), \ldots, i(l)$. If k is algebraically closed $d = \dim I = \max\{ l \colon x_{i(1)}, \ldots, x_{i(l)}$ are independent w.r.t. $I \}$.

We note that if G is a characteristic set of I then $n \geq |G| \geq n - \dim I$. For a given ordering of the variables, a characteristic set of an ideal is not necessarily unique but any two characteristic sets of an ideal must be of the same rank.

The following well-known proposition provides an algebraic characterization of characteristic sets, and motivates their use in checking geometric statements:

Proposition 7. *Let I be an ideal in $k[x_1, \ldots, x_n]$. Then the ascending set $G = \langle g_1, \ldots, g_r \rangle$ is a characteristic set of I if and only if $\mathrm{prem}(f, G) = 0$ for all $f \in I$.*

A bound on the degree of the pseudo-remainder with respect to a characteristic set of a given ideal is needed to use a probability bound. The known results ([9, 10]) are summarized in the following theorem:

Theorem 8. *Let $I = (f_1, \ldots, f_s)$ be an ideal in $k[x_1, \ldots, x_n]$, where k is an arbitrary field, and $\mathrm{Tdeg}(f_i) \leq d$, $1 \leq i \leq s$. Assume that $x_1, \ldots, x_l$ are independent variables with respect to I. Let $r = n - l = n - \dim I$. Then I has a characteristic set $G = \langle g_1, \ldots, g_r \rangle$ with respect to the ordering*

$$x_1 \prec x_2 \prec \cdots \prec x_n,$$

where for all $1 \leq j \leq r$,

1. $\text{class}(g_j) = j + l$ and $\text{Tdeg}(g_j) \leq 4(s+1)(9r)^{2r}d(d+1)^{4r^2} = O\left(s\, d^{O(r^2)}\right)$.

2. $\exists a_{j,1}, \ldots, a_{j,s} \in k[x_1, \ldots, x_n]$ such that $g_j = \sum_{i=1}^{s} a_{j,i}f_i$ for all j and
$\text{Tdeg}(a_{j,i}f_i) \leq 11(s+1)(9r)^{2r}d(d+1)^{4r^2} = O\left(s\, d^{O(r^2)}\right), 1 \leq i \leq s$.

3. If $f \in k[x_1, \ldots, x_n]$ has Tdegree less than δ and f_0 is the generalized pseudo-remainder of f with respect to G then

$$\text{Tdeg}(f_0) \leq (\delta + 1)(4(s+1)(9r)^{2r}d(d+1)^{4r^2} + 1)^r$$

Remark. Since in hypothesis of Theorem 5 the variables $x_1, \ldots, x_l$ are independent with respect to I, then $l \leq \dim I$. On the other hand it is well known by algebraic geometry that the number of polynomials defining an algebraic affine variety is at least $n - \dim I$. So $s \geq n - \dim I \geq n - l = r$.

4 Schwartz's Result

Let J be a finite set of elements from an infinite integral domain F and let $Q(x_1, \ldots, x_n)$ be a polynomial in n variables with coefficients in F.

Schwartz's paper [19] presents algorithms to test polynomial identities together with ancillary fast algorithms for calculating resultants and Sturm sequences. For the results reported in this paper we only need the following:

Theorem 9 (*Corollary 1, p. 702 in [19]*). *Let $J = J_1 = \ldots = J_n$ and let $|J| \geq c\text{Tdeg}(Q(x_1, \ldots, x_n))$. Then if Q is not identically zero, the number of elements of $J_1 \times \ldots \times J_n$ which are zeroes of Q is at most $c^{-1}|J|^n$.*

As an immediate consequence the following holds:

Corollary 10. *Let J such that $|J| \geq 2\text{Tdeg}(Q(x_1, \ldots, x_n))$ then if Q is zero for N random points in $J_1 \times \ldots \times J_n$ then the probability that Q is not identically zero is at most 2^{-N}.*

Remark. Theorem 9 follows from the more general fact shown in [19] that the number of zeroes of the polynomial $Q(x_1, \ldots, x_n)$ in n variables in $J_1 \times \ldots \times J_n$ is bounded by $|J|^{n-1}\text{Tdeg}(Q(x_1, \ldots, x_n))$. So the probability of finding a zero of $Q(x_1, \ldots, x_n)$ in $J_1 \times \ldots \times J_n$ is at most $|J|^{-1}\text{Tdeg}(Q(x_1, \ldots, x_n))$. If $\text{Tdeg}(Q(x_1, \ldots, x_n)) + 1 \leq |J|$ and Q is zero for N random points in $J_1 \times \ldots \times J_n$, then the probability that Q is not identically zero is not greater than $(|J|^{-1}\text{Tdeg}(Q(x_1, \ldots, x_n)))^N$.

5 Truth and Generic Truth of a Geometric Statement

By an *elementary geometry statement*, we mean a formula of the following kind:

$$\left(f_1 = 0 \wedge f_2 = 0 \wedge \cdots \wedge f_s = 0 \right) \Rightarrow \left(g = 0 \right), \tag{3}$$

where the f_i's and g are polynomials in $k[x_1, \ldots, x_n]$, the variables x_i's are assumed to be bound by universal quantification and their ranges are assumed to be the field k, the base field of the underlying geometry. We further assume that the base field is algebraically closed.

The conjunct $\bigwedge_i (f_i = 0)$ is called the *premise* of the geometry statement, and will be assumed to be nontrivial, i.e., the set of points in k^n satisfying the premise is nonempty. The statement $g = 0$ is its *conclusion*.

To prove a statement, then, is to show that a geometric formula is valid. Thus an algebraic statement is true iff every n-tuple $(x_1, \ldots, x_n)$ in k^n, that satisfies the premise, is also satisfying the conclusion. Thus by Hilbert's Nullstellensatz an algebraic statement is true iff $g \in \mathrm{rad}(f_1, \ldots, f_s)$, which is equivalent to show by Theorem 2, that the polynomial $f = g^q - \sum_{i=1}^r b_i f_i$ is identically zero.

One problem with the above algebraic statement is that it does not mention certain *geometric degeneracies* that are implicitly excluded: for instance, when a geometric statement mentions a *triangle*, it is conventionally assumed to mean those non-degenerate triangles whose vertices are non-collinear.

Let $G = \langle g_1, \ldots, g_r \rangle$ be a characteristic set computed from the polynomial in the premise relatively to an ordering that ranks independent variables as the lowest ones. Wu has heuristically observed that such geometric degeneracies correspond, algebraically, to the zeroes of the initials of the polynomials in G. For this reason Wu's prover checks for *generic* true of a geometric statement and also for *universal* true of a geometric statement, and in which degenerate cases the geometric statement is true. Let g_0 the pseudo-remainder of the conclusion g with respect to G. Wu's method requires to check if g_0 is identically zero. In this case, from the pseudo-remainder formula, whenever the premise is true and none of the initial polynomials is zero (i.e. we are not in presence of a degenerate case) the conclusion must be true. Observe that is not generally true the converse: g_0 could be not identically zero and the theorem could be true at least on some irreducible component of the variety determined by the premise.

6 Translating Plane Geometry into Polynomials

To apply the quantitative bounds reported in the previous section we need to bound the degrees of the polynomials f_i's and g in the premise and in the conclusion of a geometric statement. In particular it is important to have such bound in terms of the number of the geometric primitives involved in the property under consideration. To do so we need to restrict a little the kind of geometric statements considered. Any geometric statement essentially describes some property of a finite set of points. We will consider only set of points whose elements are of the following kinds:

- Points arbitrarily chosen in the plane. These points introduce two new independent variables each and no polynomial equation;
- Points that must lie on a line described by two other points previously introduced. These points introduce two new variables and a polynomial equation

of degree 2 and Tdegree at most 6, since it involves three points and its Tdegree is at most equal to the sum of the variables, that appear effectively in the polynomial.

- Points that must lie on a circle described by a center and a point on the circumference. These points introduce two new variables and a polynomial equation of degree 2 and Tdegree at most 12, since it involves three points and its Tdegree is at most equal to twice the sum of variables, that appear effectively in the polynomial.

- Points that must lie on a line through a point previously introduced and either parallel or orthogonal to a line described by two other points previously introduced. These points introduce two new variables and a polynomial equation of degree 2 and Tdegree at most 8, since it involves four points and its Tdegree is at most equal to the sum of the variables, that appear effectively in the polynomial.

- Points that must lie on two circles described by the same center and a point on each circumference, i.e. points that have proportional distances from two points previously introduced. These points introduce two new variables and a polynomial equation of degree 2 and Tdegree at most 12, since it involves three points and its Tdegree is at most equal to twice the sum of variables, that appear effectively in the polynomial.

- Points that must lie on a circle described by a center and a point on the circumference with assigned radius, i.e. points that have distance from a point previously introduced proportional to the distance of other two points previously introduced. These points introduce two new variables and a polynomial equation of degree 2 and Tdegree at most 16, since it involves four points and its Tdegree is at most equal to twice the sum of variables, that appear effectively in the polynomial.

Hence if a statement requires P points and C constructions the premise will be made of at most C polynomials of degree 2 in each variable and in at most $2P$ variables.

We will also consider only conclusion that can be expressed as a point lying on a line or on a circle or on a line either parallel or orthogonal to a line described by two other points previously introduced or points with an assigned distance from another point previously assigned: again these conditions can be expressed with a quadratic polynomial.

Remark. It is important to observe that the above estimate on the degree of the polynomials involved in the algebraic translation of geometric properties is largely conservative. In most cases big savings in the number of variables and/or polynomial equations can be obtained if some care is taking in the formalization. A saving in the number of variables and polynomials involved in a geometric statement, in practice, makes the difference between a theorem that can be realistically proved automatically and one that cannot.

Of course any new arbitrary point in a geometric construction requires the introduction of two new variables. The points introduced with straight lines or circles introduce new variables that must satisfy polynomial equations of degree at most two in each variable and Tdegree at most 16.

7 Probabilistic Estimates of Truth

Suppose that a geometric statement has been completely formalized in algebraic terms (hopefully without involving too many variables). In this section we will present a probabilistic procedure to check its truth and its generic truth.

7.1 Estimation of truth

Let $\{f_1, \ldots, f_s\}$ be the polynomials in the premise and let g be the polynomial in the conclusion. Let $f = g^q - \sum_{i=1}^r b_i f_i$ for some polynomials $b_1, \ldots, b_r$. Let $(x_1, \ldots, x_n)$ be a point corresponding to an example. It must be assumed that $x_1, \ldots, x_n$ have been obtained solving the algebraic system of the premise, i.e. the point $(x_1, \ldots, x_n)$ is a zero of the polynomials in the premise. If the property of the theorem is satisfied for this example, $(x_1, \ldots, x_n)$ is also a zero of the conclusion g. From the Nullstellensatz it follows that $(x_1, \ldots, x_n)$ is also a zero of f.

From the probabilistic point of view the probability p that a theorem is true is equal to the probability $p_{T|_H}$ of finding a zero of the conclusion conditioned by the fact that such a zero must be a zero of the premise. It follows by probability theory that p is equal to the probability $p_{H \cap T}$ of finding a zero of the premise and the conclusion divided by the probability p_H of finding a zero of the premise. An exact estimation of such probability requires some tools from measure theory. An immediate consequence of probability theory is $p = 0$, when the dimension of the set of zeroes of the premise is greater than the dimension of the set of zeroes of the premise and the conclusion.

So a different kind of probabilistic approaches shall be used in this paper.

From the results above the Tdegree of f can be bounded without any actual need to compute explicitly polynomials b_i's.

Suppose that the geometric statement under investigation involves P points and C constructions.

By Section 5 we can always assume that $d \leq 2$, $d' = 3$ and $\delta \leq 2$. Furthermore $2P \geq n \geq s = C \geq r$.

From Theorem 2 Tdegree of g^q is at most $2^4[(s+1)(n+2)(d+1)^{s+1}] = 2^4[(s+1)(n+2)(3)^{s+1}]$ and Tdegree of f is at most

$$\max(2^4[(s+1)(n+2)(d+1)^{s+1}], n[(s+1)(n+2)(d+1)^{s+1} + 2] =$$

$$\max(2^4[(s+1)(n+2)(3)^{s+1}], n[(s+1)(n+2)(3)^{s+1} + 2]$$

i.e. Tdegree of f is at most $A =$

$$\max(2^4[(C+1)(2P+2)(3)^{C+1}], 2P[(C+1)(2P+2)(3)^{C+1} + 2] =$$

$$\max(2^5[(C+1)(P+1)(3)^{C+1}], 4P[(C+1)(P+1)(3)^{C+1} + 1]$$

i.e. A is $O(cP^2C3^C)$, where c is a constant.

From Theorem 4 Tdegree of g^q is at most $2^4[3^s]$ and Tdegree of f is at most

$$\max(2^4[3^s], n[(1+\delta)(3)^s]) = \max(2^4[3^s], n[(3)^{s+1}])$$

i.e. Tdegree of f is at most $B = \max(2^4[3^C], 2P[(3)^{C+1}])$

i.e. B is $O(cP3^C)$, where c is a constant.

Hence from Theorem 9 take a set J of $2B$ elements in k and choosing at random in J the parameters necessary to completely specify an example one can successively construct N examples.

A possible way to check if f is identically zero is to choose randomly $x_1, \ldots, x_n$ in the range J in order to completely specify an example, i.e. to evaluate $f_1, \ldots, f_s, g$ at this set of values. If $f_i = 0$ for all $i = 1, \ldots, s$ and $g \neq 0$ then the theorem is not true and $f \neq 0$. If $f_1 = \ldots = f_s = g = 0$, then $f = 0$ and the theorem is true. If this is the case, then the probability of the error f not identically zero is bounded by $p = \frac{1}{|J|}\mathrm{Tdeg}(f)$. Therefore, if we choose $|J| = \frac{1}{\epsilon}\mathrm{Tdeg}(f)$, then the algorithm returns the correct answer with probability at least $1 - \epsilon$, for any $0 < \epsilon < 1$.

7.2   Estimation of generic truth

Let $G = \langle g_1, \ldots, g_r \rangle$ be a characteristic set computed from the polynomials in the premise relatively to an ordering that ranks independent variables as the lowest ones. Let g_0 be the pseudo-remainder of the conclusion g with respect to G. Let $(x_1, \ldots, x_n)$ be a point corresponding to an example. It must be assumed that $x_1, \ldots, x_{n-r}$ parameters have been chosen arbitrarily and the other r have been obtained solving the algebraic system of the premise. This can be done using exact arithmetic (algebraic numbers) as the polynomial equations describing the hypothesis have degree at most 2.

The point $(x_1, \ldots, x_n)$ is a zero of the polynomials in the premise and hence also of the g_i's. If the property of the theorem is satisfied for this example, $(x_1, \ldots, x_n)$ is also a zero of the conclusion g. From the pseudo-remainder formula follows that $(x_1, \ldots, x_n)$ is also a zero of g_0.

From the results above Tdegree of g_0 can be bounded without any actual need to compute explicitly the polynomials g_i's.

Suppose that the geometric statement under investigation involves P points and C constructions.

By Section 5 we can always assume that $d \leq 16$ and $\delta \leq 16$. Furthermore $s = C \geq r$. From Theorem 8 Tdegree of g_0 is at most

$$(\delta + 1)(4(s+1)(9r)^{2r}d(d+1)^{4r^2} + 1)^r \leq 17(4(s+1)(9r)^{2r}(16)(17)^{4r^2} + 1)^r =$$

$$17((2)^6(s+1)(9r)^{2r})(17)^{4r^2} + 1)^r = 17((2)^6(3^{4r})(r^{2r})(s+1)(17)^{4r^2} + 1)^r \leq$$

$$17(2^6(3^{4s})(s^{2s})(s+1)(17)^{4s^2} + 1)^s \leq 17(2^6(3)^{4s}(s^{2s})(s+1)(2)^{4s^2}(3)^{8s^2} + 1)^s =$$

$$17(((2)^{4s^2+6})(3)^{8s^2+4s}(s^{2s})(s+1) + 1)^s =$$

$$17((2)^{4C^2+6}(3)^{8C^2+4C}(C^{2C})(C+1)+1)^C$$

i.e. Tdegree of g_0 is at most $D = 17((2)^{4C^2+6}(3)^{8C^2+4C})(C^{2C})(C+1)+1)^C = O(c(2)^{C^3}(3)^{C^3}C^{C^2})$, where c is a constant.

Hence from Theorem 9 take a set J of $2D$ elements in k and choosing at random in J the parameters necessary to completely specify an example one can successively construct N examples.

A possible way to check if g_0 is identically zero is to choose randomly in J the parameters necessary to completely specify an example. Then one chooses the dependent variables in a manner to satisfy the hypothesis $f_1 = \ldots = f_s = 0$ at this set of values. Then the conclusion g is evaluated on these points.

If $f_1 = \ldots = f_s = g = 0$, then $g_1 = \ldots = g_r = 0$ and $g_0 = 0$ and the theorem is generically true. If this is the case, then the probability of the error g_0 not identically zero is bounded by $p = \frac{1}{|J|}\text{Tdeg}(g_0)$. Therefore, if we choose $|J| = \frac{1}{\epsilon}\text{Tdeg}(g_0)$, then the algorithm returns the correct answer with probability at least $1 - \epsilon$, for any $0 < \epsilon < 1$.

What if $f_i = 0$ for all i and $g \neq 0$ then the theorem could still be generically true, as the pseudo-remainder g_0 could be zero because the initials are zero. That is it may be that we picked a degenerate example. In this case in order to verify if g_0 is null on this point, a computation of the characteristic set is neeeded and the pseudo-division of g has to be carried out. However notice that such computations can be carried out (*after*) having instantiated the parameters. This involves a gain in efficiency as we are now computing the characteristic set of a zero-dimensional ideal (there are polynomial time algorithms for this purpose). Also the pseudo-division is carried out on a set of polynomials with much smaller number of variables.

As it is suggested in [16], p.130 this algorithm is not very practical, because $|J|$ may be very large, in which case many computational problems arise. As it is suggested still in [16] one simple solution is to evaluate the polynomials $f_1, \ldots, f_s, g$ for several sets of values instead of evaluating them for only one set of values. However we need to repeat the steps only if $f_1 = \ldots = f_s = g = 0$. The algorithm works as follows. Repeat the following steps N times (where N is a fixed number).

1. Choose randomly $x_1, \ldots, x_{n-r}$ in the range J.
2. Evaluate $x_{n-r+1}, \ldots, x_n$ such that $f_1 = \ldots = f_s = 0$.
3. Evaluate g on $x_1, \ldots, x_n$.
4. If $g(x_1, \ldots, x_n) = 0$ then return $g_0 = 0$.
5. If $g(x_1, \ldots, x_n) \neq 0$ then do the following:
   (a) Compute the characteristic set $g'_1, \ldots, g'_r$ of the polynomials f_i's with the first $n - r$ variables instantiated to $x_1, \ldots, x_{n-r}$.
   (b) Let g' be the polynomial g with the first $n - r$ variables instantiated to $x_1, \ldots, x_{n-r}$. Compute the pseudo-remainder g'_0 obtained by pseudo-division of g' by the characteristic set $g'_1, \ldots, g'_r$.
   (c) If $g'_0 = 0$ then return $g_0 = 0$ otherwise return $g_0 \neq 0$.

Now the probability that g_0 evaluates to zero, all of N times, even though $g_0 \neq 0$, is at most c^{-N}, provided that $|J| \geq c\mathrm{Tdeg}(g_0)$. By choosing $c = 2$ and N equal to the minimum integer greater than or equal to $\log(\frac{1}{\epsilon})$, we can ensure that the algorithm is correct with the probability at least $1 - \epsilon$. If g_0 has M terms, then the running time of the algorithm is bounded by $O(Nn(M + \mathrm{Tdeg}(g_0))$. In order to ensure that the probability of the error is $o(1)$ then $N = \Theta(\log(n))$, so the running time of the algorithm is $O((M + \deg(g_0))n\log(n))$ and the probability of correctness becomes $1 - (\frac{1}{n^{\Theta(1)}})$.

Provided that exact arithmetic is used one can verify for each of these examples if it satisfies the theorem. If some example fails the theorem is false and we have found a counter-example, otherwise, if all N examples satisfy the statement, then the statement is generically true with probability at least $1 - 2^{-N}$. Observe that no actual calculation of characteristic sets or of pseudo-remainder is needed unless the randomly chosen example does not satisfy the thesis. This means that either the theorem is false and a counterexample has been found or that the theorem is generically true and a degenerate example has been picked. In any case such computation is carried out only over r variables and on a zero-dimensional ideal, thus it does not involve the same complexity as the general case. [3]

Of course the computational cost of verifying, using exact arithmetic, N numerical examples could make the proposed method practically infeasible.

8 Conclusions

The probabilistic approach turns out to be interesting, since it is able to show also counter-examples to the statement, if they exist. Unfortunately the bounds D and B are very large and this fact could make the method practically infeasible, also when we decide to use the lower bound of $|J|$ given respectively by $|J| = B + 1$ and $|J| = D + 1$.

On the other hand since the distribution of the probability is binomial it can be approximated by the normal distribution when N is big enough. It is well known in statistics that the hypothesis "The probability that the theorem is true is at least $1 - (|J|^{-1}B)^{N}$" is true up to $99, 8$ per cent whenever $1 - (|J|^{-1}B)^{N} \geq N/N + 3$ and $N > 30$ (respectively "The probability that the theorem is generically true is at least $1 - (|J|^{-1}D)^{N}$" is true up to $99, 8$ per cent whenever $1 - (|J|^{-1}D)^{N} \geq N/N + 3$ and $N > 30$.

So the number of samples necessary for the evaluation of the probability that a statement is true (respectively generic true), can be evaluated a priori once B (respectively D) is known and the lower bound for $|J|$ equal to $B + 1$ (respectively $D + 1$) is used.

---

[3] Heuristically the second case is not very likely to occur, as in general the degenerate examples are few and not likely to show up in a random sample of examples. Algebraically speaking the degenerate examples constitute the algebraic set defined by the zeroes of the initial polynomials. Often such set has dimension much smaller than the universe of all examples. However this is not a rule.

9 Examples

Example 1. Let $A_1 B_1 C_1 D_1$ be a parallelogram and let E_1 be the intersection of the diagonals $A_1 C_1$ and $B_1 D_1$. Then we want to show that $A_1 E_1 = C_1 E_1$.
Let $A_1 = (0,0)$, $B_1 = (u_1, 0)$, $C_1 = (u_2, u_3)$, $D_1 = (x_2, x_1)$ and $E_1 = (x_4, x_3)$.
$h_1 = u_1 \neq 0$, i.e. $B_1 \neq A_1$
$h_2 = u_3 \neq 0$, i.e. C_1 not in $A_1 B_1$
$f_1 = u_1 x_1 - u_1 u_3 = 0$, i.e., $A_1 B_1$ parallel to $D_1 C_1$
$f_2 = u_3 x_2 - (u_2 - u_1)x_1 = 0$, i.e., $D_1 A_1$ parallel to $C_1 B_1$
$f_3 = x_1 x_4 - (x_2 - u_1)x_3 - u_1 x_1 = 0$, i.e., E_1 is on $B_1 D_1$
$f_4 = u_3 x_4 - u_2 x_3 = 0$, i.e., E_1 is on $A_1 C_1$
$\bigwedge_{i=1,2}(h_i \neq 0)$ are the *geometric non-degenerate conditions*
$\bigwedge_{i=1,\ldots,4}(f_i = 0)$ is the *premise*
$g = 2u_2 x_4 + 2u_2 x_3 - u_3^2 - u_2^2 = 0$, i.e., $A_1 E_1 = C_1 E_1$ is the *conclusion*.
$n = 7$, $l = 3$, $r = 4$, $s = 4$, $P = 5$, $C = 4$
$\deg(f_1) = \ldots = \deg(f_4) = \deg(g) = 2$, while $\mathrm{Tdeg}(f_1) = \mathrm{Tdeg}(f_4) = 3$, $\mathrm{Tdeg}(f_2) = \mathrm{Tdeg}(f_3) = 5$ and $\mathrm{Tdeg}(g) = 6$
From Theorem 2 $\mathrm{Tdeg}(g^q) \leq 2^4[(5)(9)(3)^5 = (2)^4(3)^7 5$ and
$\mathrm{Tdeg}(f) \leq \max(2^5[(5)(6)(3)^5], 40[(5)(6)(3)^5 + 1]) = (40)[(10)(3)^6 + 1]$,
i.e. $A = (40)[(10)(3)^6 + 1]$.
From Theorem 4 $\mathrm{Tdeg}(g^q) \leq (2)^4(3)^4$ and
$\mathrm{Tdeg}(f) \leq \max((2)^4(3)^4, (10)(3)^5 = (10)(3)^5$
i.e. $B = (10)(3)^5$.
$\mathrm{Tdeg}(g_0) \leq 7(4(5)(36)^8(5)(6)^{64} + 1)^4 = 7((2)^{82}(3)^{80}(5)^2 + 1)^4 = D$.
If $|J| = 2B$ (respectively $|J| = 2D$), then the hypothesis "The probability that the theorem is true is at least $1 - (2^{-N})$" (respectively "The probability that the theorem is generically true is at least $1 - (2^{-N})$" is true up to $99,8$ per cent whenever $N > 30$.
By $f_1 = u_1 x_1 - u_1 u_3 = 0$ and $h_1 = u_1 \neq 0$ it follows $x_1 = u_3$.
By $f_2 = u_3 x_2 - (u_2 - u_1)x_1 = 0$ and $h_2 = u_3 \neq 0$ it follows $x_2 = \frac{(u_2 - u_1)}{u_3} x_1 = u_2 - u_1$.
By $f_3 = u_3 x_4 - (u_2 - 2u_1)x_3 - u_1 u_3 = 0$ it follows $x_4 = \frac{(u_2 - 2u_1)}{u_3} x_3 + u_1$
By $f_4 = u_3 x_4 - u_2 x_3 = 0$ it follows $(u_2 - 2u_1)x_3 + u_1 u_3 - u_2 x_3 = -2u_1 x_3 + u_1 u_3 = 0$, i.e. $x_3 = \frac{u_3}{2}$ and then $x_4 = \frac{u_2}{2}$
By substituting the values of x_3 and x_4 we have $g = 2u_2 x_4 + 2u_2 x_3 - u_3^2 - u_2^2 = u_2^2 + u_3^2 - u_3^2 - u_2^2 = 0$
The theorem is true. In fact let J be the set of the first $2B$ even positive integers, i.e., $|J| = 2B$. Let's choose $u_1, u_2, u_3 \in J$ with $u_2 - u_1 \in J$ and $x_1 = u_3$, $x_2 = u_2 - u_1$, $x_3 = \frac{u_3}{2}$, $x_4 = \frac{u_2}{2}$. The corresponding (u_2, u_3, x_3, x_4) is a solution of the equation $g = 0$.

Example 2. Let $A_1 B_1 C_1$ be a triangle, such that $A_1 C_1 = B_1 C_1$. Then we want to show that $A_1 B_1 = A_1 C_1$.
Let $A_1 = (0,0)$, $B_1 = (u_1, 0)$, $C_1 = (x_2, x_1)$.
$h_1 = u_1 \neq 0$, i.e. $B_1 \neq A_1$
$h_2 = u_3 \neq 0$, i.e. C_1 not in $A_1 B_1$

$\bigwedge_{i=1,2}(h_i \neq 0)$ are the *geometric non-degenerate conditions*
$f_1 = -2u_1x_2 + u_1^2 = 0$, i.e., $A_1C_1 = B_1C_1$ is the *premise*.
$g = x_2^2 + x_1^2 - u_1^2 = 0$, i.e., $A_1B_1 = A_1C_1$ is the *conclusion*.
$n = 2, l = 2, r = 1, s = 1, P = 3, C = 1$
$\deg(f_1) = \deg(g) = 2$, while $\mathrm{Tdeg}(f_1) = 3, \mathrm{Tdeg}(g) = 6$
From Theorem 2 $\mathrm{Tdeg}(g^q) \leq 2^4[(2)(4)(3)^2 = (2)^7(3)^2$ and
   $\mathrm{Tdeg}(f) \leq \max(2^7[(3)^2], 3[(2)(4)(3)^2 + 2]) = (2)^7(3)^2$,
   i.e. $A = (2)^7(3)^2$.
From Theorem 4 $\mathrm{Tdeg}(g^q) \leq (2)^4(3)$ and
   $\mathrm{Tdeg}(f) \leq \max((2)^4(3), (3)(3)^2 = (2)^4(3)$
   i.e. $B = (2)^4(3)$.
   $\mathrm{Tdeg}(g_0) \leq 6(4(2)(18)^4(5)(6)^{16} + 1)^2 = 6((2)^{23}(3)(24)^{24}(5) + 1)^4 = D$.
The theorem is false. In fact if $J = \{1, \ldots, 96\}$, i.e. $|J| = 2B$, and we choose
$u_1, x_2 \in J$ with $-2x_2 + u_1 = 0$ and x_1 with $x_1^2 - 3x_2^2 = 0$, then (u_1, x_1, x_2) is a
solution of the equation $g = 0$, but such values of x_1 are not in J.

References

1. D. Brownawell: Bounds for the Degree in the Nullstellensatz. Ann. Math. **126** (1987), 577–591
2. L. Caniglia, A. Galligo, J. Heintz: Borne simple exponentielle pour le degrés dans le théoréme des zéros sur un corps de caractéristique quelconque. C.R.A.S. Paris **307** (1988), 255–258
3. G. Carrá, G. Gallo: A Procedure to Prove Geometrical Statements. Proc. AAECC-5 (Menorca) Lecture Notes in Computer Science **365** (1986), 141–150
4. S.-C. Chou: Mechanical Geometry Theorem Proving. D. Reidel Publishing Company, Kluwer Academic Publishers Group, Dordrecht Boston (1988)
5. S.-C. Chou: Proving and Discovering Theorems in Elementary Geometries Using Wu's Method. PhD. Thesis, Department of Mathematics, University of Texas at Austin (1985)
6. M.K. Deng: The Parallel Numerical Method of Proving the Constructive Geometric Theorems. Chinese Sci. Bull. **34** (1989), 1066–1070
7. N. Fitchas, A. Galligo: Nulstellensatz effective et conjecture de Serre (Theoreme de Quillen-Suslin) pour le calcul formel. Math. Nachr. **149** (1990), 231–253
8. G. Gallo: La Dimostrazione Automatica in Geometria e Questioni di Complessita' Correlate. Tesi di Dottorato di Ricerca, Catania (1989)
9. G. Gallo, B. Mishra: Efficient Algorithms and Bounds for Wu-Ritt Characteristic Sets. Progress in Math. **94**, Proc. of MEGA 90, Birkhäuser (1990), 119–142
10. G. Gallo, B. Mishra: Wu-Ritt Characteristic Sets and Their Complexity. DIMACS series in Discrete Mathematics and Theoretical Computer Science **6** (1991), 111–136
11. J.W. Hong: Proving by Example and Gap Theorem. 27th Annual Symposium on Foundations of Computer Science, IEEE Computer Society Press (1986), 107–116
12. D. Kapur: Using Gröbner Bases to Reason about Geometry Problems. Journal of Symbolic Computation **2** (1986), 399–408
13. D. Kapur, T. Saxena, L. Yang: Algebraic and Geometric Reasoning using Dixon Resultants. Proc. ISSAC-94 (Oxford, July 20-22, 1994), 97–107

14. J. Kollar: Sharp Effective Nullstellensatz. J. Amer. Math. Soc. **1** (1988), 963–975
15. B. Kutzler, S. Stifter: Automated Geometry Theorem Proving Using Buchberger's Algorithm. Proceedings of the 1986 Symposium on Symbolic and Algebraic Computation (Waterloo, July 21-23, 1986), 209–214
16. B. Mishra: Algorithmic Algebra. Springer Verlag, New York (1993)
17. J.F. Ritt: Differential Equations from an Algebraic Standpoint. AMS Colloquium Publications **XIV**, New York (1932)
18. J.F. Ritt: Differential Algebra. AMS Colloquium Publications **XXXII**, New York (1950)
19. J.T. Schwartz: Fast Probabilistic Algorithms for Verification of Polynomial Identities. Journal of ACM **27** (1980), 701–717
20. D. Wang: Proving-by-Examples Method and Inclusion of Varieties. Kexue Tongbao **33**(4) (1988), 2015–2018
21. D. Wang: Elimination Procedures for Mechanical Theorem Proving in Geometry. Ann. Math. Artif. Intell. **3** (1995), 1–24
22. D. Wang: Geometry Machines: From AI to SMC. Proc. AISMC-3 (Steyr, Sept. 23–25) Lecture Notes in Computer Science **1138** (1996), 213–239
23. F. Winkler: A Geometric Decision Algorithm Based on the Gröbner Bases Algorithm. Proc. ISSAC-88 (Rome) Lecture Notes in Computer Science **358** (1988), 356–363
24. W-T. Wu: On the Decision Problem and the Mechanization of Theorem Proving in Elementary Geometry. Scientia Sinica **21** (1978), 157–179
25. W-T. Wu: Basic Principles of Mechanical Theorem Proving in Elementary Geometries. Journal of Sys. Sci. and Math. Sci. **4** (1984), 207–235; also in Journal of Automated Reasoning **4** (1986), 221–252
26. W-T. Wu: Some Recent Advances in Mechanical Theorem-Proving of Geometries. Automated Theorem Proving: After 25 Years, Contemporary Mathematics. American Mathematical Society. **29** (1984), 235–242
27. W-T. Wu: Mechanical Theorem Proving in Geometries: Basic Principles (translated from the Chinese by Xiaofan Jin and Dongming Wang). Springer-Verlag, Wien New York (1994)
28. L. Yang: A New Method of Automated Theorem Proving. The Mathematical Revolution Inspired by Computing. Oxford Univ. Press, New York (1991), 115–126
29. L. Yang, J.Z. Zhang, C.Z. Li: A Prover for Parallel Numerical Verification of a Class of Constructive Geometry Theorems. Proc. IWMM 92 (Beijing, July 16-18, 1992), 244–250
30. L. Yang, J.Z. Zhang: Searching Dependency between Algebraic Equations: An Algorithm Applied to Automated Reasoning. Artificial Intelligence in Mathematics, Oxford University Press, Oxford (1994), 147–156
31. J.Z. Zhang, L. Yang, M.K. Deng: The Parallel Numerical Methods in Mechanical Theorem Proving. Theoret. Comput. Sci. **74** (1990), 253–271

Computational Synthetic Geometry
with Clifford Algebra

Timothy F. Havel

Biological Chemistry and Molecular Pharmacology,
Harvard Medical School, Boston, MA 02115

Abstract. *Computational synthetic geometry* is an approach to solving
geometric problems on a computer, in which the quantities appearing in
the equations are all covariant under the corresponding group of transformations, and hence possess intrinsic geometric meanings. The natural
covariant algebra of metric vector spaces is called *Clifford algebra*, and it
includes Gibbs' vector algebra as a special case. As a preliminary essay
demonstrating that one can develop practical computer programs based
on this approach for solving problems in Euclidean geometry, we have implemented a MAPLE package, called *Gibbs*, for the elementary expansion
and simplification of expressions in Gibbs' *abstract* vector algebra. We
also show how to translate any origin-independent scalar-valued expression in the algebra into an element of the corresponding invariant ring,
which we have christened the *Cayley-Menger ring*. Finally, we illustrate
the overall approach by using it to derive a new kinematic parametrization of the conformation space of the molecule *cyclohexane*.

1   Background and Introduction

Clifford algebra provides a natural, coordinate-free language for metric geometry.
It has numerous applications in the physical sciences, and includes Gibbs' vector
algebra as a special case. A variety of symbolic algebra packages for performing
calculations in general Clifford algebras are available, but all of these require
one to choose a basis and expand everything in terms of coordinates (several
examples may be found in [1]). A closely related geometric language is obtained
by imposing a metric structure on the exterior calculus, and a coordinate-free
package for exterior calculus, EXCALC, has been developed for the REDUCE
system, but it too relies upon a basis in order to specify the metric tensor to
be used [2]. Even in calculations within reach of pencil and paper, expansion
in terms of coordinates often renders the intermediate expressions prohibitively
complicated. Such expansions may be needed for numerical calculations, but
they should be optional rather than obligatory. Unfortunately, packages that
can operate directly in these higher, abstract algebras are relatively difficult to
implement.

The remainder of this section will briefly introduce Clifford algebra, its subal-
gebra of invariant geometric quantities, and explain how these apply to everyday
Euclidean geometry. These results are classical, but they are not widely known.
The following section will demonstrate the possibility and practical advantages

to be had from working entirely at abstract level, by describing an implementation of Gibbs' vector algebra in the MAPLE symbolic language [3]. This package, called *Gibbs*, was written to derive a system of equations characterizing the solutions to a distance constraint problem that arises in the study of molecular conformation, which existing coordinate-based implementations could not handle [4, 5]. The third and final section will illustrate the overall approach by deriving a new, closed-form solution to a particular instance of this problem, namely the molecule *cyclohexane*.

This approach falls within the scope of *computational synthetic geometry*, as advocated by Sturmfels and others [6, 7]. Most work on this subject has dealt with projective geometry, using the Grassmann algebra and its ring of invariants, the *bracket ring* [8, 9]. An analogous theory for metric vector spaces dates back to Hermann Weyl's work on the classical groups, where it was shown that the invariant ring of the orthogonal group $O(p, q)$ is generated by the inner products of vectors [10]; in the case of $SO(p, q)$ one must include the oriented volumes spanned by all ordered n-tuples as well ($n = p + q$). The vanishing of all $(n + 1) \times (n + 1)$ principal minors of the metric tensor (matrix of inner products) then generates the *syzygy ideal* consisting of all algebraic relations among these invariants. For a modern proof of this fact, see [11], and for a study of the underlying combinatorial structure of these algebraic dependencies, see [12]. It seems appropriate to call the corresponding coordinate ring the *Weyl ring*.

The natural covariant algebra of metric vector spaces is *Clifford algebra* [13, 14], while the corresponding geometries are (projectively viewed) the *Cayley-Klein* geometries, which include the classical elliptic and hyperbolic geometries. One can construct a Clifford algebra from a Grassmann algebra together with a (not necessarily definite) vector inner product, by defining the *Clifford product* of a k-vector with an ℓ-vector according to the formula

$$(\mathbf{x}_k \wedge \ldots \wedge \mathbf{x}_1)(\mathbf{y}_1 \wedge \ldots \wedge \mathbf{y}_\ell) =$$

$$\sum_{m=0}^{\min(k,\ell)} \sum_{\substack{(i_1,\ldots,i_k) \\ \in S(m,k)}} \sum_{\substack{(j_1,\ldots,j_\ell) \\ \in S(m,\ell)}} \sigma(i_1,\ldots,i_k)\sigma(j_1,\ldots,j_\ell) \tag{1}$$

$$\det \begin{pmatrix} \mathbf{x}_{i_1} \cdot \mathbf{y}_{j_1} & \cdots & \mathbf{x}_{i_1} \cdot \mathbf{y}_{j_m} \\ \vdots & \ddots & \vdots \\ \mathbf{x}_{i_m} \cdot \mathbf{y}_{j_1} & \cdots & \mathbf{x}_{i_m} \cdot \mathbf{y}_{j_m} \end{pmatrix} (\mathbf{x}_{i_k} \wedge \ldots \wedge \mathbf{x}_{i_{k-m+1}} \wedge \mathbf{y}_{j_{\ell-m+1}} \wedge \ldots \wedge \mathbf{y}_{j_\ell})$$

(where $S(m, k)$ denotes the set of "shuffles", i.e. permutations of $(1, \ldots, k)$ that preserve the order in the first m and last $k - m$ places, and $\sigma(i_1, \ldots, i_k)$ is the sign of the permutation). Note this product is indicated by the juxtaposition of symbols, with no intervening operator symbol.

The Clifford product (left-hand side of eq. (1)) enables us to express geometric relations very concisely, while expansion in terms of inner and outer products (right-hand side of eq. (1)) enables us to spell out these relations in greater detail. Clifford algebra also gives us a coordinate-free language for describing geometric operations in metric vector spaces. For example, *reflections* are represented in the Clifford algebra by

$$-\mathbf{u}^2(\mathbf{u}\,\mathbf{x}\,\mathbf{u}) \;=\; \mathbf{x} - 2\,\mathbf{u}\cdot\mathbf{x} \qquad (\mathbf{u}^2 = \pm 1)\,. \tag{2}$$

Since the product of any two reflections is a rotation and (by the Cartan-Dieudonné theorem) any rotation is a product of an even number of reflections, this shows that we can represent rotations by products of the form $\mathbf{R}^\dagger\,\mathbf{x}\,\mathbf{R}$, where $\mathbf{R} = \mathbf{r}_1\cdots\mathbf{r}_{2k}$ is the product of an even number of unit vectors and $\mathbf{R}^\dagger = \mathbf{r}_{2k}\cdots\mathbf{r}_1$ is its *reverse*.

The situation for Euclidean space, and metric affine spaces more generally, differs in surprising ways from the above. The problem is that translations of the Cartesian coordinates are not *linear* transformations, and the usual trick of linearizing them by going over to homogeneous coordinates does *not* result in a transvection invariant inner product. Instead, the squared *distances* $D(i,j)$ among Euclidean points constitute a complete system of invariants for Euclidean geometry. The squared distances among a set of n-dimensional Euclidean points, in turn, are equal to the inner products of a set of null vectors in a *Minkowski* space two dimensions higher, and which have been normalized so that their inner product with an additional fixed null vector is unity. This additional null vector corresponds to the classical "point at infinity", and its stablizer in $O(1, n+1)$ is the n-dimensional Euclidean group. In other words, this construction not only enables us to linearize the translations, but to actually represent them by Lorentz transformations in $O(1, n+1)$ [15]. For an account, see [16, 17] and references therein.

The minors of the corresponding metric tensor that include the row/column of the point at infinity are known as *Cayley-Menger bideterminants*, i.e.

$$D(i_1,\ldots,i_k;j_1,\ldots,j_k) \;=\; \frac{(-1)^k}{2^{k-1}}\det\begin{bmatrix} 0 & 1 & \cdots & 1 \\ 1 & D(i_1,j_1) & \cdots & D(i_1,j_k) \\ \vdots & \vdots & \ddots & \vdots \\ 1 & D(i_k,j_1) & \cdots & D(i_k,j_k) \end{bmatrix}. \tag{3}$$

We also abbreviate $D(i_1,\ldots,i_k;i_1,\ldots,i_k)$ as $D(i_1,\ldots,i_k)$, and call it a Cayley-Menger determinant. The constant $2(-1/2)^k$ ensures that these determinants are equal to $(k!)^2$ times the squared volume of the simplex spanned by the points, and in particular that $D(i_1,i_2)$ is just the squared distance between the points. Given that the $(n+1)\times(n+1)$ principal minors of the metric tensor generate the syzygy ideal of metric vector spaces, it is not surprising that the $(n+1)$-point Cayley-Menger bideterminants generate the syzygy ideal of Euclidean spaces [18]. We shall call the corresponding coordinate ring the *Cayley-Menger ring*.

Figure 1 illustrates the hierarchy of methods that can be used to study the classical geometries. The point we wish to make is that each level of these parallel hierarchies has both advantages and disadvantages, so it is important to be able to scale them at will. The advantages to working at the bottom-most levels are that existing computer algebra programs are best at doing polynomial algebra,

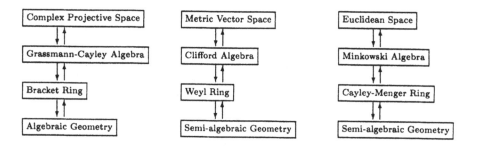

Fig. 1. The Hierarchies of the Classical Geometries

and that this algebra is widely known and understood. The disadvantages are, first, that the use of coordinates obscures the geometry, and second, the size of the expressions involved tends to be much greater when everything is expanded in terms of a basis. We shall now describe a package for the elementary expansion and simplification of expressions in *abstract* vector algebra, which generally produces simpler expressions.

2 The Gibbs Package

The *Gibbs* package performs elementary expansion and simplification of expressions in Gibbs' vector algebra, wherein each "vector" is represented by a single abstract symbol rather than by its 3-D coordinates. It also contains procedures whereby one can go from the abstract algebra level down to the Cayley-Menger ring, and thence to polynomials in the Cartesian coordinates. This enables MAPLE's builtin polynomial manipulation procedures to be used to derive additional results, or to solve the resulting equations by numerical methods when symbolic solutions prove too difficult.

To understand how this package relates to the forgoing, it is necessary to understand how Gibbs' vector algebra relates to 3-D Euclidean Clifford algebra. The inner product is, of course, the same, whereas the outer product is related to the cross product by duality, i.e.

$$\mathbf{x} \times \mathbf{y} \;=\; (\mathbf{x} \wedge \mathbf{y})\iota^{\dagger} , \tag{4}$$

where $\iota = \mathbf{e}_1\mathbf{e}_2\mathbf{e}_3$ for any orthonormal basis is the *unit pseudo-scalar* of the algebra. Similarly, the triple product of vectors is the dual of their outer product:

$$\mathbf{x} \cdot (\mathbf{y} \times \mathbf{z}) \;=\; (\mathbf{x} \wedge \mathbf{y} \wedge \mathbf{z})\iota^{\dagger} \tag{5}$$

Thus one can do pretty much everything in Gibbs' vector algebra that one could do in the Clifford algebra of three dimensions, although the latter generally provides a more compact notation (and avoids the use of vectors whose physical dimensions are area!).

Our decision to implement Gibbs' vector algebra rather than the 3-D Clifford algebra was simply a matter of expedience: Only two types are needed in order to produce an abstract implementation of Gibbs' vector algebra, whereas four types (and many more operations) would be needed to implement the Clifford algebra. Thus the package uses the following simple "lexical" type conventions:

– All Greek letters and numeric symbols are scalars.
– All other symbols are of type vector.

The validity and type of an expression can be checked using a pair of coroutines, called type_scalar and type_vector, which are based on the rules described below for combining "atomic" scalars and vectors into compound expressions.

Because of this type convention, and because scalar multiplication and vector addition are associative and commutative, the ordinary '+' and '*' operators in MAPLE can be used to represent linear combinations of vectors; the type of these operators is vector or scalar as both or one of their operands is vector or scalar, respectively. The inner and cross products were implemented by means of the MAPLE define command as commutative and anticommutative neutral operators called '&.' and '&x', respectively; the '&.' operator is considered to be of type scalar, whereas '&x' is of type vector. These neutral operators evaluate to themselves, save that the order of their arguments is standardized and the sign changed if need be in the case of '&x'. Therefore, they are true symbolic expressions of the corresponding vector algebra operations.

In order to compare different expressions, it is desirable to be able to fully expand them and thereby enable the builtin MAPLE simplification procedure to automatically cancel terms that differ only by sign. Unfortunately, the MAPLE expand function does not know, and cannot be taught, to expand the dot and cross products of arbitrary linear combinations of vectors. It was therefore necessary to write a pair of coroutines that can perform this vital function, namely expand_scalar and expand_vector, which recursively call each other as they descend the expression tree, according as the type of the current subexpression is scalar or vector, respectively. The following example illustrates the use of expand_scalar:

```
mapleV> expand_scalar( (b - a) &. ((c - a) &x (d - a))) );
```
$$(b \mathbin{\&.} (c \mathbin{\&x} d)) - (a \mathbin{\&.} (b \mathbin{\&x} c)) + (a \mathbin{\&.} (b \mathbin{\&x} d)) -$$
$$(a \mathbin{\&.} (c \mathbin{\&x} d)) + (a \mathbin{\&.} (a \mathbin{\&x} d)) - (a \mathbin{\&.} (a \mathbin{\&x} c))$$

It will be observed that it is possible to simplify the above expression by cancelling triple products with a repeated factor (those containing a × a were canceled automatically by the '&x' procedure). While it is clear that the latter is indeed a simpler form, the general question of what it means to be "simpler" is a difficult one that is open to interpretation. There is also the closely related problem of finding a normal form for expressions in Gibbs' vector algebra, so that one can tell if two expressions are the same or not (i.e. equal for all values of their component scalars and vectors). In multivariate polynomial algebra over a field of characteristic zero, a normal form is obtained simply by fully expanding the polynomial, then sorting the factors in the monomials together with monomials themselves according to the lexicographic order induced by a fixed ordering of the indeterminates. We have therefore chosen to "simplify" expressions in *Gibbs* by expanding them to polynomials in the inner products of atomic vectors to the maximum extent possible. The metric vector space syzygies described above, however, ensure that even an expanded and sorted polynomial in the inner products is not a normal form.

Thus we have written an additional pair of coroutines, `simplify_scalar` and `simplify_vector`, that take a fully expanded expression and attempt to further expand the individual terms as inner products. First, any iterated cross products are replaced by linear combinations of vectors using the usual relation:

$$\mathbf{a} \times (\mathbf{b} \times \mathbf{c}) = (\mathbf{a} \cdot \mathbf{c})\mathbf{b} - (\mathbf{a} \cdot \mathbf{b})\mathbf{c} . \tag{6}$$

Second, the factors in any triple products are ordered in a standard way (determined by MAPLE's internal variable ordering), and in particular triple products containing repeated factors are cancelled. Third, any powers of triple products are replaced by the corresponding determinant of a metric tensor, e.g.

$$(\mathbf{a} \cdot (\mathbf{b} \times \mathbf{c}))^2 = \det \begin{bmatrix} \mathbf{a} \cdot \mathbf{a} & \mathbf{a} \cdot \mathbf{b} & \mathbf{a} \cdot \mathbf{c} \\ \mathbf{a} \cdot \mathbf{b} & \mathbf{b} \cdot \mathbf{b} & \mathbf{b} \cdot \mathbf{c} \\ \mathbf{a} \cdot \mathbf{c} & \mathbf{b} \cdot \mathbf{c} & \mathbf{c} \cdot \mathbf{c} \end{bmatrix} \tag{7}$$

An analogous transformation could be applied to arbitrary algebraic products of triple products, but this has not so far proved necessary in the applications we have pursued. Together, these simplifications enable us to reduce any scalar-valued expression to a polynomial in the dot and triple products of vectors, wherein each term is at most linear in the triple products.

Similarly, it is possible to expand the products of triple products with cross products as

$$(\mathbf{a} \cdot (\mathbf{b} \times \mathbf{c}))(\mathbf{d} \times \mathbf{e}) = \det \begin{bmatrix} \mathbf{a} \cdot \mathbf{d} & \mathbf{a} \cdot \mathbf{e} & \mathbf{a} \\ \mathbf{b} \cdot \mathbf{d} & \mathbf{b} \cdot \mathbf{e} & \mathbf{b} \\ \mathbf{c} \cdot \mathbf{d} & \mathbf{c} \cdot \mathbf{e} & \mathbf{c} \end{bmatrix} , \tag{8}$$

but we have not implemented this since we have been concerned primarily with scalar-valued expressions. A more useful simplification, requiring however the scanning of all triples of terms, would be

$$(\mathbf{a} \cdot \mathbf{n}_1)(\mathbf{n}_2 \times \mathbf{n}_3) + (\mathbf{a} \cdot \mathbf{n}_2)(\mathbf{n}_3 \times \mathbf{n}_1) + (\mathbf{a} \cdot \mathbf{n}_3)(\mathbf{n}_1 \times \mathbf{n}_2) \tag{9}$$
$$= \mathbf{a} \left(\mathbf{n}_1 \cdot (\mathbf{n}_2 \times \mathbf{n}_3) \right) .$$

The following examples indicate some of the present capabilities of `simplify_scalar` and `simplify_vector`:

```
mapleV> simplify_scalar( (a &x b) &. ((b &x c) &x (c &x a)) );
          (a &. (b &x c))²
```

```
mapleV> simplify_scalar( " );
```

$$2(a \;\&.\; b)(a \;\&.\; c)(b \;\&.\; c) + (a \;\&.\; a)(b \;\&.\; b)(c \;\&.\; c) -$$
$$(a \;\&.\; a)(b \;\&.\; c)^2 - (b \;\&.\; b)(a \;\&.\; c)^2 - (c \;\&.\; c)(a \;\&.\; b)^2$$

```
mapleV> simplify_vector( a &x (b &x c) + b &x (c &x a) +
         c &x (a &x b) );
       0
```

To illustrate how *Gibbs* can prove geometric theorems, let us verify that the incenter of the tetrahedron whose vertices are a, b, c, d is given by

$$z = \frac{\alpha a + \beta b + \gamma c + \delta d}{\alpha + \beta + \gamma + \delta} \tag{10}$$

where $\alpha, \beta, \gamma, \delta$ are the areas of the faces opposite their coefficient vertices. Clearly it suffices to show that the distance to any two of these faces is the same, e.g. that

$$(z - a) \cdot ((b - a) \times (c - a))/(2\delta) \tag{11}$$
$$= (z - a) \cdot ((d - a) \times (b - a))/(2\gamma)$$

Clearing denominators by multiplying through with $4\delta\gamma(\alpha + \beta + \gamma + \delta)$ and expanding yields an equation with 18 terms on each side, but both sides simplify to

$$2\delta\gamma (b \cdot (c \times d) - a \cdot (c \times d) + a \cdot (b \times d) - a \cdot (b \times c)) \tag{12}$$
$$= 6\delta\gamma V(a, b, c, d)$$

This shows that the distance of the incenter to each face is thrice the volume $V(a, b, c, d)$ of the tetrahedron divided by its total surface area.

Finally, we note that it is often desirable to avoid any origin-dependent terms by computing entirely with interpoint vectors. For this reason *Gibbs* defines an antisymmetric table, e.g. $T(a, b) = -T(b, a) = b - a$. If all the vectors are expressed in terms of T, these polynomials may then be further converted into polynomials in the squared distances D and oriented volumes V of the tetrahedra spanned by the quadruples of points, by means of the substitutions

$$T(a, b) \cdot T(c, d) = \tfrac{1}{2} (D(a, d) - D(a, c) + D(b, c) - D(b, d)) \tag{13}$$

and

$$T(a, b) \cdot (T(c, d) \times T(e, f)) \tag{14}$$
$$= 6 (V(a, b, c, e) - V(a, b, c, f) - V(a, b, d, e) + V(a, b, d, f)) \; .$$

These polynomials represent elements of what might be called the *oriented Cayley-Menger ring*, including oriented volumes as well as squared distances, and having $36 V^2(a, b, c, d) = D(a, b, c, d)$ as a syzygy. The forgoing constitutes a constructive proof that any scalar-valued origin-independent vector algebra expression can be rewritten as a polynomial in these invariants, thereby also proving (at least in three dimensions) that they constitute a complete system of invariants for the *proper* Euclidean group.

3   A Chemical Example

As a considerably more challenging example, we shall now derive a new parametrization for the conformation space of the kinematic framework corresponding to the molecule *cyclohexane*. In this problem, the distances between consecutive pairs $(i, i + 1)$ of bonded atoms (mod 6) around the six-atom ring are fixed at unity, while the distances between alternate pairs $(i, i + 2)$ are fixed at $\sqrt{8/3}$ (which ensures that the angles between consecutive bonds are fixed at the tetrahedral value of 109.5°). These twelve constraints are generically sufficient to hold the framework rigid in three dimensions [19], but due to the symmetry of the framework there exists a topological circle of conformations related by an internal motion known among chemists as a "pseudo-rotation" [20] (see Fig. 2). This problem is actually a special case of the local deformation problem mentioned above [4, 5], for which we have been able to explicitly solve the equations.

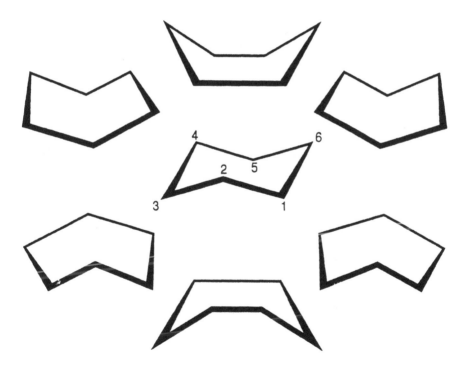

Fig. 2. The molecule cyclohexane can assume a rigid "chair" conformation (center) as well as a circle (one-parameter cyclic family) of "boat" conformations. The latter are arranged about the chair in the order in which they occur physically in the course of such a "pseudo-rotation". In the text, the atoms are referred to by their numerical indices, which occur consecutively around the ring as shown on the chair conformation.

These equations are based on the *stereographic* representation of rotations, i.e.

$$\mathbf{u} \mapsto \mathbf{R}^\dagger \mathbf{u} \mathbf{R} = \frac{(1 - \rho \iota r)\,\mathbf{u}\,(1 + \rho \iota r)}{1 + \rho^2}, \tag{15}$$

where $\mathbf{r}$ is a unit vector along the axis of rotation and $\rho = \tan(\phi/2)$ is the tangent of the half-angle of the rotation. With a bit of work, this can be rewritten solely in terms of the operations of vector algebra as:

$$\mathbf{R}^\dagger \mathbf{u} \mathbf{R} = \frac{\mathbf{u} + 2\rho(\mathbf{r} \times \mathbf{u}) + \rho^2[2(\mathbf{r} \cdot \mathbf{u})\mathbf{r} - \mathbf{u}]}{1 + \rho^2} \tag{16}$$

With a bit more work, one can further show that the product of two rotations $\mathbf{R}$ and $\mathbf{S} = 1 + \sigma \iota s$ in this representation is

$$\mathbf{R}^\dagger \mathbf{S}^\dagger \mathbf{u} \mathbf{S} \mathbf{R} = \mathbf{T}^\dagger \mathbf{u} \mathbf{T} = \frac{(1 - \tau \iota t)\mathbf{u}(1 + \tau \iota t)}{1 + \tau^2}, \tag{17}$$

where

$$\mathbf{t} = \frac{\mathbf{r} + \mathbf{s} + \mathbf{r} \times \mathbf{s}}{1 - \mathbf{r} \cdot \mathbf{s}} \tag{18}$$

$$1 + \tau^2 = \frac{(1 + \rho^2)(1 + \sigma^2)}{(1 - \mathbf{r} \cdot \mathbf{s})^2}.$$

Using Gibbs' together with a bit of pencil and paper (to apply a few simplifications that Gibbs' cannot yet handle, e.g. eq. (9)), one can substitute $\mathbf{t}$ and τ in eq. (18) for $\mathbf{r}$ and ρ in eq. (16), then expand and simplify the result to obtain:

$$(1 + \rho^2)(1 + \sigma^2)\mathbf{T}^\dagger \mathbf{u} \mathbf{T} = \mathbf{u} + \tag{19}$$
$$2(\mathbf{r} \times \mathbf{u})\rho +$$
$$2(\mathbf{s} \times \mathbf{u})\sigma +$$
$$2(\mathbf{r} \cdot \mathbf{u})\mathbf{r} - \mathbf{u})\rho^2 +$$
$$2(\mathbf{s} \cdot \mathbf{u})\mathbf{s} - \mathbf{u})\sigma^2 +$$
$$4((\mathbf{r} \cdot \mathbf{u})\mathbf{s} - (\mathbf{s} \cdot \mathbf{u})\mathbf{r})\rho\sigma +$$
$$2((\mathbf{u} \cdot (\mathbf{r} \times \mathbf{s}))\mathbf{r} - 2(\mathbf{s} \times \mathbf{u}))\rho^2\sigma +$$
$$2((\mathbf{s} \cdot \mathbf{u})(\mathbf{r} \times \mathbf{s}) - 2(\mathbf{r} \times \mathbf{u}))\rho\sigma^2 +$$
$$2(2(\mathbf{r} \cdot \mathbf{s})(\mathbf{s} \cdot \mathbf{u})\mathbf{r} - (\mathbf{r} \cdot \mathbf{u})\mathbf{r} - (\mathbf{s} \cdot \mathbf{u})\mathbf{s})\rho^2\sigma^2$$

We now specialize this formula to the case of cyclohexane. It is easily verified that the coordinates of the chair conformation

$$[\mathbf{x}_0, \ldots, \mathbf{x}_5] = \begin{bmatrix} 0 & \frac{\sqrt{6}}{3} & \frac{\sqrt{6}}{3} & 0 & -\frac{\sqrt{6}}{3} & -\frac{\sqrt{6}}{3} \\ \frac{\sqrt{8}}{3} & \frac{\sqrt{2}}{3} & -\frac{\sqrt{2}}{3} & -\frac{\sqrt{8}}{3} & -\frac{\sqrt{2}}{3} & \frac{\sqrt{2}}{3} \\ \frac{1}{6} & -\frac{1}{6} & \frac{1}{6} & -\frac{1}{6} & \frac{1}{6} & -\frac{1}{6} \end{bmatrix} \tag{20}$$

are a solution to the problem. We will take this as our reference conformation, and measure the rotation angles $\psi_{i,i+1 \bmod 6}$ about the bonds around the ring as differences from the dihedral angles between the planes defined by each triple of consecutive atoms in the chair, which are all $\pm 60°$. We shall further define the interatomic vectors $\mathbf{v}_{ij} = \mathbf{x}_j - \mathbf{x}_i$ for $0 \le i < j \le 5$, and set $\mathbf{r} = \mathbf{v}_{01}$, $\mathbf{s} = \mathbf{v}_{12}$ and $\mathbf{u} = \mathbf{v}_{23}$ in the above equations. Then the squared distance from $\mathbf{x}_3$ to $\mathbf{x}_5$ is:

$$D_{35}(\rho,\sigma) = \| -\mathbf{v}_{05} + \mathbf{v}_{01} + \mathbf{R}^\dagger \mathbf{v}_{12}\mathbf{R} + \mathbf{T}^\dagger \mathbf{v}_{23}\mathbf{T}\|^2 \tag{21}$$

On multiplying this by $(1 + \rho^2)(1 + \sigma^2)$, then expanding and simplifying the right-hand side using Gibbs', we obtain a (fairly complicated) polynomial in ρ and σ, which is only *quadratic* in each of these variables separately.

Next, we assign the coordinate differences from eq. (20) to the abstract vectors $\mathbf{v}_{ij}$ above, and redefine the neutral operators '&.' and '&x' to be MAPLE procedures that return the the dot and cross-products of their coordinate vector arguments. Then if we subtract $(1 + \rho^2)(1 + \sigma^2)D_{35}(\rho,\sigma)$ from its desired value of $(1 + \rho^2)(1 + \sigma^2)8/3$, the coefficients in the polynomial evaluate to numerical quantities, which are (up to a constant factor):

$$\Omega^+(\rho,\sigma) = -\sqrt{27}\rho + \sqrt{27}\sigma + 9\rho^2 + 9\sigma^2 + 6\rho\sigma - \sqrt{75}\rho\sigma^2 + \sqrt{75}\rho^2\sigma + 8\rho^2\sigma^2 \tag{22}$$

A slightly different polynomial is obtained if we shift the coordinate indices by one (mod 6), namely

$$\Omega^-(\rho,\sigma) = \sqrt{27}\rho - \sqrt{27}\sigma + 9\rho^2 + 9\sigma^2 + 6\rho\sigma + \sqrt{75}\rho\sigma^2 - \sqrt{75}\rho^2\sigma + 8\rho^2\sigma^2 . \tag{23}$$

The polynomials obtained by further shifts are all the same as these two polynomials. Note these polynomials have no constant term, so that $\rho, \sigma = 0$ (the chair form!) is a root. Another solution is given by $\rho, \sigma = \pm\sqrt{3}$, which corresponds to the mirror image of the coordinates in eq. (20).

Now let $\chi_{01} = \tan(\psi_{01}/2), \ldots, \chi_{45} = \tan(\psi_{45}/2), \chi_{05} = \tan(\psi_{05}/2)$ be the six tangents of the half rotation angles around each bond of the ring relative to the chair conformation. Substituting each consecutive pair of these tangents for ρ and σ alternately in eqs. (22) and (23) yields a total of six equations, which we will now proceed to solve. The strategy will be to eliminate our way up both sides of the ring starting from a pair of equations sharing a common unknown, obtaining two equations in two unknowns when these two chains of eliminations meet. Up to a constant factor, the resultants of the first two pairs of equations are

$$\begin{aligned}
\Upsilon(\chi_{01},\chi_{23}) &= Res_{12}(\Omega^+(\chi_{01},\chi_{12}), \Omega^-(\chi_{12},\chi_{23})) \\
&= (\chi_{01} - \chi_{23})^2(9 - \sqrt{432}\chi_{01} - \sqrt{432}\chi_{23} + 9\chi_{01}^2 \\
&\quad + 9\chi_{23}^2 - \sqrt{48}\chi_{01}\chi_{23}^2 - \sqrt{48}\chi_{01}^2\chi_{23} + 25\chi_{01}^2\chi_{23}^2)
\end{aligned} \tag{24}$$

and

$$\begin{aligned}
\Upsilon(\chi_{01},\chi_{45}) &= Res_{05}(\Omega^-(\chi_{01},\chi_{05}), \Omega^+(\chi_{05},\chi_{45})) \\
&= (\chi_{01} - \chi_{45})^2(9 - \sqrt{432}\chi_{01} - \sqrt{432}\chi_{45} + 9\chi_{01}^2 \\
&\quad + 9\chi_{45}^2 - \sqrt{48}\chi_{01}\chi_{45}^2 - \sqrt{48}\chi_{01}^2\chi_{45} + 25\chi_{01}^2\chi_{45}^2)
\end{aligned} \tag{25}$$

Due to its D_{3d} symmetry, the first factor on the right-hand side of these equations vanishes in the chair form and its mirror image, which we already have, and thus we use the other factor to compute the next pair of resultants, namely

$$\Xi(\chi_{01}, \chi_{34}) = Res_{23}(\Upsilon_2(\chi_{01}, \chi_{23}), \Omega^+(\chi_{23}, \chi_{34})) \tag{26}$$
$$= (\sqrt{3} - \chi_{01} + \chi_{34} + \sqrt{3}\chi_{01}\chi_{34})^2(-\sqrt{27}\chi_{01} + \sqrt{27}\chi_{34}$$
$$+ 9\chi_{01}^2 + 9\chi_{34}^2 - \sqrt{75}\chi_{01}\chi_{34}^2 + \sqrt{75}\chi_{01}^2\chi_{34} + 8\chi_{01}^2\chi_{34}^2)$$
$$= Res_{45}(\Upsilon_2(\chi_{01}, \chi_{45}), \Omega^-(\chi_{45}, \chi_{34})),$$

where Υ_2 denotes the second factor in eqs. (24) and (25).

The fact that these last two resultants are the same shows that we may choose one of our tangents freely, e.g. χ_{01}, providing that the discriminants of the quadratic factors with respect to the other variables are nonnegative in each of $\Omega^+(\chi_{01}, \chi_{12})$, $\Omega^-(\chi_{01}, \chi_{05})$, $\Upsilon(\chi_{01}, \chi_{23})$, $\Upsilon(\chi_{01}, \chi_{45})$ and $\Xi(\chi_{01}, \chi_{34})$. Up to the usual constant factor, these discriminants all turn out to be

$$- (71\chi_{01}^2 - (36\sqrt{2} + 56\sqrt{3})\chi_{01} + 15 - 12\sqrt{6}) \tag{27}$$
$$(71\chi_{01}^2 + (36\sqrt{2} - 56\sqrt{3})\chi_{01} + 15 + 12\sqrt{6}).$$

The second factor has no real roots, while the roots of the first factor are:

$$\frac{1}{71}\left(18\sqrt{2} + 28\sqrt{3} \pm \sqrt{1935 + 1860\sqrt{6}}\right) \tag{28}$$

Translating these bounds on the tangents back into the corresponding angles and offsetting them by the chair dihedral angle of $-60°$ shows that the dihedral angle about the $0-1$ bond (and hence all bonds by symmetry) lies in the interval $[-70.643717, 70.643717]$ in degrees.

The quadratic factor in $\Xi(\chi_{01}, \chi_{34})$ vanishes in the chair conformation and its mirror image. We now solve the first factor of $\Xi(\chi_{01}, \chi_{34})$ for χ_{34}, obtaining

$$\chi_{34} = \frac{\chi_{01} - \sqrt{3}}{1 + \sqrt{3}\chi_{01}}. \tag{29}$$

To correct for the fact that the $0-1$ and $3-4$ dihedral angles in the reference chair conformation differ by 120°, we make the substitution $\chi_{01} = (\bar{\chi}_{01} - \sqrt{3})/(1 + \sqrt{3}\bar{\chi}_{01})$, where $\bar{\chi}_{01} = \tan(\psi_{01}/2 + \pi/3)$, and find that $\chi_{34} = \bar{\chi}_{01}$. This shows that, if this solution obtains, the two dihedral angles across the ring are the same (which is consistent with the observed C_2 symmetry of all the boat conformations).

Thus we can cover the one-dimensional manifold of the boat conformations of cyclohexane with two coordinate patches, both using the tangents of the half angle χ_{01} as the parameter. The tangents χ_{12} and χ_{23} can be computed by solving quadratic equations (in which two of the four possible combinations of roots are allowed), and finally computing the rest of the tangents via eq. (29). The only other exact parametrization of which we are aware, which was derived by eliminating squared distances among the Cayley-Menger determinants around the ring, involves solving a cubic equation [20, 21].

4 Summary and Prospects

In this paper we have outlined the four levels at which one could use computer algebra systems to solve problems in Euclidean geometry. We have also described an eminently practical implementation of Gibbs vector algebra, which demonstrates that one can actually perform computations in an entirely coordinate-free fashion at the abstract algebra level. It seems rather ironic that even though the advantages of a coordinate-free approach to geometry are widely recognized, most of the computer algebra procedures geometers have written to aid them in their work rely exclusively on coordinates. We hope our demonstration may help change this situation.

Our demonstration, of course, simplified matters considerably by restricting itself to a signature of $(3,0)$. More general implementations able to operate on arbitrary signatures (without introducing a basis) will need to be able to specify the signs of the squares of the vectors. In MAPLE, this could be done with an extension of the `assume` command, e.g. the command "`assume( a^2 > 0, b^2 < 0 )`" would guarantee that 'a', 'b' spanned a hyperbolic plane. This style of programming would clearly benefit from using a computer algebra system with strong typing and more highly developed rules-based programming methods. Further work along these lines is envisioned.

Acknowledgements

This work was supported by grant BIR-9511892 from the National Science Foundation of the U.S.A.

References

1. R. Ablamowicz, P. Lounesto, and J. M. Parra, editors. *Clifford Algebras with Numeric and Symbolic Computations*. Birkhäuser, Boston, MA, 1996.
2. E. Schrufer, F. W. Hehl, and J. D. McCrea. Exterior calculus on the computer: The REDUCE-package EXCALC applied to general relativity and to the Poincare gauge theory. *General Relativity and Gravitation*, 19:197–218, 1987.
3. B. W. Char, G. J. Fee, K. O. Geddes, G. H. Gonnet, and M. B. Monagan. A tutorial introduction to maple. *J. Symb. Comput.*, 2:179–200, 1986.
4. T. F. Havel and I. Najfeld. Applications of geometric algebra to the theory of molecular conformation. Part 2. The local deformation problem. *J. Mol. Struct. (TheoChem)*, 336:175–189, 1995.
5. T. F. Havel and I. Najfeld. A new system of equations, based on geometric algebra, for ring closure in cyclic molecules. In *Applications of Computer Algebra in Science and Engineering*, pages 243–259. World Scientific, Singapore, 1995.
6. J. Bokowski and B. Sturmfels. *Computational Synthetic Geometry*. Lect. Notes Math. 1355. Springer-Verlag, Berlin, F.R.G., 1989.
7. B. Sturmfels. *Algorithms in Invariant Theory*. Springer-Verlag, New York, NY, 1993.

8. M. Barnabei, A. Brini, and G.-C. Rota. On the exterior calculus of invariant theory. *J. Algebra*, 96:120–160, 1985.

9. N. L. White. A tutorial on Grassmann-Cayley algebra. In N. L. White, editor, *Invariant Methods in Discrete and Computational Geometry*, pages 93–106. Kluwer Academic, Amsterdam, Holland, 1995.

10. H. Weyl. *The Classical Groups.* Princeton University Press, Princeton, NJ, 1939.

11. D. R. Richman. The fundamental theorems of vector invariants. *Adv. Math.*, 73:43–78, 1989.

12. A. W. M. Dress and T. F. Havel. Some combinatorial properties of discriminants in metric vector spaces. *Adv. Math.*, 62:285–312, 1986.

13. D. Hestenes and G. Sobczyk. *Clifford Algebra to Geometric Calculus.* D. Reidel Pub. Co., Dordrecht, Holland, 1984.

14. D. Hestenes and R. Ziegler. Projective geometry with Clifford algebra. *Acta Appl. Math.*, 23:25–63, 1991.

15. D. Hestenes. The design of linear algebra and geometry. *Acta Appl. Math.*, 23:65–93, 1991.

16. A. W. M. Dress and T. F. Havel. Distance geometry and geometric algebra. *Found. Phys.*, 23:1357–1374, 1993.

17. T. F. Havel. Geometric algebra and Möbius sphere geometry as a basis for euclidean invariant theory. In *Invariant Methods in Discrete and Computational Geometry*, pages 245–256. Kluwer Academic, Amsterdam, Holland, 1995.

18. J. P. Dalbec. Straightening Euclidean invariants. *Ann. Math. Artif. Intel.*, 13:97–108, 1995.

19. J. Graver, B. Servatius, and H. Servatius. *Combinatorial Rigidity*, volume 2 of *Graduate Studies in Mathematics.* Amer. Math. Soc., Providence, RI, 1993.

20. G. M. Crippen and T. F. Havel. *Distance Geometry and Molecular Conformation.* Research Studies Press, Taunton, England, 1988.

21. A. W. M. Dress. Vorlesungen über kombinatorische Geometrie. Univ. Bielefeld, Germany, 1982.

Clifford Algebraic Calculus for Geometric Reasoning

with Application to Computer Vision

Dongming Wang

LEIBNIZ–IMAG, 46, avenue Félix Viallet, 38031 Grenoble Cedex, France

Abstract. In this paper we report on our recent study of Clifford algebra for geometric reasoning and its application to problems in computer vision. A general framework is presented for construction and representation of geometric objects with selected rewrite rules for simplification. It provides a mechanism suitable for devising methods and software tools for geometric reasoning and computation. The feasibility and efficiency of the approach are demonstrated by our preliminary experiments on automated theorem proving in plane Euclidean geometry. We also explain how non-commutative Gröbner bases can be applied to geometric theorem proving. In addition to several well-known geometric theorems, two application examples from computer vision are given to illustrate the practical value of our approach.

1 Introduction

Much of the recent research on automated geometric deduction has focused on developing high-level coordinate-free techniques [1, 2, 3, 5, 10–14]. One of the techniques is using vector representation and calculus which was advocated early by R. Wong [18] and investigated recently by S.-C. Chou, X.-S. Gao and J.-Z. Zhang [2], and S. Stifter [14]. Two different methods are proposed in [2] and [14] respectively for proving certain classes of geometric theorems: the former proceeds by eliminating constructed points from the conclusion of each theorem using some basic equalities on inner and cross products of vectors, while the latter employs Gröbner bases in vector spaces for the involved computation. On the other hand, H. Li and M.-t. Cheng observed that Clifford algebra is much more expressive than vector algebra for geometric problems. They have studied how to use this algebra to represent notions and relations in various geometries and proposed a general hybrid approach that combines Clifford algebraic representation, parametrization and coordinate representation for theorem proving [10, 11, 12]. The approach is an analogy to the well-known method of Wu [20] and also makes use of Wu's method when Clifford algebraic calculus does not succeed. Our interest in Clifford algebra for geometric reasoning has been motivated by Li and Cheng's work.

In this report, we describe a general framework of geometric constructions with Clifford algebraic representation for calculus and reasoning. By means of representing each newly constructed point in terms of the points previously constructed using Clifford algebraic operators, we propose a simple yet efficient approach. The idea of considering this kind of linear constructions goes back to the early work of Wu [19] and the author [16]. As a matter of fact, the class of theorems and problems that can be dealt with using linear constructions is somewhat restricted, but it is large enough to be of interest. In particular, the class covers more problems in the case of Clifford algebra representation than in coordinate formulation. This can be seen from our generalization of Steiner's theorem and other examples given later.

Although the method of Li and Cheng appears more general than ours, the scope of theirs is not necessarily much broader if coordinates are not introduced. This may be observed from the pseudo-reduction with respect to a vector triangular form which is effective only if every vector polynomial in the triangular form is linear in its leading variable. In our approach, it is made clearer when parameters need be introduced in Clifford algebraic representation, the proving procedure is considerably simplified, and concrete rewrite rules for simplifying Clifford algebraic expressions are presented. The last is a key issue that has not yet been well addressed. The method presented in this paper and that of Chou, Gao and Zhang have some common features yet use different constructions, representations and deduction mechanisms. In contrast to point elimination in [2], simple substitution and simplification using rewrite rules are the main reduction strategies in our method. In addition to the general approach, the applications of non-commutative Gröbner bases in Grassmann algebra to geometric theorem proving and Clifford-algebra-based methods to problems in computer vision are also discussed in the paper.

In what follows, a point P is also regarded as the vector from the origin O to P, $Q - P$ as the vector from point P to Q, and $P + Q$ as the vector from O to the opposite vertex of the parallelogram formed by OP and OQ. The outer product "$\wedge$" and the inner product "$\cdot$" among vectors will be introduced in Section 2 to represent geometric relations. For example,

- line AB is perpendicular to line CD if and only if $(A - B) \cdot (C - D) = 0$;
- point C lies on line AB if and only if $(C - A) \wedge (B - A) = 0$.

Both products are associative, and distributive with addition. The inner product is commutative, while the outer product is anticommutative, i.e.,

$$\mathbf{a} \wedge \mathbf{b} = -\mathbf{b} \wedge \mathbf{a},$$

and thus $\mathbf{a} \wedge \mathbf{a} = 0$ for any vectors $\mathbf{a}$ and $\mathbf{b}$. To show the simplicity of proving geometric theorems using Clifford algebra, we give two examples first.

Example 1 (Centroid). Let ABC be an arbitrary triangle and A_1, B_1, C_1 be the midpoints of the three sides BC, CA, AB. Then the three lines AA_1, BB_1, CC_1 are concurrent.

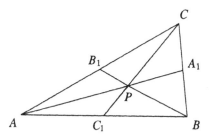

Let A be located at the origin; then $A = 0$. Hence, we have

$$A_1 = \frac{B+C}{2}, \quad B_1 = \frac{C}{2}, \quad C_1 = \frac{B}{2}.$$

Let P denote the intersection point of BB_1 and CC_1. It follows that

$$h_1 = (P - B_1) \wedge (P - B) = \frac{C \wedge B}{2} - P \wedge B - \frac{C \wedge P}{2} = 0, \qquad (1)$$
$$\text{\% col}(P, B, B_1)$$
$$h_2 = (P - C_1) \wedge (P - C) = \frac{B \wedge C}{2} - P \wedge C - \frac{B \wedge P}{2} = 0, \qquad (2)$$
$$\text{\% col}(P, C, C_1)$$

where $\text{col}(P, B, B_1)$ means that the three points P, B, B_1 are collinear, and similarly for others. The conclusion of the theorem is

$$g = (P - A_1) \wedge (P - A) = \frac{P \wedge B}{2} + \frac{P \wedge C}{2} = 0. \qquad \text{\% col}(P, A, A_1)$$

This is proved simply by adding (1) and (2).

Example 2 (Orthocenter). The three altitudes of an arbitrary triangle ABC intersect at a common point O.

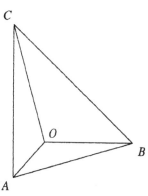

Let O be the origin, i.e., $O = 0$. Then we have the following hypotheses

$$h_1 = (A - B) \cdot C = 0, \qquad \text{\% } AB \perp OC$$
$$h_2 = (B - C) \cdot A = 0. \qquad \text{\% } BC \perp OA$$

We want to prove that

$$g = (A - C) \cdot B = 0. \qquad \text{% } AC \perp OB$$

This is true simply because $g = h_1 + h_2$.

The above proofs are surprisingly simple. In general, the proofs may become somewhat more complicated when difficult theorems are considered. One objective of this work is to develop the above techniques into a systematic method.

2 Clifford Algebra and Its Operators

Let $\mathbb{K}$ be a field of characteristic $\neq 2$ and V an n-dimensional vector space with base $e_1, \ldots, e_n$. Then any vector $\mathbf{v} \in V$ can be represented uniquely as

$$\mathbf{v} = \sum_{i=1}^{n} v_i e_i, \quad v_i \in \mathbb{K}.$$

Let Q be a quadratic form on V written explicitly as

$$Q = Q(\mathbf{v}) = Q(v_1, \ldots, v_n) = \sum_{1 \leq i \leq n} q_i v_i^2 + \sum_{1 \leq i < j \leq n} q_{ij} v_i v_j.$$

Construct a 2^n-dimensional vector space C over $\mathbb{K}$ with base

$$1, e_i, e_i e_j \ (i < j), \ldots, e_1 e_2 \cdots e_n.$$

Then, any $\mathbf{C} \in C$ is of the form

$$\mathbf{C} = c + \sum_{1 \leq i \leq n} c_i e_i + \sum_{1 \leq i < j \leq n} c_{ij} e_i e_j + \cdots + c_{1 \cdots n} e_1 \cdots e_n, \qquad (3)$$

$$c, c_{ij}, \ldots, c_{1 \cdots n} \in \mathbb{K}.$$

We want to define multiplication among the elements of C. For this purpose, let

$$e_i e_i = q_i, \quad 1 \leq i \leq n. \qquad (4)$$

It follows that

$$e_i e_j + e_j e_i = q_{ij}, \quad 1 \leq i < j \leq n. \qquad (5)$$

Applying the relations (4) and (5), one can always write the product $e_i e_j e_k \cdots$ for arbitrary i, j, k etc. in the form (3). For example,

$$e_1 e_4 e_1 e_3 = -q_1 q_{34} + q_{14} e_1 e_3 + q_1 e_3 e_4.$$

The product of any two elements of the form (3) is defined to be the sum of the products of all the terms, according to the distribution law. It is easy to show that the multiplication is associative; under it, C forms an associative algebra, called the *Clifford algebra* associated to Q. The elements of C are called *multivectors* or *Clifford numbers*. The product so defined for multivectors is also called the *geometric product*.

The following properties may be easily verified:

– For any $\mathbf{u}, \mathbf{v} \in \mathcal{V}$,

$$\mathbf{u}^2 = Q(\mathbf{u}), \quad \mathbf{u}\mathbf{v} + \mathbf{v}\mathbf{u} = B(\mathbf{u}, \mathbf{v}),$$

where

$$B(\mathbf{u}, \mathbf{v}) = Q(\mathbf{u} + \mathbf{v}) - Q(\mathbf{u}) - Q(\mathbf{v})$$

is the symmetric bilinear form associated with Q.

– If $\mathbf{e}_1, \ldots, \mathbf{e}_n$ is an orthogonal base relative to Q, then $q_{ij} = 0$; in this case

$$\mathbf{e}_i \mathbf{e}_j = -\mathbf{e}_j \mathbf{e}_i, \quad \mathbf{e}_i^2 = q_i, \quad i, j = 1, \ldots, n, \ i \neq j.$$

In particular, $\mathcal{C}$ is called *exterior algebra* or *Grassmann algebra* when $Q = 0$.

The above definition of Clifford algebra is taken from [15, §143]. Clifford algebra can also be defined without using the base of $\mathcal{V}$: Denote the tensor algebra over $\mathcal{V}$ by $\mathfrak{T}(\mathcal{V})$ and the tensor multiplication by $\otimes$. Let $\mathfrak{J}(Q)$ be the two-sided ideal of $\mathfrak{T}(\mathcal{V})$ generated by the elements $\mathbf{v} \otimes \mathbf{v} - Q(\mathbf{v})$, $\mathbf{v} \in \mathcal{V}$. Then the resulting associative quotient algebra $\mathfrak{T}(\mathcal{V})/\mathfrak{J}(Q)$ is the *Clifford algebra* associated to the quadratic form Q. This definition involves the concept of tensor algebra which need some technical explanations, while the constructive definition using the base of $\mathcal{V}$ seems easier to be understood.

Clifford algebra provides a rich and unified language for geometric calculus. This has been well developed, in particular, by D. Hestenes and his co-workers in their study of classical mechanics and mathematical physics [6, 7]. Here we introduce some of the Clifford algebraic operators for late use. Any vector of $\mathcal{V}$ can be represented as a directed line segment in an n-dimensional space. For any two vectors $\mathbf{a}, \mathbf{b} \in \mathcal{V}$ the sum of $\mathbf{a}$ and $\mathbf{b}$ is also a vector of $\mathcal{V}$, viz.,

$$\mathbf{a} + \mathbf{b} \in \mathcal{V}.$$

It is also called the *geometric sum* of $\mathbf{a}$ and $\mathbf{b}$ and has the familiar geometric meaning. The *inner product* $\mathbf{a} \cdot \mathbf{b}$ of two vectors $\mathbf{a}$ and $\mathbf{b}$ is defined by

$$\mathbf{a} \cdot \mathbf{b} = \frac{1}{2} B(\mathbf{a}, \mathbf{b}) \in \mathbb{K}.$$

If $\mathbf{e}_1, \ldots, \mathbf{e}_n$ are taken as an orthogonal base with $q_i = -1, 0$ or 1 for $1 \leq i \leq n$ (i.e., an *orthonormal base*), the above definition leads to the familiar inner product of two vectors in Euclidean space. That is, the inner product of $\mathbf{a}$ and $\mathbf{b}$ is the length of the line segment obtained by dilating the projection of $\mathbf{a}$ on $\mathbf{b}$ by the length of $\mathbf{b}$. Using the commonly known notations, we have

$$\mathbf{a} \cdot \mathbf{b} = |\mathbf{a}||\mathbf{b}| \cos \theta,$$

where θ is the angle between $\mathbf{a}$ and $\mathbf{b}$. In the following sections, we shall fix $q_i = 1$ and $q_{ij} = 0$ for $0 \leq i < j \leq n$ (and thus an orthonormal base $\mathbf{e}_1, \ldots, \mathbf{e}_n$) without further indication.

The *outer product* of two vectors **a** and **b**, denoted by **a** ∧ **b**, is defined to be the difference of **ab** and **a** · **b**:

$$\mathbf{a} \wedge \mathbf{b} = \mathbf{ab} - \mathbf{a} \cdot \mathbf{b}.$$

It is a bivector corresponding to the parallelogram obtained by sweeping the vector **a** along the vector **b**. The parallelogram obtained by sweeping **b** along **a** differs only in orientation from that obtained by sweeping **a** along **b**. This is simply expressed as

$$\mathbf{b} \wedge \mathbf{a} = -\mathbf{a} \wedge \mathbf{b}.$$

The reader may refer to the first two chapters in [6] for nice geometric interpretations of sum, inner and outer products defined above. The definitions can be naturally extended to arbitrary multivectors, i.e., elements of the Clifford algebra $\mathcal{C}$ (see [7]). When the quadratic form $Q = 0$, the geometric product is identical to the outer product.

Starting from geometric sum, inner and outer products, one can define various other operators like the *cross product*, *meet* and *dual operator* for multivectors. The dual operator for $n = 2$ will be used repeatedly in later sections; we explain it as follows. Let $n = 2$, e_1, e_2 be an orthonormal base of the vector space $\mathcal{V}$ and

$$I = e_1 \wedge e_2.$$

The *dual* $\mathbf{C}^\sim$ of any multivector $\mathbf{C} \in \mathcal{C}$ with respect to I is defined to be $\mathbf{C}I$, the geometric product of $\mathbf{C}$ and I. Geometrically, let e_1 and e_2 form an orthonormal frame of the Euclidean plane. If **v** is a vector, then $\mathbf{v}^\sim$ is the vector obtained by rotating **v** 90° anticlockwise in the plane. The duality considered in this paper is always for $n = 2$ and with respect to the fixed I. A general introduction to the dual operator involves a couple of technical terms and is omitted here. The reader may consult [7, 11].

We shall refer to the geometric sum and product, the inner product ·, the outer product ∧, the dual operator ~ and other operators such as the cross product × and meet ∨ not introduced here as *Clifford algebraic operators*. These operators obey some basic calculational laws and relate to each other; some of the laws and relations will be given in Section 5. The operators combined together lead to a powerful language for expressing geometric notions and statements. Extensive investigations on representing concepts and relations in different geometries have been carried out by Li [10, 11].

3   Geometric Construction and Representation

One of the typical actions in studying geometry is to construct new objects from already constructed ones. In fact, many geometric problems are/may be stated in terms of sequential constructions. In this section, we give a few illustrative examples of such constructions with the corresponding Clifford algebraic representation, which serve to substantiate our general framework for geometric calculus and reasoning.

Although the following constructions are given mainly for plane Euclidean geometry, we can work with any geometry of concrete dimension n, for instance, plane non-Euclidean or 3-dimensional Euclidean geometry. The method explained in later sections may also be generalized for other geometries without major difficulty. We take points, denoted by capital letters like A, B, P, Q, as the most fundamental geometric objects. Other geometric objects are represented by means of points. For example, a straight line is represented by two distinct points on the line, a plane is represented by three non-collinear points in the plane, and a circle is represented by three non-collinear points on the circle or by its center and another point on the circle. The vector $Q - P$ from P to Q also represents the line PQ passing through P and Q.

We consider three kinds of constructions.

C1. Construct arbitrary points, lines, planes, circles, etc. This is a simple case: take a finite number of free points. There is no constraint among these points.

C2. From already constructed points $P_1, \ldots, P_s$, construct a new point X that is neither arbitrary nor completely determined by $P_1, \ldots, P_s$. In other words, X has some freedom, but it is not a free point. In this case, determine the *degree* d of freedom of X according to the dimension n and the number r of independent constraints for X. Normally, $d = n - r$. This will be seen from the example constructions. Then, introduce d scalar parameters or free points $\mu_1, \ldots, \mu_d$ and represent the point X in terms of $P_1, \ldots, P_s$ and $\mu_1, \ldots, \mu_d$

$$X = f(P_1, \ldots, P_s; \mu_1, \ldots, \mu_d)$$

using Clifford algebraic operators. The following examples are given for Euclidean geometry of dimension n (≥ 2), where the degree of freedom for X is 1.

C2.1. Take a point X on the line $P_1 P_2$ determined by two points P_1 and P_2 already constructed. Introduce one scalar parameter u. The point X may be represented as

$$X = \text{on\_line}(P_1, P_2) = u P_1 + (1 - u) P_2.$$

Note that $X = P_2$ when $P_1 = P_2$. In the following two constructions, $X = P_3$ when $P_1 = P_2$.

C2.2. From three points P_1, P_2, P_3 already constructed, construct a point X such that the line $P_3 X$ is parallel to $P_1 P_2$. Introduce one scalar parameter u. The point X may be represented as

$$X = \text{par}(P_1, P_2, P_3) = P_3 + u(P_2 - P_1).$$

C2.3. From three points P_1, P_2, P_3 already constructed, construct a point X such that the line $P_3 X$ is perpendicular to $P_1 P_2$. Introduce one scalar parameter u. The point X may be represented as

$$X = \text{per}(P_1, P_2, P_3) = P_3 + u(P_2 - P_1)^{\sim}.$$

C2.4. Take a point X on the circle passing through three non-collinear points P_1, P_2, P_3 already constructed.

Let $Y = \text{cir\_ctr}(P_1, P_2, P_3)$ denote the circumcenter of triangle $P_1 P_2 P_3$ which will be constructed in C3.4. Introduce one free point U. Let $M = \text{per\_ft}(U, Y, P_1)$ denote the perpendicular foot of the line $P_1 M$ to UY which will be constructed in C3.3. Then the point X may be represented as

$$X = \text{on\_cir}(P_1, P_2, P_3) = 2M - P_1 = 2\,\text{per\_ft}(U, \text{cir\_ctr}(P_1, P_2, P_3), P_1) - P_1.$$

This example shows how geometric problems involving circles can be formulated by using linear constructions. Similarly, one can take a point X on the circle with center O and passing through P:

$$X = \text{on\_cir}^*(O, P) = 2\,\text{per\_ft}(U, O, P) - P,$$

where U is a free point introduced.

C3. From already constructed points $P_1, \ldots, P_s$, construct a new point X that is completely determined by $P_1, \ldots, P_s$. In other words, X has no freedom when the P_i are fixed. In this case, represent the point X in terms of $P_1, \ldots, P_s$

$$X = f(P_1, \ldots, P_s)$$

using Clifford algebraic operators.

C3.1. Construct the midpoint X of two points P_1 and P_2 already constructed. The point X may be represented as

$$X = \text{midp}(P_1, P_2) = \frac{P_1 + P_2}{2}.$$

C3.2. From four points $P_1, \ldots, P_4$ already constructed with lines $P_1 P_2$ and $P_3 P_4$ in the same plane but not parallel, construct the intersection point X of $P_1 P_2$ and $P_3 P_4$. The point X may be represented as

$$X = \text{int}(P_1, P_2, P_3, P_4) = \frac{[(P_2 - P_1) \cdot (P_3 \wedge P_4)]^\sim - [(P_4 - P_3) \cdot (P_1 \wedge P_2)]^\sim}{[(P_2 - P_1) \wedge (P_4 - P_3)]^\sim}.$$

Since $P_1 P_2$ and $P_3 P_4$ are assumed to be non-parallel, the denominator of the above expression is a non-zero scalar. In other words, the non-degeneracy condition

$$(P_2 - P_1) \wedge (P_4 - P_3) \neq 0, \qquad \%\ P_1 P_2 \nparallel P_3 P_4$$

is introduced in the way of representing X in terms of $P_1, \ldots, P_4$. The situation for the following constructions is very similar and will not be discussed again.

C3.3. From three points P_1, P_2, P_3 already constructed with $P_1 \neq P_2$, construct the foot X of the perpendicular drawn from point P_3 to line $P_1 P_2$. The point X may be represented as

$$X = \text{per\_ft}(P_1, P_2, P_3) = \frac{P_3 \cdot (P_2 - P_1)(P_2 - P_1) - (P_2 - P_1) \cdot (P_1 \wedge P_2)}{(P_2 - P_1) \cdot (P_2 - P_1)}.$$

When P_3 lies on $P_1 P_2$, $X = P_3$.

C3.4. From three non-collinear points P_1, P_2, P_3 already constructed, construct the circumcenter X of triangle $P_1 P_2 P_3$. The point X is represented as

$$X = \mathsf{cir\_ctr}(P_1, P_2, P_3)$$
$$= \frac{(P_2 \cdot P_2 - P_1 \cdot P_1)(P_3 - P_1)^\sim - (P_3 \cdot P_3 - P_1 \cdot P_1)(P_2 - P_1)^\sim}{2\left[(P_2 - P_1) \wedge (P_3 - P_1)\right]^\sim}.$$

Similarly, the centroid and orthocenter X of triangle $P_1 P_2 P_3$ may be constructed respectively:

$$X = \mathsf{ctr}(P_1, P_2, P_3) = \mathsf{int}(\mathsf{midp}(P_1, P_3), P_2, \mathsf{midp}(P_2, P_3), P_1) = \frac{P_1 + P_2 + P_3}{3},$$
$$X = \mathsf{ort\_ctr}(P_1, P_2, P_3) = \mathsf{int}(\mathsf{per\_ft}(P_2, P_3, P_1), P_1, \mathsf{per\_ft}(P_1, P_3, P_2), P_2).$$

The following constructions generalize C3.2 and C3.3.

C3.5. From five points $P_1, \ldots, P_5$ in the same plane already constructed with $P_1 P_2 \nparallel P_3 P_4$, construct the intersection point X of line $P_3 P_4$ and the line passing through P_5 and parallel to $P_1 P_2$. The point X may be represented as

$$X = \mathsf{par\_int}(P_1, P_2, P_3, P_4, P_5)$$
$$= \frac{[(P_2 - P_1) \cdot (P_3 \wedge P_4)]^\sim - \{(P_4 - P_3) \cdot [P_5 \wedge (P_2 - P_1)]\}^\sim}{[(P_2 - P_1) \wedge (P_4 - P_3)]^\sim}.$$

C3.6. From five points $P_1, \ldots, P_5$ in the same plane already constructed with lines $P_1 P_2$ and $P_3 P_4$ not perpendicular, construct the intersection point X of line $P_3 P_4$ and the line passing through P_5 and perpendicular to $P_1 P_2$. The point X may be represented as

$$X = \mathsf{per\_int}(P_1, P_2, P_3, P_4, P_5) = \frac{P_5 \cdot (P_2 - P_1)(P_4 - P_3) - (P_2 - P_1) \cdot (P_3 \wedge P_4)}{(P_2 - P_1) \cdot (P_4 - P_3)}.$$

C3.7. From three points P_1, P_2, P_3 already constructed with $P_1 \neq P_2$ and a non-negative scalar d, construct a point X such that $X P_3$ is perpendicular to $P_1 P_2$ and the perpendicular distance from X to $P_1 P_2$ is equal to d. The point X may be represented as

$$X = \mathsf{per\_dis}(P_1, P_2, P_3, d) = \mathsf{per\_ft}(P_1, P_2, P_3) \pm d \frac{(P_2 - P_1)^\sim}{\sqrt{(P_2 - P_1) \cdot (P_2 - P_1)}}.$$

This construction takes into account of distance. See Example 7 for one of its nice applications.

In any case of C2 and C3, the constructed point X is called a *dependent point*. The correctness of the above expressions for X can be easily verified by taking coordinates for the involved points (see the δ-rules in Section 5). Some of the expressions are derived from the solutions of vector equations given in [10].

4 Geometric Calculus and Theorem Proving

In this section we are concerned with a particular class of problems in any geometry of concrete dimension $n \geq 2$, in which each problem may be formulated constructively as follows.

Suppose that a finite number of scalar parameters $u_1, \ldots, u_e$ and free points $U_1, \ldots, U_h$ are given first. Starting with these parameters and points, construct finitely many new points step by step according to the three kinds of constructions explained in Section 3. Some of the new points may be completely free; let them be denoted $U_{h+1}, \ldots, U_l$ ($l \geq h$). In the way of construction, some scalar parameters $u_{e+1}, \ldots, u_d$ ($d \geq e$) and other free points denoted by $U_{l+1}, \ldots, U_m$ ($m \geq l$) may be introduced so that the remaining points $X_1, \ldots, X_r$ are not completely free and each dependent X_i can be represented in terms of the scalar parameters $u_1, \ldots, u_d$, free points $U_1, \ldots, U_m$, and the previously constructed dependent points $X_1, \ldots, X_{i-1}$:

$$X_i = f_i(u_1, \ldots, u_d; U_1, \ldots, U_m; X_1, \ldots, X_{i-1}), \quad i = 1, \ldots, r,$$

using Clifford algebraic operators. In the above representation of X_i, the denominator d_i of each f_i is assumed to be non-zero. Usually, $d_i \neq 0$ corresponds to some non-degeneracy condition of the geometric problem. The original geometric problem we want to solve is reduced to calculating the value of a geometric quantity given as a Clifford algebraic expression

$$\Delta = g(u_1, \ldots, u_d; U_1, \ldots, U_m; X_1, \ldots, X_r)$$

or to verifying whether Δ is equal to 0 when the X_i are substituted by the corresponding expressions. Examples of the geometric quantity Δ among the given and constructed points are the distance between two points, the area of a triangle, and any expression that has geometric meaning or corresponds to some geometric property.

The calculation of Δ is straightforward. Substituting the expression of X_1, $X_1 = f_1^* = f_1$, into f_2, one obtains

$$X_2 = f_2^*(u_1, \ldots, u_d; U_1, \ldots, U_m)$$

which involves the variables u_i and U_j only. Similarly, substituting $X_1 = f_1^*$ and $X_2 = f_2^*$ into f_3, one obtains

$$X_3 = f_3^*(u_1, \ldots, u_d; U_1, \ldots, U_m),$$

and so on. In this way, we shall get

$$X_i = f_i^*(u_1, \ldots, u_d; U_1, \ldots, U_m), \quad 1 \leq i \leq r,$$

where each f_i^* contains only the scalar parameters and free points. Finally, substituting all $X_i = f_i^*$ into g we obtain

$$\Delta = g^*(u_1, \ldots, u_d; U_1, \ldots, U_m)$$

under the non-degeneracy conditions $d_i \neq 0$, $1 \leq i \leq r$. The Clifford algebraic expression g^* in terms of u_i and U_j is what we want to derive.

For geometric theorem proving, one may consider the expressions

$$X_1 - f_1 = 0, \ldots, X_r - f_r = 0$$

as the hypotheses and $g = 0$ as the conclusion of a theorem to be proved. Then, under the conditions $d_i \neq 0$, the theorem is true if and only if $g^* = 0$ is an identity.

Example 3. From a point D draw three perpendiculars to the three sides of an arbitrary triangle ABC. Calculate the area Δ of the triangle formed by the three perpendicular feet P, Q and R.

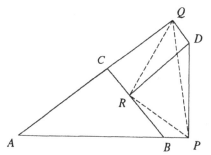

For simplicity and without loss of generality, let A be located at the origin, i.e., $A = 0$. The points P, Q, R may be constructed and calculated as follows:

$$P = \mathsf{per\_ft}(A, B, D) = \frac{B \cdot D\,B}{B \cdot B},$$

$$Q = \mathsf{per\_ft}(C, A, D) = \frac{C \cdot D\,C}{C \cdot C},$$

$$R = \mathsf{per\_ft}(B, C, D) = \frac{(C-B) \cdot D\,(C-B) - B \cdot C\,(C+B) + B \cdot B\,C + C \cdot C\,B}{C \cdot C - 2\,B \cdot C + B \cdot B}.$$

The denominators of the above expressions are non-zero when any two vertices of $\triangle ABC$ do not coincide. For any $\mathbf{a}, \mathbf{b} \in V$, let

$$|\mathbf{a} \wedge \mathbf{b}| = |\mathbf{a}||\mathbf{b}| \sin \theta$$

denote the magnitude of the bivector $\mathbf{a} \wedge \mathbf{b}$, where θ is the angle between $\mathbf{a}$ and $\mathbf{b}$. Then the area Δ in terms of P, Q, R is represented as

$$\Delta = \frac{1}{2}|(P - Q) \wedge (P - R)|.$$

Substituting the expressions of P, Q, R into Δ, we obtain

$$\Delta = \frac{|\Gamma||B \wedge C|}{2\,B \cdot B\,C \cdot C\,(C \cdot C - 2\,B \cdot C + B \cdot B)}.$$

with some simplification (see Section 5), where

$$\Gamma = -2\,B \cdot D\,B \cdot C\,C \cdot D + C \cdot C\,[(B \cdot D)^2 - B \cdot B\,B \cdot D + B \cdot C\,B \cdot D]$$
$$+ B \cdot B\,[(C \cdot D)^2 - C \cdot C\,C \cdot D + B \cdot C\,C \cdot D].$$

Example 4 (Euler Line). The orthocenter, the circumcenter and the centroid of an arbitrary triangle are collinear.

Let one vertex of the triangle be at the origin and the two other vertices be A and B. Then the centroid X, the circumcenter Y, and the orthocenter Z of the triangle may be expressed as

$$X = \mathsf{ctr}(A, B, 0) = \frac{A + B}{3},$$

$$Y = \mathsf{cir\_ctr}(A, B, 0) = \frac{A \cdot A\,B^\sim - B \cdot B\,A^\sim}{2\,A \cdot B^\sim},$$

$$Z = \mathsf{ort\_ctr}(A, B, 0) = \frac{A \cdot B\,(B^\sim - A^\sim)}{A \cdot B^\sim}.$$

The conclusion of the theorem to be proved is

$$(Y - X) \wedge (Z - X) = 0.$$

Substituting the expressions of X, Y, Z into the left-hand side, one obtains its numerator Δ as follows

$$\Delta = 3\,A \cdot B^\sim [A \cdot A\,(A + B) \wedge B^\sim - B \cdot B\,A \wedge (A^\sim + B^\sim)$$
$$+ 2\,A \cdot B\,(A \wedge A^\sim - B \wedge B^\sim)] + 9\,A \cdot B\,(A \cdot A - B \cdot B)A \wedge B.$$

The theorem is true under the non-degeneracy condition

$$A \cdot B^\sim \neq 0 \qquad \% \text{ the triangle does not degenerate to a line}$$

if and only if $\Delta = 0$ is an identity. In the next section, we shall explain how to prove identities of this type by simplification. In fact, the above Δ can be easily simplified to 0 when some of the basic rules α–γ given there are applied.

For the sake of efficiency, we may also reverse the substitution steps to reduce the conclusion expression to 0 as follows. Substituting $X_r = f_r$ into g, one obtains

$$g = \frac{g_{r-1}(u_1, \ldots, u_d; U_1, \ldots, U_m; X_1, \ldots, X_{r-1})}{h_{r-1}(u_1, \ldots, u_d; U_1, \ldots, U_m; X_1, \ldots, X_{r-1})}.$$

Next, substituting $X_{r-1} = f_{r-1}$ into g_{r-1} yields

$$g_{r-1} = \frac{g_{r-2}(u_1, \ldots, u_d; U_1, \ldots, U_m; X_1, \ldots, X_{r-2})}{h_{r-2}(u_1, \ldots, u_d; U_1, \ldots, U_m; X_1, \ldots, X_{r-2})}.$$

This process is continued by substituting $X_{r-2} = f_{r-2}$ into g_{r-2} and so forth; we shall finally obtain

$$g_{-1} = \frac{g_0(u_1, \ldots, u_d; U_1, \ldots, U_m)}{h_0(u_1, \ldots, u_d; U_1, \ldots, U_m)}.$$

The theorem is true under the subsidiary conditions $d_i \neq 0$ if and only if $g_0 = 0$ is an identity.

5 Clifford Algebraic Simplification

As shown in the preceding section, in both cases of calculating a geometric quantity and proving a theorem, we finally arrive at an expression g^* or g_0 which only involves the scalar parameters u_i and free points U_j. Nonetheless, the expression is composed by means of the Clifford algebraic operators $\wedge, \cdot$ and $\sim$ as well as the usual sum and product among scalars and between scalars and multivectors. It can be complex. How to effectively simplify such Clifford algebraic expressions is one of the principal tasks in our approach. It is necessary to introduce appropriate rules which relate the different operators and use them to simplify Clifford algebraic expressions.

The framework proposed in the previous two sections is general and applies to any geometry. But the adoption and design of simplification rules may depend on the concrete geometry under consideration. Our general idea is to select a few groups of rules with a careful study of Clifford algebra for each geometry and develop a term rewriting system based on these rules. In this way, the mathematical power of Clifford algebra will be combined with well-developed techniques and tools of term rewriting. However, we cannot get very far in this paper. In what follows, we concentrate on Euclidean geometry and see how the approach works in proving non-trivial theorems. Some of the rules are given only for the case $n = 2$.

Consider any Clifford algebraic expression g in u_i and U_j. Assume that the Clifford algebraic operators other than $\wedge, \cdot$ and $\sim$ in g have already been eliminated by using some Clifford algebraic relations. This can be done in most cases as far as theorem proving in Euclidean geometry is of concern. Here we do not discuss further how to eliminate such operators and assume that g is an algebraic expression composed by $\wedge, \cdot, \sim$ and the usual sum and product.

The *grade* of any element of $\mathbb{K}$ is defined to be 0. The *grade* $\operatorname{gr}(\mathbf{C})$ of any $\mathbf{C} \in \mathcal{C} \setminus \mathbb{K}$ is defined to be k if there exist $\mathbf{v}_1, \ldots, \mathbf{v}_k \in \mathcal{V}$ such that $\mathbf{C} = \mathbf{v}_1 \wedge \cdots \wedge \mathbf{v}_k$, and -1 otherwise. Computationally, we have

$$
\operatorname{gr}(\mathbf{C}) = \begin{cases}
0 & \text{if } \mathbf{C} \in \mathbb{K}, \\
1 & \text{if } \mathbf{C} \in \mathcal{V}, \\
|k_1 - k_2| & \text{if } \mathbf{C} = \mathbf{C}_1 \cdot \mathbf{C}_2 \text{ and } k_i = \operatorname{gr}(\mathbf{C}_i) \geq 1 \text{ for } i = 1, 2, \\
k_1 + k_2 & \text{if } \mathbf{C} = \mathbf{C}_1 \wedge \mathbf{C}_2, k_i = \operatorname{gr}(\mathbf{C}_i) \geq 0 \text{ and } k_1 + k_2 \leq n, \\
0 & \text{if } \mathbf{C} = \mathbf{C}_1 \cdot \mathbf{C}_2 \text{ and } \operatorname{gr}(\mathbf{C}_1) = 0 \text{ or } \operatorname{gr}(\mathbf{C}_2) = 0; \\
& \text{or } \mathbf{C} = \mathbf{C}_1 \wedge \mathbf{C}_2, k_i = \operatorname{gr}(\mathbf{C}_i) \geq 0 \text{ and } k_1 + k_2 > n, \\
-1 & \text{if } \not\exists \mathbf{v}_1, \ldots, \mathbf{v}_k \in \mathcal{V} \text{ such that } \mathbf{C} = \mathbf{v}_1 \wedge \cdots \wedge \mathbf{v}_k \\
& \text{for any } k \geq 1.
\end{cases}
$$

$\mathbf{C}$ is called a *k-vector* if $\operatorname{gr}(\mathbf{C}) = k \geq 0$.

Assume for the moment that an ordering has been introduced for Clifford algebraic terms and thus expressions; the introduction of such an ordering needs more technical exploration and will be discussed briefly below. We propose to use the following four kinds of rewrite rules to simplify the expression g.

α. *Trivial Rules*

α.1. For any $\mathbf{C} \in \mathcal{C}$:

$$0 \wedge \mathbf{C} \to 0, \ \mathbf{C} \wedge 0 \to 0; \ \ 0 \cdot \mathbf{C} \to 0, \ \mathbf{C} \cdot 0 \to 0;$$
$$0 + \mathbf{C} \to 0, \ \mathbf{C} + 0 \to 0; \ \ 0^{\sim} \to 0.$$

α.2. For any $a, b \in \mathbb{K}$ and $\mathbf{A}, \mathbf{B} \in \mathcal{C}$:

$$(a\mathbf{A}) \wedge \mathbf{B} \to a\mathbf{A} \wedge \mathbf{B}, \ \ \mathbf{A} \wedge (a\mathbf{B}) \to a\mathbf{A} \wedge \mathbf{B};$$
$$(a\mathbf{A}) \cdot \mathbf{B} \to a\mathbf{A} \cdot \mathbf{B}, \ \ \mathbf{A} \cdot (a\mathbf{B}) \to a\mathbf{A} \cdot \mathbf{B};$$
$$(a + b)\mathbf{A} \to a\mathbf{A} + b\mathbf{A}; \ \ (a\mathbf{A})^{\sim} \to a\mathbf{A}^{\sim}.$$

α.3 (Distributivity). For any $\mathbf{A}, \mathbf{B}, \mathbf{C} \in \mathcal{C}$:

$$\mathbf{A} \wedge (\mathbf{B} + \mathbf{C}) \to \mathbf{A} \wedge \mathbf{B} + \mathbf{A} \wedge \mathbf{C}, \ \ (\mathbf{A} + \mathbf{B}) \wedge \mathbf{C} \to \mathbf{A} \wedge \mathbf{C} + \mathbf{B} \wedge \mathbf{C};$$
$$\mathbf{A} \cdot (\mathbf{B} + \mathbf{C}) \to \mathbf{A} \cdot \mathbf{B} + \mathbf{A} \cdot \mathbf{C}, \ \ (\mathbf{A} + \mathbf{B}) \cdot \mathbf{C} \to \mathbf{A} \cdot \mathbf{C} + \mathbf{B} \cdot \mathbf{C};$$
$$(\mathbf{A} + \mathbf{B})^{\sim} \to \mathbf{A}^{\sim} + \mathbf{B}^{\sim}.$$

α.4 (Associativity). For any $\mathbf{A}, \mathbf{B}, \mathbf{C} \in \mathcal{C}$:

$$\mathbf{A} \wedge (\mathbf{B} \wedge \mathbf{C}) \to (\mathbf{A} \wedge \mathbf{B}) \wedge \mathbf{C}, \ \ \mathbf{A} \wedge \mathbf{B} \wedge \mathbf{C} \to (\mathbf{A} \wedge \mathbf{B}) \wedge \mathbf{C};$$
$$\mathbf{A} \cdot (\mathbf{B} \cdot \mathbf{C}) \to (\mathbf{A} \cdot \mathbf{B}) \cdot \mathbf{C}, \ \ \mathbf{A} \cdot \mathbf{B} \cdot \mathbf{C} \to (\mathbf{A} \cdot \mathbf{B}) \cdot \mathbf{C}.$$

β. *Elementary Rules*

β.1. For any $\mathbf{v} \in \mathcal{V}$:
$$\mathbf{v} \wedge \mathbf{v} \to 0; \ \ \mathbf{v} \cdot \mathbf{v}^{\sim} \to 0 \text{ if } n = 2.$$

β.2. For any $a \in \mathbb{K}$ and $\mathbf{C} \in \mathcal{C}$:
$$a \wedge \mathbf{C} \to a\mathbf{C}, \ \ a \cdot \mathbf{C} \to 0; \ \ a^{\sim} \to a\mathbf{I} \text{ if } n = 2.$$

β.3. For any $m > n$ and $\mathbf{v}_1, \ldots, \mathbf{v}_m \in \mathcal{V}$:
$$\mathbf{v}_1 \wedge \cdots \wedge \mathbf{v}_m \to 0.$$

β.4. For $n = 2$ and any $\mathbf{a}, \mathbf{b} \in \mathcal{V}$:
$$(\mathbf{a}^{\sim})^{\sim} \to -\mathbf{a}, \ \ (\mathbf{a} \wedge \mathbf{b})^{\sim} \to \mathbf{a} \cdot \mathbf{b}^{\sim};$$
$$\mathbf{a} \wedge \mathbf{b}^{\sim} \to \mathbf{b} \wedge \mathbf{a}^{\sim} \text{ if } \mathbf{a} \succ \mathbf{b}.$$

β.5. For any $\mathbf{A}, \mathbf{B} \in \mathcal{C}$ with $k_1 = \mathrm{gr}(\mathbf{A}) \geq 0$ and $k_2 = \mathrm{gr}(\mathbf{B}) \geq 0$: if $\mathbf{A} \succ \mathbf{B}$ then

$$\mathbf{A} \wedge \mathbf{B} \to (-1)^{k_1 k_2} \mathbf{B} \wedge \mathbf{A},$$
$$\mathbf{A} \cdot \mathbf{B} \to (-1)^{k_1 k_2 - \min(k_1, k_2)} \mathbf{B} \cdot \mathbf{A}.$$

γ. Advanced Rules

γ.1. For any $\mathbf{a}, \mathbf{b} \in \mathcal{V}$ and $\mathbf{C} \in \mathcal{C}$ with $\mathrm{gr}(\mathbf{C}) \geq 0$:

$$\mathbf{a} \cdot (\mathbf{b} \wedge \mathbf{C}) \to \mathbf{a} \cdot \mathbf{b}\, \mathbf{C} - \mathbf{b} \wedge (\mathbf{a} \cdot \mathbf{C}).$$

γ.2. For $n = 2$ and any $\mathbf{a}, \mathbf{b}, \mathbf{c} \in \mathcal{V}$: if $\mathbf{a} \succ \mathbf{b}$ then

$$\mathbf{b} \cdot \mathbf{c}\, \mathbf{a} \wedge \mathbf{c}^{\sim} \to \mathbf{a} \cdot \mathbf{c}\, \mathbf{b} \wedge \mathbf{c}^{\sim}, \quad \mathbf{c} \cdot \mathbf{b}\, \mathbf{a} \wedge \mathbf{c}^{\sim} \to \mathbf{c} \cdot \mathbf{a}\, \mathbf{b} \wedge \mathbf{c}^{\sim}.$$

γ.3. For $n = 2$ and any $\mathbf{a}, \mathbf{b}, \mathbf{c}, \mathbf{d} \in \mathcal{V}$: if $\mathbf{a} \succ \mathbf{c}$ and $\mathbf{b} \succ \mathbf{d}$ then

$$\mathbf{a} \cdot \mathbf{b}\, \mathbf{c} \wedge \mathbf{d} \to \mathbf{a} \cdot \mathbf{d}\, \mathbf{c} \wedge \mathbf{b} - \mathbf{a} \cdot \mathbf{c}\, \mathbf{d} \wedge \mathbf{b}.$$

γ.4. For $n = 2$ and any $\mathbf{a}, \mathbf{b} \in \mathcal{V}$:

$$\mathbf{a} \wedge \mathbf{b} \to -\mathbf{a} \cdot \mathbf{b}^{\sim} \boldsymbol{I}, \quad \mathbf{a} \cdot \boldsymbol{I} \to \mathbf{a}^{\sim},$$
$$(\mathbf{a} \cdot \mathbf{b}^{\sim})^2 \to \mathbf{a} \cdot \mathbf{a}\, \mathbf{b} \cdot \mathbf{b} - (\mathbf{a} \cdot \mathbf{b})^2.$$

γ.5. $\boldsymbol{I} \cdot \boldsymbol{I} \to -1;\ \boldsymbol{I}^{\sim} \to -1.$

δ. Coordinate Rules

δ.1. For any $\mathbf{a}, \mathbf{b} \in \mathcal{V}$ with respective coordinates $(a_1, \ldots, a_n)$ and $(b_1, \ldots, b_n)$:

$$\mathbf{a} \cdot \mathbf{b} \to a_1 b_1 + \cdots + a_n b_n.$$

δ.2. For $n = 2$ and any $\mathbf{a}, \mathbf{b} \in \mathcal{V}$ with respective coordinates (a_1, a_2) and (b_1, b_2):

$$\mathbf{a} \wedge \mathbf{b} \to (a_1 b_2 - a_2 b_1) \boldsymbol{I}.$$

δ.3. For $n = 2$ and any $\mathbf{v} \in \mathcal{V}$ with coordinates (v_1, v_2): $\mathbf{v}^{\sim} \to (-v_2, v_1)$.

The above rules are classified so that the first two groups of rules can be applied along with the constructions; normally, they do not make the expressions more complicated than expanded. Application of the last two groups of rules may make the expressions more complex yet more canonical. To simplify the expression g composed by means of $\wedge, \cdot, ^{\sim}$ and the usual sum and product, we apply the rules from α to δ successively and recursively until the resulting expression cannot be further simplified. If $g = 0$ is an identity, g will always be simplified to 0 before or at the end when the coordinate rules δ are applied.

The correctness of the rewrite rules is based on the known properties and relations among Clifford algebraic operators (see [6, 7]). Some of the rules can be easily verified by using coordinates.

Before entering into other technical discussions, let us show a few non-trivial examples of geometric theorem proving. The calculations for these and other examples have been carried out using an experimental implementation of the described method in Maple. The detailed proof of each theorem may be provided as one wishes by the substitution steps sketched in Section 4 and a sequence of rewrite steps which simplify the expression g^* or g_0 to 0 using the rules α–δ.

Example 5 (Pappus Theorem). Let A, B, C and A_1, B_1, C_1 be two sets of points respectively on two distinct lines. Then the points P_1, P_2, P_3 of intersection are collinear.

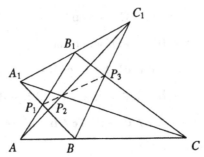

Take the intersection point of the two lines AB and $A_1 B_1$ as the origin. The points A and A_1 are arbitrarily taken and the other points are constructed as follows

$$B = u_1 A, \quad C = u_2 A, \quad B_1 = u_3 A_1, \quad C_1 = u_4 A_1,$$
$$P_1 = \text{int}(A, B_1, A_1, B), \quad P_2 = \text{int}(A, C_1, A_1, C), \quad P_3 = \text{int}(B, C_1, B_1, C).$$

The concrete expression for each P_i may be copied from that in C3.2 with simple renaming of points. The conclusion expression

$$g = (P_1 - P_2) \wedge (P_1 - P_3)$$

is simplified to 0 by substitution and application of the rules α and β. In addition to the assumption

$$A \cdot A_1^{\sim} \neq 0, \qquad\qquad \text{% the two lines } AB \text{ and } A_1 B_1 \text{ do not coincide}$$

the non-degeneracy conditions for the theorem to be true are

$$u_1 u_3 - 1 \neq 0, \qquad \text{% } AB_1 \nparallel A_1 B$$
$$u_2 u_4 - 1 \neq 0, \qquad \text{% } AC_1 \nparallel A_1 C$$
$$u_1 u_3 - u_2 u_4 \neq 0. \qquad \text{% } BC_1 \nparallel B_1 C$$

These conditions guarantee the existence of the intersection points P_1, P_2, P_3.

In our implementation, the generated proof steps may be printed automatically in plain text as well as LATEX format. They are not copied here because more work is needed on structuring and presenting the proofs for brevity and readability. In fact, tracing all the detailed substitution and rewrite steps is not our current concern.

Example 6 (Simson Theorem). From a point D draw three perpendiculars to the three sides of any triangle ABC. Then the three perpendicular feet P, Q and R are collinear if and only if D lies on the circumscribed circle of $\triangle ABC$.

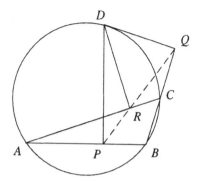

Now A, B, C are free points. For proving the "if" direction, the constructions may be as follows

$$O = \text{cir\_ctr}(A, B, C),$$
$$D = 2\,\text{per\_ft}(O, U, A) - A,$$
$$P = \text{per\_ft}(A, B, D),$$
$$Q = \text{per\_ft}(B, C, D),$$
$$R = \text{per\_ft}(C, A, D).$$

During the construction, a free point U is introduced. The algebraic expressions for the dependent points may be obtained easily from those given in Section 3. To simplify calculation, we take A as the origin, i.e., $A = 0$. The theorem is then proved by substituting the expressions of P, Q, R, etc. to the conclusion expression

$$g = (P - Q) \wedge (P - R)$$

and simplifying the result to 0 with application of the rules α–δ. Production of the detailed rewrite steps and non-degeneracy conditions is omitted.

To see "if and only if," we recall the area Δ computed in Example 3. Assume that $\triangle ABC$ does not degenerate to a line, so that $B \wedge C \neq 0$. Thus, $\Delta = 0$ if and only if $\Gamma = 0$.

On the other hand, D lies on the circumscribed circle of $\triangle ABC$ if and only if

$$\text{per\_ft}(A, D, \text{cir\_ctr}(A, B, C)) - \text{midp}(A, D) = \frac{\Gamma^* G}{2\,B \cdot C^\sim G \cdot G} = 0,$$

where

$$\Gamma^* = -B \cdot B\,C \cdot G^\sim + C \cdot C\,B \cdot G^\sim - G \cdot G\,B \cdot C^\sim.$$

Application of the simplification rules γ shows that

$$B \cdot C^\sim \Gamma^* = \Gamma.$$

Hence, Γ and Γ^* are equivalent when $B \wedge C \neq 0$. The theorem is now completely proved.

In the above proof of Simson's theorem, the coordinate rules are also used. The following example provides an easy proof of a difficult theorem without using the coordinate rules. For any three points A, B, C, let $|AB|$ denote the distance between A and B, and $|C(AB)|$ the (perpendicular) distance from C to the line AB.

Example 7 (Steiner Theorem Generalized). Take three points C', A' and B' respectively on the three perpendicular bisectors of AB, BC and CA of any $\triangle ABC$ such that

$$|C'(AB)| = t|AB|, \quad |A'(BC)| = t|BC|, \quad |B'(CA)| = t|CA|,$$

where t is an arbitrary non-negative number. Then the three lines AA', BB' and CC' are concurrent.

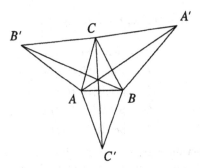

Let the vertex C of the triangle be located at the origin and I be the intersection of AA' and BB'. The points may be constructed as follows

$$A' = \mathsf{per\_dis}(C, B, \mathsf{midp}(C, B), t|CB|) = \frac{B}{2} + tB^{\sim},$$

$$B' = \mathsf{per\_dis}(A, C, \mathsf{midp}(A, C), t|AC|) = \frac{A}{2} - tA^{\sim},$$

$$C' = \mathsf{per\_dis}(B, A, \mathsf{midp}(B, A), t|BA|) = \frac{A + B}{2} + t(A^{\sim} - B^{\sim}),$$

$$I = \mathsf{int}(A, A', B, B') = \frac{[(A' - A) \cdot (B \wedge B')]^{\sim} - [(B' - B) \cdot (A \wedge A')]^{\sim}}{[(A' - A) \wedge (B' - B)]^{\sim}}.$$

It is easy to simplify the conclusion expression

$$(I - C) \wedge (C' - C) = I \wedge C'$$

to 0 by substitution and application of the rules α, β and γ. Hence, the theorem is proved to be true under the non-degeneracy condition

$$(A' - A) \wedge (B' - B) = -[t^2 A \cdot B^{\sim} + t(A \cdot B - B \cdot B - A \cdot A) + \frac{3}{4} A \cdot B^{\sim}]\mathbf{I} \neq 0,$$

i.e., the two lines AA' and BB' are not parallel. $AA' \parallel BB'$ happens, for example, when $t = 1/2$, and A and B are situated at $(1, 0)$ and $(1, -1)$ respectively. In this case A' coincides with A, and B' lies on BC.

The centroid theorem in Example 1, the well-known Steiner theorem, and the orthocenter theorem in Example 2 are special cases of the above-proved theorem with $t = 0$, $\sqrt{3}/2$ and infinity, respectively. Proving Steiner's theorem using coordinates may encounter the reducibility problem and is not easy. The proof given here is very simple. We have not looked up the geometry literature to find out where appeared the theorem in this generalized form we discovered independently in January 1994.

The above examples demonstrate that the simple approach we have suggested may be used to prove difficult geometric theorems automatically. Nevertheless, there are several theoretical and practical questions, in particular for simplifying Clifford algebraic expressions, that remain to be answered. We mention some of them briefly below. They will be clarified and addressed in our future research.

The first question is about the termination of the simplification procedure. For most of the rules given above, the termination is obvious. But it is not the case for the rules in which ordering is involved. To ensure the termination for these rules, Clifford algebraic terms (i.e., the inner/outer/usual power products of scalars and vectors) and expressions need be properly ordered. It is not easy to introduce a suitable ordering because, for example, checking whether two Clifford algebraic expressions are equal falls into the same simplification problem. However, in practice it is not difficult to implement some heuristical ordering to force the procedure to terminate, or give up applying some of the rules when there is doubt about the termination. For our experiment, we have used an ordering according to the order of the vectors and the addresses of expressions in the memory. The suitability of this ordering has yet to be studied.

The second question is what other rules need be added to make the rewriting system confluent, so that any Clifford algebraic expression that is identically 0 will always be simplified to 0, without using coordinate rules. And, what is the minimal set of such rules? These questions are beyond the reach of this paper. Here we are satisfied by only listing some of the rules and showing how difficult theorems may be easily proved by using them.

At the practical level, different variations and strategies may be developed and implemented. An expression might be simplified to 0 heuristically by a simple application of one or two rules, while it may get more complicated if the simplification rules are applied in a different way. This is quite like the case of proving trigonometric identities where tricky use of existing relations may lead to short and elegant proofs. Our initial attempt here is to adopt a few elementary rules as the basis of simplification. Developing heuristics and incorporating sophisticated rules as lemmas may enhance the proving power but have not been investigated yet. It is obvious that shorter and more readable proofs can be produced if more rules are implemented as lemmas or propositions. Moreover, provers based on our approach can be easily designed for "self-learning:" the conclusion expression of any true theorem will be simplified, with application of the rules α–γ, to 0 or to an expression h that is identically equal to 0; thus, in the latter case $h \to 0$ may be proved by using the coordinate rules δ and added automatically as a new rule to the rewriting system.

6 Non-commutative Gröbner Bases Applied

The theory and method of non-commutative Gröbner bases have been developed by several researchers, in particular for non-commutative polynomial rings of solvable type. Here we explain how the algorithm proposed in [8] for solving the (radical) ideal membership problem in Grassmann algebra (i.e., the Clifford algebra associated to the null quadratic form) can be applied to geometric theorem proving [17]. This section is independent of the general approach described in the previous three sections.

Let $\mathbb{K}$ be a (commutative) field and $\mathbb{R} = \mathbb{K}[x_1,\ldots,x_n]$ the (commutative) polynomial ring in the indeterminates x_i over $\mathbb{K}$. A solvable polynomial ring $\mathbb{K}\{x_1,\ldots,x_n\}$ is an ordinary commutative polynomial ring $\mathbb{R}$ equipped with a new non-commutative multiplication $*$ satisfying some conditions. See [8] for the definition as well as the concepts of *left, right* and *two-sided ideals, reduction,* and *Gröbner bases* in solvable polynomial rings. According to [8], there is an algorithm which computes a left, right or two-sided Gröbner basis of the left, right or two-sided ideal generated by any given set of polynomials in $\mathbb{K}\{x_1,\ldots,x_n\}$.

Now, let $\mathbb{P} = \mathbb{K}\langle x_1,\ldots,x_n\rangle$ denote the ring of polynomials in non-commutative indeterminates $x_1,\ldots,x_n$ that commute with the elements of $\mathbb{K}$, i.e., the free associative algebra over $\mathbb{K}$ generated by $\{x_1,\ldots,x_n\}$. Fix an admissible ordering $\prec$ on the set of commutative terms in $x_1,\ldots,x_n$. A *commutation system* for $(\mathbb{P},\prec)$ is a set of polynomials of the form

$$\mathcal{Q} = \{x_j * x_i - c_{ij}x_i * x_j - p_{ij}|\ 1 \le i < j \le n\},$$

where $0 \ne c_{ij} \in \mathbb{K}$, $p_{ij} \in \mathbb{K}[x_1,\ldots,x_n]$ and $p_{ij} \prec x_i * x_j$. Let $\mathfrak{I}(\mathcal{Q})$ denote the two-sided ideal generated by $\mathcal{Q}$ in $\mathbb{P}$. Assume that $\mathfrak{I}(\mathcal{Q})$ does not contain any non-zero commutative polynomial. The following result is established in [8].

Theorem. *The quotient* $\mathbb{P}/\mathfrak{I}(\mathcal{Q})$ *is isomorphic to a solvable polynomial ring* $\mathbb{R} = \mathbb{K}\{y_1,\ldots,y_n\}$ *with respect to* $\prec$ *and the multiplication of* $\mathbb{R}$ *under an isomorphism fixing* $\mathbb{K}$ *pointwise and mapping the residue class of* x_i *modulo* $\mathfrak{I}(\mathcal{Q})$ *onto* y_i.

Now, consider Grassmann algebra and let $\mathbf{v}_1,\ldots,\mathbf{v}_m$ be m vectors of $\mathcal{V}$. Construct formally the polynomial ring $\mathbb{K}[\mathbf{v}_1,\ldots,\mathbf{v}_m]$ and take the commutation system

$$\mathcal{Q} = \{\mathbf{v}_j \wedge \mathbf{v}_i + \mathbf{v}_i \wedge \mathbf{v}_j|\ 1 \le i < j \le n\}.$$

It is easy to show that $\mathfrak{I}(\mathcal{Q})$ does not contain any non-zero commutative polynomial. Therefore,

$$\mathbb{R} = \mathbb{K}\{\mathbf{v}_1,\ldots,\mathbf{v}_m\} = \mathbb{K}\langle\mathbf{v}_1,\ldots,\mathbf{v}_m\rangle/\mathfrak{I}(\mathcal{Q})$$

is a polynomial ring of solvable type. Let $\mathcal{I}$ be the two-sided ideal in $\mathbb{R}$ generated by

$$\mathcal{P} = \{\mathbf{v}_1 \wedge \mathbf{v}_1,\ldots,\mathbf{v}_m \wedge \mathbf{v}_m\}.$$

It is not difficult to verify that $\mathcal{P}$ is a two-sided Gröbner basis in $\mathbb{R}$. Hence, the ideal membership problem in the quotient algebra $\mathbb{R}/\mathcal{I}$ can be treated in the following manner: Given the generating set $\mathcal{F}$ of a two-sided (left, right) ideal $\mathcal{J}$ and a polynomial f in $\mathbb{R}/\mathcal{I}$, construct a (left, right) Gröbner basis $\mathcal{G}$ of $\mathcal{I}+\mathcal{J}$ from $\mathcal{P} \cup \mathcal{F}$ and compute a (left, right) normal form h of f with respect to $\mathcal{G}$. Then $f \in \mathcal{J}$ if and only if $h = 0$.

Remark that our construction of the quotient algebra $\mathbb{R}/\mathcal{I}$ is different from the Grassmann algebra formed in [8], where the base vectors $e_1, \ldots, e_n$ were taken as non-commutative indeterminates. In that way, every element of the Grassmann algebra has to be represented as a non-commutative polynomial in the base vectors. This is equivalent to using coordinates to represent k-vectors and thus inadequate for the coordinate-free approach we want.

Example 1*. Refer to Example 1 about the centroid of a triangle. Computing a two-sided non-commutative Gröbner basis $\mathcal{G}$ of

$$\{h_1, h_2\} \cup \{B \wedge B, C \wedge C, P \wedge P\}$$

with $B \prec C \prec P$, one finds that

$$\mathcal{G} = \{B \wedge B, C \wedge C, B \wedge P - \frac{1}{3} B \wedge C, C \wedge P + \frac{1}{3} B \wedge C, P \wedge P\}.$$

A (left) normal form of g with respect to $\mathcal{G}$ is 0, so the theorem is proved to be true. As a side effect, the third and fourth elements in the Gröbner basis $\mathcal{G}$ indicate that the area of $\triangle ABP$ is one-third of the area of $\triangle ABC$, and so is the area of $\triangle ACP$. In consequence, $\triangle ABP$ and $\triangle ACP$ have the same area, and

$$|PC_1| = \frac{1}{3}|CC_1|, \quad |PB_1| = \frac{1}{3}|BB_1|.$$

The non-commutative Gröbner bases in this and other examples were computed, with respect to the inverse lexicographical term ordering, by using functions implemented in the MAS algebra system. This system is developed by H. Kredel and his colleagues at the University of Passau, Germany.

Example 8 (Gauss Line). The midpoints M_1, M_2, M_3 of the three diagonals A_1B_1, A_2B_2, A_3B_3 of any complete quadrilateral are collinear.

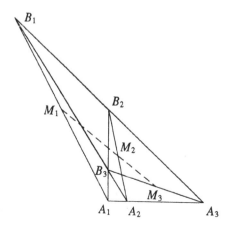

The hypothesis of the theorem may be expressed as

$$
\begin{aligned}
h_1 &= (A_1 - A_2) \wedge (A_1 - A_3) = 0, &&\text{\% col}(A_1, A_2, A_3) \\
h_2 &= (A_1 - B_2) \wedge (A_1 - B_3) = 0, &&\text{\% col}(A_1, B_2, B_3) \\
h_3 &= (B_1 - A_2) \wedge (B_1 - B_3) = 0, &&\text{\% col}(B_1, A_2, B_3) \\
h_4 &= (B_1 - B_2) \wedge (B_1 - A_3) = 0, &&\text{\% col}(B_1, B_2, A_3) \\
h_5 &= 2M_1 - (A_1 + B_1) = 0, &&\text{\% } M_1 = \text{midp}(A_1, B_1) \\
h_6 &= 2M_2 - (A_2 + B_2) = 0, &&\text{\% } M_2 = \text{midp}(A_2, B_2) \\
h_7 &= 2M_3 - (A_3 + B_3) = 0. &&\text{\% } M_3 = \text{midp}(A_3, B_3)
\end{aligned}
$$

Under the variable ordering $A_1 \prec A_2 \prec A_3 \prec B_1 \prec B_2 \prec B_3 \prec M_1 \prec M_2 \prec M_3$, a two-sided Gröbner basis $\mathcal{G}$ of $\{h_1, \ldots, h_7\} \cup \{A_1 \wedge A_1, \ldots, M_3 \wedge M_3\}$ consists of $A_1 \wedge A_1, A_2 \wedge A_2, A_3 \wedge A_3, B_1 \wedge B_1, B_2 \wedge B_2, B_3 \wedge B_3$ and

$$
\begin{aligned}
&A_2 \wedge A_3 - A_1 \wedge A_3 + A_1 \wedge A_2, \\
&B_1 \wedge B_2 - A_3 \wedge B_2 + A_3 \wedge B_1, \\
&B_1 \wedge B_3 - A_2 \wedge B_3 + A_2 \wedge B_1, \\
&B_2 \wedge B_3 - A_1 \wedge B_3 + A_1 \wedge B_2, \\
&M_1 - \frac{1}{2}B_1 - \frac{1}{2}A_1, \\
&M_2 - \frac{1}{2}B_2 - \frac{1}{2}A_2, \\
&M_3 - \frac{1}{2}B_3 - \frac{1}{2}A_3.
\end{aligned}
$$

The conclusion of the theorem is

$$
g = (M_1 - M_2) \wedge (M_1 - M_3) = -M_1 \wedge M_3 - M_2 \wedge M_1 + M_2 \wedge M_3 = 0.
$$

It is easy to verify that a (left) normal form of g with respect to $\mathcal{G}$ is 0, so the theorem is proved to be true.

However, without further extension the above method can prove only a small class of geometric theorems. The reason seems that the mere Gröbner bases computation is not complete for the decision problem in Grassmann algebra. We have thought about some possible extensions for the method. These include pre-processing the set of hypothesis expressions using Clifford algebraic operations and adding special identities to the hypothesis set. We hope that investigations on these possibilities will be conducted in the future.

7 Application to Computer Vision

Constructive methods based on Clifford algebra have potential applications in computer vision and other geometry engineering areas such as geometric modeling and computer-aided geometric design. This can be seen from the methods' geometric features and the indispensability of dealing with 3-dimensional objects in the above-mentioned areas. For example, analyzing 3-dimensional scenes and constructing realistic models from sets of different images are two key tasks in computer vision. For this, pertinent information about characteristic objects like points, lines, planes and circles must be extracted, and relationships among these objects in the images and in the scenes have to be established. This amounts to manipulating and reasoning about the involved geometric objects, for which automated methods and tools are desirable. Typical computer vision problems to which Clifford-algebra-based methods may be applied include image understanding and analysis, computing geometric invariants, generating and verifying geometric properties and relations, and deriving and solving geometric constraints. In fact, vector representation has been used extensively for formal calculation and derivation treating these problems in the computer vision literature. The computational tasks should be made easier and more systematic when Clifford algebraic methods are developed, implemented and used.

The application of Wu's method to problems in computer vision has been studied and reported in a few papers, for example, [4, 9]. In this section, we present two examples to illustrate how the problems can also be handled by using Clifford algebraic calculus. The examples are about deriving geometric constraints and relations in projective and perspective viewing.

Example 9 (Parallel Lines). Let two parallel lines P_1P_2 and P_3P_4 in a scene remain parallel in an image plane (i.e., $Q_1Q_2 \parallel Q_3Q_4$). Find possible constraints.

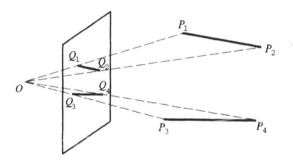

Let the viewpoint O be located at the origin. The geometric conditions can be expressed as

$$h_1 = (P_4 - P_3) \wedge (P_2 - P_1) = 0, \qquad \text{\% } P_1P_2 \parallel P_3P_4$$
$$h_{i+1} = P_i \wedge Q_i = 0, \quad i = 1, \dots, 4, \qquad \text{\% col}(O, P_i, Q_i)$$
$$h_6 = (Q_1 - Q_2) \wedge (Q_1 - Q_3) \wedge (Q_1 - Q_4) = 0,$$
$$\qquad \text{\% } Q_1Q_2 \text{ and } Q_3Q_4 \text{ are coplanar}$$
$$h_7 = (Q_4 - Q_3) \wedge (Q_2 - Q_1) = 0. \qquad \text{\% } Q_1Q_2 \parallel Q_3Q_4$$

Let $\mathcal{Q} = \{P_1 \wedge P_1, \dots, P_4 \wedge P_4, Q_1 \wedge Q_1, \dots, Q_4 \wedge Q_4\}$. With respect to the variable ordering $P_1 \prec \cdots \prec P_4 \prec Q_1 \prec \cdots \prec Q_4$, a two-sided Gröbner basis of

$\{h_1, \ldots, h_7\} \cup \mathcal{Q}$ is

$$\mathcal{G} = \begin{Bmatrix} P_2 \wedge P_4 - P_1 \wedge P_4 - P_2 \wedge P_3 + P_1 \wedge P_3, \\ P_1 \wedge Q_1, \\ P_2 \wedge Q_2, \\ P_1 \wedge P_4 \wedge Q_2 - P_1 \wedge P_3 \wedge Q_2, \\ P_3 \wedge Q_3, \\ P_4 \wedge Q_2 \wedge Q_3 - P_4 \wedge Q_1 \wedge Q_3, \\ P_2 \wedge P_3 \wedge Q_4 - P_1 \wedge P_3 \wedge Q_4, \\ P_4 \wedge Q_4, \\ P_2 \wedge Q_1 \wedge Q_4 - P_2 \wedge Q_1 \wedge Q_3, \\ Q_2 \wedge Q_4 - Q_1 \wedge Q_4 - Q_2 \wedge Q_3 + Q_1 \wedge Q_3 \end{Bmatrix} \cup \mathcal{Q}.$$

Let plane(A, B, C) stand for the plane determined by the three points A, B, C. From the Gröbner basis $\mathcal{G}$, the following relations may be observed:

$$P_1 \wedge (P_4 - P_3) \wedge Q_2 = 0 \iff P_3 P_4 \text{ lies in or is parallel to } \mathsf{plane}(O, P_1, Q_2);$$
$$P_4 \wedge (Q_2 - Q_1) \wedge Q_3 = 0 \iff Q_1 Q_2 \text{ lies in or is parallel to } \mathsf{plane}(O, P_4, Q_3);$$
$$(P_2 - P_1) \wedge P_3 \wedge Q_4 = 0 \iff P_1 P_2 \text{ lies in or is parallel to } \mathsf{plane}(O, P_3, Q_4);$$
$$P_2 \wedge Q_1 \wedge (Q_4 - Q_3) = 0 \iff Q_3 Q_4 \text{ lies in or is parallel to } \mathsf{plane}(O, P_2, Q_1).$$

Hence, either

- the lines $Q_1 Q_2, Q_3 Q_4, P_1 P_2, P_3 P_3$ are all coplanar, and thus $Q_1 Q_2$ coincides with $Q_3 Q_4$, or
- $P_1 P_2$ is parallel to the image plane.

The above conclusion was derived by D. Kapur and J. L. Mundy in [9] using Wu's method, where coordinate representation is used. We do not know how they got the polynomial equations interpreted geometrically. Here the expressions are given as outer products of vectors; their geometric interpretation can be figured out easily.

Example 10. Determine geometric relations among points $P_1, P_2, P_3, \ldots$ in a scene plane in terms of their image projections $Q_1, Q_2, Q_3, \ldots$.

This problem was investigated by K. Deguchi [4] using Wu' method. The derivation of geometric constraints and relations of length and angle correspondences between 3-dimensional target and its projected image plane was considered for different cases. Here is given a simple and general setting for all the relations established in [4]. Let the viewpoint O be the origin and the scene plane α be given by

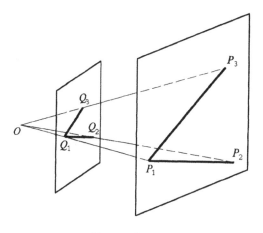

$$\alpha: \quad \mathbf{X} \cdot \mathbf{n} + k = 0,$$

where $\mathbf{n} = (p, q, -1)$. Then the geometric conditions may be expressed as

$$P_i \wedge Q_i = 0, \qquad \quad \% \ \mathrm{col}(O, Q_i, P_i)$$
$$P_i \cdot \mathbf{n} + k = 0, \qquad \% \ P_i \ \text{lies in } \alpha$$
$$i = 1, 2, \ldots.$$

Solving the above equations for P_i according to [10], we obtain

$$P_i = -\frac{k}{Q_i \cdot \mathbf{n}} Q_i, \quad i = 1, 2, \ldots.$$

Therefore, any geometric quantity among the P_i can be calculated as a function of the Q_i. For examples, by substituting the expressions of P_i into the right-hand side of the equations

$$|P_1 P_2| = |P_1 - P_2| = \sqrt{(P_1 - P_2) \cdot (P_1 - P_2)},$$
$$\cos \angle P_2 P_1 P_3 = \frac{(P_2 - P_1) \cdot (P_3 - P_1)}{|P_2 - P_1||P_3 - P_1|},$$

one establishes the distance between P_1 and P_2, and the cosine of the angle formed by the lines $P_1 P_2$ and $P_1 P_3$ in terms of the image points Q_1, Q_2, Q_3.

Acknowledgments. This research is supported by AFCRST under Project PRA M94-1 and by CEC under Reactive LTR Project 21914 – CUMULI.

References

1. Chou, S.-C., Gao, X.-S., Zhang, J.-Z.: Automated production of traditional proofs for constructive geometry theorems. In: Proc. 8th IEEE Symp. LICS (Montreal, June 19–23, 1993), pp. 48–56.
2. Chou, S.-C., Gao, X.-S., Zhang, J.-Z.: Automated geometry theorem proving by vector calculation. In: Proc. ISSAC '93 (Kiev, July 6–8, 1993), pp. 284–291.

3. Chou, S.-C., Gao, X.-S., Zhang, J.-Z.: Machine proofs in geometry. World Scientific, Singapore (1994).

4. Deguchi, K.: An algebraic framework for fusing geometric constraints of vision and range sensor data. In: Proc. IEEE Int. Conf. MFI '94 (Las Vegas, October 2–5, 1994), pp. 329–336.

5. Havel, T. F.: Some examples of the use of distances as coordinates for Euclidean geometry. J. Symb. Comput. 11: 579–593 (1991).

6. Hestenes, D.: New foundations for classical mechanics. D. Reidel, Dordrecht Boston Lancaster Tokyo (1987).

7. Hestenes, D., Sobczyk, G.: Clifford algebra to geometric calculus. D. Reidel, Dordrecht Boston Lancaster Tokyo (1984).

8. Kandri-Rody, A., Weispfenning, V.: Non-commutative Gröbner bases in algebras of solvable type. J. Symb. Comput. 9: 1–26 (1990).

9. Kapur, D., Mundy, J. L.: Wu's method and its application to perspective viewing. Artif. Intell. 37: 15–26 (1988).

10. Li, H.: New explorations on mechanical theorem proving of geometries. Ph.D thesis, Beijing University, China (1994).

11. Li, H.: Clifford algebra and area method. Math. Mech. Res. Preprints 14: 37–69 (1996).

12. Li, H., Cheng, M.-t.: Proving theorems in elementary geometry with Clifford algebraic method. Preprint, MMRC, Academia Sinica, China (1995).

13. Richter-Gebert, J.: Mechanical theorem proving in projective geometry. Ann. Math. Artif. Intell. 13: 139–172 (1995).

14. Stifter, S.: Geometry theorem proving in vector spaces by means of Gröbner bases. In: Proc. ISSAC '93 (Kiev, July 6–8, 1993), pp. 301–310.

15. van der Waerden, B. L.: Algebra, vol. II. Springer, Berlin Heidelberg (1959).

16. Wang, D.: On Wu's method for proving constructive geometric theorems. In: Proc. IJCAI '89 (Detroit, August 20–25, 1989), pp. 419–424.

17. Wang, D.: Geometry machines: From AI to SMC. In: Proc. AISMC-3 (Steyr, September 23–25, 1996), LNCS 1138, pp. 213–239.

18. Wong, R.: Construction heuristics for geometry and a vector algebra representation of geometry. Tech. Memo. 28, Project MAC, MIT, Cambridge, USA (1972).

19. Wu, W.-t.: Toward mechanization of geometry — Some comments on Hilbert's "Grundlagen der Geometrie". Acta Math. Scientia 2: 125–138 (1982).

20. Wu, W.-t.: Mechanical theorem proving in geometries: Basic principles. Springer, Wien New York (1994).

Area in Grassmann Geometry

Desmond Fearnley-Sander[1] and Tim Stokes[2]

[1] Department of Mathematics, University of Tasmania
GPO Box 252C, Hobart, Tasmania 7001, Australia
URL: http://www.maths.utas.edu.au/People/dfs/dfs.html
EMAIL: dfs@hilbert.maths.utas.edu.au
[2] School of Mathematical and Physical Sciences
Murdoch University, Murdoch, WA 6150, Australia
EMAIL: stokes@prodigal.murdoch.edu.au

Abstract. In this survey paper we give the basic properties of Grassmann algebras, present a generalised theory of area from a Grassmann algebra perspective, present a version for Grassmann algebras of the Buchberger algorithm, and give examples of computation and deduction in Grassmann geometry.

The automatic proof of geometry theorems using the powerful algorithms of Wu and Buchberger is the most impressive achievement to date in automated theorem proving. It is our view, though, that progress in the automation of geometry requires something more than the invention and refinement of algorithms. What we have in mind is the creation of algebraic structures that internalize the rich variety of geometric concepts in ways that are amenable to computation. In this survey paper we present such a structure, along with its variant of the Buchberger algorithm.

Grassmann algebras are appropriate many-sorted algebraic structures for affine geometry, well-suited to deduction and to computation of quantities such as areas. The basic geometric objects are points. Vectors and less familiar geometric entities such as bipoints and bivectors are generated as we will describe. To illustrate the invariant coordinate-free flavour of Grassmann algebra we consider a general notion of 'area' enclosed by a closed curve in a space of arbitrary dimension, which in two-dimensional spaces reduces to a scale-free version of the familiar concept. The general notion supports the extension of many plane geometry theorems involving area to higher dimensions.

White [28], Sturmfels and Whitely [27] and others have developed a similar approach to projective geometry theorem proving, based on the Cayley Algebra, and using Cayley factorization as the basic algorithmic tool. Hestenes and Ziegler in [17] present an extensive study of the projective model of affine geometry using Grassmann-Cayley algebra. Also interesting is the work of Chou, Gao and Zhang ([5], [6] and [7]) in which theorems of two and three dimensional geometry are proved using a formalism based on areas and volumes which allows higher level interpretation of the resultant proofs. Our way is in a sense more elementary and its development has been partly motivated by the desire to transparently incorporate elementary geometric reasoning and computation in systems with wider reasoning capabilities.

1   Grassmann Algebras

A *Grassmann algebra* $\Omega[\mathcal{K}, \mathcal{P}]$ is a ring (associative and with unit element 1) which is generated by (the union of) disjoint distinguished subsets $\mathcal{K}$ and $\mathcal{P}$ such that

GA1 $\mathcal{K}$ is a field (under the ring operations);
GA2 $\mathcal{P}$ is an affine space over $\mathcal{K}$ (under the ring operations);
GA3 $aA = Aa$ for every $a \in \mathcal{K},\, A \in \mathcal{P}$;
GA4 $BA = -AB$ for every $A, B \in \mathcal{P}$.
GA5 for $P_1, P_2, \ldots, P_k$ in $\mathcal{P}$,

$$P_1 P_2 \cdots P_k = 0 \Rightarrow P_1, P_2, \ldots, P_k \text{ dependent (over } \mathcal{K}).$$

The meaning of **GA2** is that $\mathcal{P}$ is closed under *affine combinations*:

$$A, B \in \mathcal{P}, a, b \in \mathcal{K} \text{ and } a + b = 1 \Rightarrow aA + bB \in \mathcal{P}.$$

For the real case, all geometric interpretations of expressions and equations between expressions follow from the single fundamental *semantic rule*:

if A and B are points and a and b positive real numbers with $a + b = 1$ then $aA + bB$ may be interpreted as the point P which divides the line segment from A to B in the ratio b to a.

Elements of $\mathcal{K}$ are called *numbers*, elements of $\mathcal{P}$, *points*, and elements of the set $V = \mathcal{P} - \mathcal{P}$, *vectors*. These sets are disjoint, except that 0 is both a number and a vector. Throughout this papaer we use the letters $A, \ldots, P$ for points, and $U, \ldots, Z$ for vectors.

The implication **GA5** is in fact an equivalence:

$$P_1 P_2 \cdots P_k = 0 \iff P_1, P_2, \ldots, P_k \text{ are dependent.}$$

For if $P_1, P_2, \ldots, P_k$ are dependent then one of the P_j is expressible as a linear combination of the others and hence, clearly, the product $P_1 P_2 \cdots P_k$ is zero. This condition is often easier to handle than the standard definition of (linear) dependence. The equivalence continues to hold, as one may show, even if some or all of the P_j are replaced by vectors.

Theorem 1 (*The Instantiation Theorem*). *For points* $P_1, \ldots, P_k$ *in a Grassmann algebra,*

$$P_1 P_2 \cdots P_k = 0 \iff \begin{aligned} &\text{either } P_1 P_2 \cdots P_{k-1} = 0 \\ &\text{or } P_k \text{ is spanned by } P_1, P_2, \ldots, P_{k-1}. \end{aligned}$$

and this remains valid if any or all of the points are replaced by vectors.

Proof. Suppose that $P_1 P_2 \cdots P_k = 0$. According to **GA5** there exist scalars $p_1, p_2, \ldots, p_k$, not all zero, such that

$$p_1 P_1 + p_2 P_2 + \cdots p_k P_k = 0.$$

If p_k is non-zero then the second of the stated alternatives holds. Otherwise one of $P_1, P_2, \ldots, P_{k-1}$ is a linear combination of the rest and hence $P_1 P_2 \cdots P_{k-1} = 0$. The converse implication is obvious. $\square$

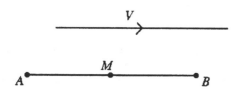

Fig. 1. $V = B - A$ and $AM + MB = AB$.

Note that $AB = 0 \iff B = A$ and $AV = 0 \iff V = 0$. Also, if we define $\mathsf{midpoint}(A, B) = \frac{1}{2}(A + B)$ then

$$AM = MB \iff M = \mathsf{midpoint}(A, B),$$

giving a precise interpretation to traditional notation.

For $k = 3$ the *Instantiation Theorem* is:

$$ABP = 0 \iff \text{either } AB = 0 \text{ or } P = aA + bB \text{ for some } a, b.$$

Hence for non-coincident points A and B we define

$$\mathsf{collinear}(A, B, C) \iff ABC = 0.$$

Similarly, for non-collinear points A, B and C

$$\mathsf{coplanar}(A, B, C, D) \iff ABCD = 0.$$

Theorem 2 (*The Boundary Theorem*). *Let P be a point and $V_1, V_2, \ldots, V_k$ vectors in a Grassmann algebra. Then*

$$PV_1 \cdots V_k = 0 \iff V_1 \cdots V_k = 0$$

Proof. Use the *Instantiation Theorem*, together with the fact that a point P cannot be a linear combination of vectors. $\square$

Theorem 3 (*The Exchange Theorem*). *If*

$$A_1 \cdots A_n P_j = 0 \text{ for } j = 1 \text{ to } n + 1,$$

then either $A_1 \cdots A_n = 0$ or $P_1 \cdots P_{n+1} = 0$.

Proof. Suppose that $A_1 \cdots A_n P_j = 0$ for $j = 1$ to $n + 1$ and that $A_1 \cdots A_n$ is not 0. According to the *Instantiation Theorem*, each P_j is expressible as a linear combination of $A_1, \ldots, A_n$. Now when the product $P_1 \cdots P_{n+1}$ is expanded each term in the resultant sum is a product of $n + 1$ elements of the set $\{A_1, \ldots, A_n\}$ and hence vanishes. □

The *Exchange Theorem* clearly remains valid if some or all of the A_j and P_j are replaced by vectors. It is the Grassmann geometry analogue of the "no zero divisors" property for fields.

2 Existence and Uniqueness

Let Ω be a Grassmann algebra. Either there exists a finite maximal set of independent points $O_0, O_1, \ldots, O_n$ in Ω or not.

If there does then we say that Ω is *finite-dimensional* and that $(O_0, O_1, \ldots, O_n)$ is an *affine coordinate system* for Ω and that $(O, X_1, \ldots, X_n)$, where

$$O = O_0, X_1 = O_1 - O_0, \ldots, X_n = O_n - O_0,$$

is the corresponding *Cartesian coordinate system*. And then, also, both $(O_0, O_1, \ldots, O_n)$ and $(O, X_1, \ldots, X_n)$ are bases for the linear space Ω_0 spanned by $\mathcal{P}$, and $(X_1, \ldots, X_n)$ is a basis for $\mathcal{V}$.

Theorem 4 (*The Dimension Theorem*). *For a finite-dimensional Grassmann algebra, all coordinate systems Ω have the same number of elements.*

The number of elements in a coordinate system of a finite-dimensional Grassmann algebra, less one, is called its *dimension*.

The next theorem pins down the structure of a finite-dimensional Grassmann algebra Ω. For each natural number k, let Ω_k be the linear subspace of Ω that is spanned by $\mathcal{P}^{k+1}$; let $\Omega_{-1} = K$. Elements of Ω_k are called *chains of dimension k* (or of *degree $k + 1$*), or, briefly, *k-chains*. In particular, -1-chains are numbers, and 0-chains are points, multiples of points or vectors.

Theorem 5 (*The Structure Theorem*). *Let Ω be an n-dimensional Grassmann algebra. As a linear space Ω is the direct sum of the subspaces $\Omega_{-1}, \Omega_0, \ldots, \Omega_n$; moreover, for each k, Ω_k has dimension $\binom{n+1}{k+1}$ and (hence) Ω has dimension 2^{n+1}.*

Using the 2^{n+1} basis elements $1, O_0, O_1, \ldots, O_n, O_0 O_1, \ldots, O_0 O_n, O_1 O_2, \ldots, O_1 O_n, \ldots, O_{n-1} O_n, \ldots, O_0 O_1 \cdots O_n$ corresponding to an affine coordinate system $O_0, \ldots, O_n$, one obtains a multiplication table for the n-dimensional algebra Ω. Since an algebra is determined by its dimension together with the multiplication table for a basis, the uniqueness (up to isomorphism) of the Grassmann algebra of dimension n is established.

If $O, X_1, \ldots, X_n$ is the cartesian coordinate system associated with $(O_0, O_1, \ldots, O_n)$ (meaning that $O = O_0, X_1 = O_1 - O, \ldots, X_n = O_n - O$) then it is easy to see that $1,\ O,\ X_1,\ \ldots,\ X_n,\ OX_1,\ \ldots, OX_n,\ X_1 X_2,\ \ldots,\ X_1 X_n,\ \ldots,\ OX_1 \cdots X_n$ is also a basis for Ω, with the property that its elements of dimension k form a basis for Ω_k.

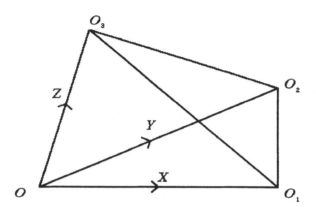

Fig. 2. A coordinate system.

To take an important example, in the 3-dimensional Grassmann algebra Ω every element ϕ of is uniquely expressible in the form

$$\phi = \phi_{-1} + \phi_0 + \phi_1 + \phi_2 + \phi_3$$

where each ϕ_i is a chain of dimension i. Moreover if (O, X, Y, Z) is a cartesian coordinate system for Ω, these components are themselves uniquely expressible, respectively, as

$$\phi_{-1} = k,$$
$$\phi_0 = oO + xX + yY + zZ,$$
$$\phi_1 = aOX + bOY + cOZ + uXY + vYZ + wZX$$
$$\phi_2 = eOXY + fOYZ + gOZX + tXYZ$$
$$\phi_3 = dOXYZ,$$

the coefficients $k, o, x, \ldots, d$ being numbers.

Theorem 6 (*The Existence Theorem*). *For each natural number n there exists a Grassmann algebra of dimension n over $\mathcal{K}$.*

Proofs of the above theorems are given in [11].

3 Multipoints and Multivectors

A product AB of two distinct points is called a *bipoint*, a product ABC of three non-collinear points is called a *tripoint* and, in general, a non-zero product

$A_1 \cdots A_k$ of points is called a *multipoint* (of *degree k*) — it is the *exterior product* of the points.

The following easily proved equivalence shows how a bipoint is to be interpreted.

$$CD = kAB \iff \text{collinear}(A, B, C), \text{collinear}(A, B, D), D - C = k(B - A).$$

In particular the bipoint from C to D equals the bipoint from A to B if and only if all four points are collinear and the vector from C to D equals the vector from A to B. Accordingly a bipoint is sometimes called a *line vector*.

One easily sees that $kAB = AP$ for some point P. Hence the set of all bipoints is closed under multiplication by non-zero scalars. We say that two bipoints are *projectively equivalent* if one is a non-zero scalar multiple of the other. This is an equivalence relation. Writing

$$[AB] = \{kAB : \ k \text{ non-zero}\},$$

we see that there is a one-to-one correspondence between the set of such equivalence classes of bipoints (modulo multiplication by non-zero scalars) and the set of all lines. Accordingly we may unambiguously view a line as an algebraic object $[AB]$ rather than as a set of points. In a similar fashion, a plane may be viewed as an equivalence class of tripoints $[ABC]$. This is the Grassmann algebra approach to projective geometry.

A non-zero product of k vectors is called a *multivector* of *degree k*; in particular, a multivector of degree 2 is called a *bivector* and one of degree 3 is called a *trivector*. Thus, if U, V and W are independent vectors then VW is a bivector and UVW is a trivector.

For independent points A, B, C, D the 1-chain $AB + CD$ cannot be a bipoint or a bivector, since squares of bipoints and bivectors are obviously 0.

Two multivectors are said to be *parallel* if they are dependent; this just means that each is a non-zero multiple of the other. By the *Structure Theorem*, parallel multivectors must have the same degree. Parallelism is an equivalence relation on the set of all multivectors.

For vectors V and W we define

$$\text{parallel}(V, W) \iff VW = 0.$$

Theorem 7. *Bivectors VW and $V'W'$ are parallel if and only if $\text{span}(V, W) = \text{span}(V', W')$.*

This extends in the obvious way to multivectors of arbitrary degree. Thus there is a one-to-one correspondence between the projective equivalence classes of multivectors and the linear subspaces of V; these are the points, lines, planes and hyperplanes at infinity.

4 Linear Maps

Let Ω and Λ be Grassmann algebras (over the same field $\mathcal{K}$) with point spaces $\mathcal{P}$ and $\mathcal{Q}$ and vector spaces $\mathcal{V}$ and W respectively.

A ring homomorphism $\mathbf{T} : \Omega \to \Lambda$ is called a *linear map* (or *Grassmann algebra homomorphism*), if it leaves numbers fixed and maps points to points. A linear map $\mathbf{T}$ preserves lines and ratios of distances along lines, because for any a, b in $\mathcal{K}$ and A, B in $\mathcal{P}$,

$$\mathbf{T}(aA + bB) = a\mathbf{T}(A) + b\mathbf{T}(B).$$

Also $\mathbf{T}$ preserves vectors, because, for any A, B in $\mathcal{P}$,

$$\mathbf{T}(B - A) = \mathbf{T}(B) - \mathbf{T}(A),$$

and $\mathbf{T}$ acts linearly on vectors.

One easily shows that

$$\mathsf{parallelogram}(A, B, C, D) \Rightarrow \mathsf{parallelogram}(\mathbf{T}(A), \mathbf{T}(B), \mathbf{T}(C), \mathbf{T}(D));$$

that

$$M = \mathsf{centroid}(A, B, C) \Rightarrow \mathbf{T}(M) = \mathsf{centroid}(\mathbf{T}(A), \mathbf{T}(B), \mathbf{T}(C));$$

and that $\mathbf{T}$ preserves parallelism of multivectors.

To specify a linear map on a finite-dimensional Grassmann algebra one need only give its action on points, as the following theorem shows.

Theorem 8. *Let* $\mathbf{T} : \mathcal{P} \to \mathcal{Q}$ *be a function such that*

$$\mathbf{T}(aA + bB) = a\mathbf{T}(A) + b\mathbf{T}(B)$$

for all A, B *in* $\mathcal{P}$ *and* a, b *in* $\mathcal{K}$ *with* $a + b = 1$. *Then* $\mathbf{T}$ *extends uniquely to a linear map* $\mathbf{T} : \Omega \to \Lambda$.

Proof. Let $(O_0, O_1, \ldots, O_n)$ be an affine coordinate system for Ω. Every element of Ω is uniquely expressible as a linear combination of products of the form $O_i O_j \cdots O_k$ with $i < j < \cdots < k$; extend $\mathbf{T}$ to Ω by first defining $\mathbf{T}(1) = 1$ and

$$\mathbf{T}(O_i O_j \cdots O_k) = \mathbf{T}(O_i)\mathbf{T}(O_j) \cdots \mathbf{T}(O_k)$$

for these basic products and then extending linearly to arbitrary elements. Obviously the extended $\mathbf{T}$ preserves points and leaves numbers fixed, and it is straightforward to check that it is a ring homomorphism. And uniqueness follows from the fact that any other such extension must agree with $\mathbf{T}$ on the basic products and hence on arbitrary elements. $\qquad\square$

The *translation* $\mathbf{T} : \mathcal{P} \to \mathcal{P}$ defined by a vector U is given by

$$\mathbf{T}(P) = P + U \text{ for every } P \in \mathcal{P}.$$

$\mathbf{T}$ preserves affine combinations of points, since for $a + b = 1$ and A, B in P

$$\begin{aligned}
\mathbf{T}(aA + bB) &= aA + bB + U \\
&= a(A + U) + b(B + U) \\
&= a\mathbf{T}(A) + b\mathbf{T}(B).
\end{aligned}$$

Hence $\mathbf{T}$ extends to a linear map $\mathbf{T} : \Omega \to \Omega$. Note that $\mathbf{T}$ leaves all vectors (and hence all multivectors) fixed:

$$\mathbf{T}(B - A) = \mathbf{T}(B) - \mathbf{T}(A) = (B + U) - (A + U) = B - A.$$

A linear map is determined by its action on a single point A, together with its action on vectors.

If ABC and PQR are tripoints in a two-dimensional Grassmann algebra Ω then $PQR = kABC$ for some number k. Hence, applying the linear map $\mathbf{T} : \Omega \to \Omega$ to this equation

$$\mathbf{T}(PQR) = k\mathbf{T}(ABC),$$

and we write

$$\frac{\mathbf{T}(PQR)}{\mathbf{T}(ABC)} = \frac{PQR}{ABC},$$

or, equivalently,

$$\frac{\mathbf{T}(PQR)}{PQR} = \frac{\mathbf{T}(ABC)}{ABC}.$$

This ratio is a number (since $\mathbf{T}(ABC) = \mathbf{T}(A)\mathbf{T}(B)\mathbf{T}(C)$) is a multiple of ABC) that depends only on $\mathbf{T}$; it is the *determinant* of $\mathbf{T}$:

$$\det(\mathbf{T}) = \frac{\mathbf{T}(ABC)}{ABC}.$$

From this one may give basis-free proofs of the properties of determinants. In particular, if $\mathbf{T} : \Omega \to \Omega$ and $\mathbf{S} : \Omega \to \Omega$ be linear maps then $\det(\mathbf{TS}) = \det(\mathbf{T})\det(\mathbf{S})$. The definition and properties of the determinant function $\det$ extend easily to higher dimensions.

There is a standard construction, not needed for our purposes, by which any finite-dimensional linear space may be embedded in a Grassmann algebra as the space spanned by its points. From this it follows, for example, that one consequence of the *Dimension Theorem* is that all bases of a finite-dimensional linear space have the same number of elements. This may be proved without using the Grassmann multiplication, but the extra structure makes the proof simpler.

The new kinds of quantities discussed above are not only of geometric interest. In physics, the invariance properties of angular momentum, for example, are different from those of momentum (see [14], pp. 52-5) and it is incorrect to represent both as being vectors, though that is what is usually done — angular momentum is, more correctly, a bivector. Similarly, a force which is constrained to act along a line is properly represented as what we have called a bipoint or line vector. Hestenes [16] advocates a Clifford algebra approach to physics in which such matters are treated correctly.

5 Closed Curves

A polygon has area, a polyhedron has volume. Area and volume have simple properties by which they may be characterized. Of these the most important, perhaps, is additivity — if a polygon is made by glueing together two other polygons then its area is the sum of their areas. In this section, we consider these matters. But first we must pin down what the entities are that have area or volume.

We define an (*oriented*) *edge* to be an ordered pair of distinct points (A, B); A is called its *initial vertex* and B its *final vertex*. An (*oriented*) *closed curve* (or *closed polygonal arc*) is a finite (non-empty) set of edges with the property that each of its points occurs once as the initial vertex of an edge and once as the final vertex of an edge. A closed curve is called *minimal* if no proper subset of it is a closed curve.

A closed curve may be viewed as a permutation of a finite set of points which leaves no point fixed. (All permutations would be obtained if we allowed degenerate edges (P, P).) Minimal curves are the cyclic permutations, and corresponding to the fact that every permutation is a composite of unique cyclic permutations, every closed curve is the union of a unique family of minimal closed curves called its *components*.

We usually denote a minimal closed curve

$$\{(A_1, A_2), (A_2, A_3), \ldots, (A_{k-1}, A_k), (A_k, A_1)\}$$

by $\langle A_1, A_2, \ldots, A_k \rangle$. We call $\langle A_k, A_{k-1}, \ldots, A_1 \rangle$ the *opposite* of $\langle A_1, A_2, \ldots, A_k \rangle$ and extend this notion to general closed curves in the obvious way.

A curve is called a *segment* if it has 2 vertices and an (*oriented*) *triangle* if it has 3 vertices. While there are two oriented triangles associated with a set of 3 points $\{A, B, C\}$, namely $\langle A, B, C \rangle$ and its opposite $\langle C, B, A \rangle$, a segment and its opposite are identical.

To motivate what follows, readers will find it instructive to compute the value of the quantity $AB + BC + CD + DA$ for a square $\langle A, B, C, D \rangle$, where $B = A + X, C = B + Y, D = C - X$ — it turns out to be $2XY$. An equally simple calculation shows that the value of the quantity $AB + BC + CA$ for a triangle $\langle A, B, C \rangle$, where $A = O, B = O + aX, C = O + cX + dY$ is $adXY$; note that in a rectangular coordinate system, a is the base length of the triangle, and d is its height.

With a minimal closed curve $\langle A_1, A_2, \ldots, A_k \rangle$ we may associate the quantity

$$\mathsf{m}\langle A_1, A_2, \ldots, A_k \rangle = A_1 A_2 + A_2 A_3 + \cdots + A_{k-1} A_k + A_k A_1,$$

called its (generalized, oriented) *area*. We extend m to general closed curves additively.

Observe that the area of the opposite of a curve is the negative of its area and that $m\langle A, B \rangle = 0$.

One easily shows that for any point P

$$m\langle A_1, A_2, \ldots, A_k \rangle = (A_1 - P)(A_2 - P) + \cdots$$
$$+ (A_{k-1} - P)(A_k - P) + (A_k - P)(A_1 - P).$$

Hence the quantity $m\langle A_1, A_2, \ldots, A_k \rangle$ is *vectorial*, meaning that it is expressible in terms of just vectors. For example,

$$m\langle A, B, C, D \rangle = AB + BC + CD + DA = (A - C)(B - D),$$

and so the area of the quadrilateral $\langle A, B, C, D \rangle$ is zero if and only if $\text{parallel}(A - C, B - D)$.

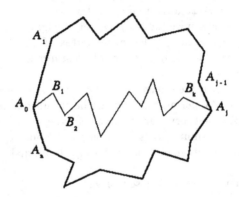

Fig. 3. The area property **A1**.

Although m is dimension-free it shares the key properties of one-dimensional oriented area. In particular, a trivial algebraic identity implies that m is *additive* and *translation-invariant* in the following sense.

Theorem 9.

A1 $m\langle A_0, A_1, \ldots, A_n \rangle =$
$\quad m\langle A_0, \ldots, A_j, B_k, \ldots, B_1 \rangle + m\langle A_0, B_1, \ldots, B_k, A_j, \ldots, A_n \rangle$;
A2 $m\langle A_0 + V, A_1 + V, \ldots, A_n + V \rangle = m\langle A_0, A_1, \ldots, A_n \rangle$.

We also have the characteristic properties of area of triangles:

Theorem 10.

T1 $m\langle A + a(C - B), B, C \rangle = m\langle A, B, C \rangle$;

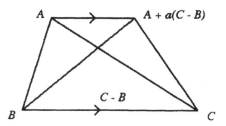

Fig. 4. The area property **T1**.

T2 $\mathsf{m}\langle A, B, B + c(C - B)\rangle = c\,\mathsf{m}\langle A, B, C\rangle$;
T3 $\mathsf{m}\langle A, B, C\rangle = 0 \iff \mathrm{collinear}(A, B, C)$;

The area identity

$$\mathsf{m}\langle A, B, C\rangle + \mathsf{m}\langle C', B', A'\rangle$$
$$+\,\mathsf{m}\langle A, A', B', B\rangle + \mathsf{m}\langle B, B', C', C\rangle + \mathsf{m}\langle C, C', A', A\rangle = 0;$$

is what lies behind the following theorem. Note that if any one of the hypotheses is swapped with the conclusion, then another valid theorem is obtained.

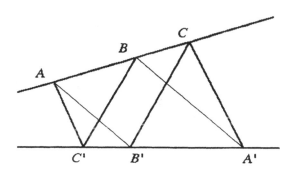

Fig. 5. The Parallel Pappus Theorem.

Geometry Theorem 1 (*The Parallel Pappus Theorem*).

hyp$_1$ collinear(A, B, C)
hyp$_2$ collinear(A', B', C')
hyp$_3$ parallel$(B - C', C - B')$
hyp$_4$ parallel$(C - A', A - C')$
conc parallel$(A - B', B - A')$

6 Plane Area

We define an (*oriented*) *polygon* (or *polygonal curve*) to be a minimal curve whose vertices are coplanar. It is called an (*oriented*) *quadrilateral* if it has 4 vertices, a *pentagon* if it has 5, and so on.

Standard oriented area, as usually defined, has the properties **T1**, **T2** and **T3**, and hence, according to the following, must be an oriented area in our invariant sense, unique up to a scalar multiple.

Theorem 11. *Let* m *and* m′ *be area functions for triangles in a plane. Then there exists a nonzero number* k *such that* m′ = k m.

Proof. Let $\langle A, B, C \rangle$ be a fixed non-degenerate triangle in the plane and let

$$k = \frac{m'\langle A, B, C \rangle}{m\langle A, B, C \rangle}.$$

Let $\langle A', B', C' \rangle$ be any non-degenerate triangle in the plane. There exist numbers b and c such that

$$B' = A + b(B - A) + c(C - A) = B_1 + c(C - A),$$

where $B_1 = A + b(B - A)$. Hence, using properties **T1** and **T2** of m and m′, we have

$$\frac{m'\langle A, B', C \rangle}{m'\langle A, B, C \rangle} = \frac{m'\langle A, B', C \rangle}{m'\langle A, B_1, C \rangle} \frac{m'\langle A, B_1, C \rangle}{m'\langle A, B, C \rangle}$$
$$= b$$
$$= \frac{m\langle A, B', C \rangle}{m\langle A, B_1, C \rangle} \frac{m\langle A, B_1, C \rangle}{m\langle A, B, C \rangle}$$
$$= \frac{m\langle A, B', C \rangle}{m\langle A, B, C \rangle}.$$

Using the analagous equations involving A' and C' instead of B', we have

$$\frac{m'\langle A', B', C' \rangle}{m'\langle A, B, C \rangle} = \frac{m'\langle A', B', C' \rangle}{m'\langle A', B', C \rangle} \frac{m'\langle A', B', C \rangle}{m'\langle A', B, C \rangle} \frac{m'\langle A', B, C \rangle}{m'\langle A, B, C \rangle}$$
$$= \frac{m\langle A', B', C' \rangle}{m\langle A', B', C \rangle} \frac{m\langle A', B', C \rangle}{m\langle A', B, C \rangle} \frac{m\langle A', B, C \rangle}{m\langle A, B, C \rangle}$$
$$= \frac{m\langle A', B', C' \rangle}{m\langle A, B, C \rangle}.$$

and so

$$m'\langle A', B', C' \rangle = k \, m\langle A', B', C' \rangle;$$

moreover, property **T3** ensures that this holds also for degenerate triangles $\langle A', B', C' \rangle$. □

Let (O, O_1, O_2) be an affine coordinate system for our plane and (O, X, Y) the associated cartesian coordinate system. Note that

$$m\langle O, O_1, O_2 \rangle = OO_1 + O_1O_2 + O_2O = XY.$$

Consider a polygon $\langle A_1, A_2, \ldots, A_k \rangle$ in the plane. Observing that $(A_i - O)(A_{i+1} - O)$, being a product of vectors spanned by X and Y is a multiple of XY, we see that

$$
\begin{aligned}
m\langle A_1, A_2, \ldots, A_k \rangle &= A_1 A_2 + \cdots + A_{k-1}A_k + A_k A_1 \\
&= (A_1 - O)(A_2 - O) + \cdots \\
&\quad + (A_{k-1} - O)(A_k - O) + (A_k - O)(A_1 - O) \\
&= 2aXY, \\
&= 2a\, m\langle O, O_1, O_2 \rangle
\end{aligned}
$$

for some number a, the basis-dependent *scalar oriented area* of the polygon. We call the absolute value of a the *scalar area* of the polygon and we call its sign the *orientation* of the polygon relative to the basis $\langle X, Y \rangle$. If $m(\sigma) = 0$ then we say that σ has orientation 0. Here we must assume that the underlying field is equipped with an absolute value function (like the real numbers or the complex numbers). Scalar area and orientation are separately translation invariant.

Typically of Grassmann geometry, the following theorem generalizes in an obvious way from paired triangles to paired closed curves.

Geometry Theorem 2.

hyp$_1$ parallel$(A - B', A' - B)$
hyp$_2$ parallel$(B - C', B' - C)$
hyp$_3$ parallel$(C - A', C' - A)$
conc $m\langle A', B', C' \rangle = m\langle A, B, C \rangle$

The invariant treatment of area we are advocating renders some affine geometry theorems transparently provable by algebraic arguments that are easily automated.

The proof of the following in [8] (p. 55) uses two auxiliary points and two lemmas. Note too that our formulation contains more information than a traditional one, since it asserts that two orientations are the same, as well as two scalar areas.

Geometry Theorem 3.

hyp$_1$ collinear(A, D, P)
hyp$_2$ collinear(B, C, P)
hyp$_3$ $M = \text{midpoint}(A, C)$
hyp$_4$ $N = \text{midpoint}(B, D)$
conc $m\langle A, B, C, D \rangle = 4\, m\langle M, N, P \rangle$

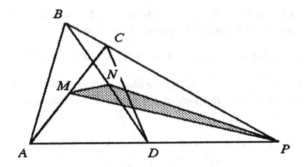

Fig. 6. Geometry Theorem 3.

Proof. We have:

$$4(PM + MN + NP)$$
$$= 2P(A+C) + (A+C)(B+D) + 2(B+D)P$$
$$= 2(AD + DP + PA) - 2(BC + CP + PB) + (AB + BC + CD + DA)$$
$$= AB + BC + CD + DA.$$

□

Suppose that the vertices of a polygon $\langle A_1, \ldots, A_n \rangle$ are given by

$$A_i = O + x(t_i)X + y(t_i)Y,$$

where x and y are functions from the field $\mathcal{K}$ to itself. Then it is straightforward to show that the scalar oriented area of the polygon relative to $\langle X, Y \rangle$ is

$$\frac{1}{2} \sum (x(t_i)y'(t_i) - y(t_i)x'(t_i))(t_{i+1} - t_i),$$

where $x'(t_i)) = x(t_{i+1}) - x(t_i)/(t_{i+1} - t_i)$ (and $\sum$ is a cyclic sum over $i = 1$ to n). In the limit this becomes the line integral formula for area.

Consider a triangle $\langle A, B, C \rangle$ in a plane having scalar oriented area a relative to $\langle X, Y \rangle$. From

$$AB + BC + CA = 2aXY,$$

we see, using the fact that $AXY = OXY$ (because A lies in the plane of (O, X, Y)), that

$$ABC = 2aOXY,$$

and hence (unambiguously)

$$\mathrm{m}\langle A, B, C \rangle = \frac{ABC}{2OXY}.$$

Thus we have an alternative representation of area for triangles in a plane as a ratio of tripoints, and

$$\frac{\mathrm{m}\langle A', B', C' \rangle}{\mathrm{m}\langle A, B, C \rangle} = \frac{A'B'C}{ABC}.$$

Grassmann geometry gives new meaning to traditional Euclidean geometry notations. For example, the following propositions have obvious interpretions in terms of absolute area and orientation.

Geometry Theorem 4.

hyp parallelogram(A, B, C, D)
conc $ACD = ABC$

Geometry Theorem 5.

hyp parallel$(B - A, C - D)$
conc $ABC = ABD$

Since

$$\det(\mathbf{T}) = \frac{\mathbf{T}(A)\mathbf{T}(B)\mathbf{T}(C)}{ABC}$$
$$= \frac{\mathrm{m}\langle\mathbf{T}(A), \mathbf{T}(B), \mathbf{T}(C)\rangle}{\mathrm{m}\langle A, B, C\rangle}$$

the determinant of a linear transformation $\mathbf{T}$ of the plane is the (constant) ratio by which $\mathbf{T}$ transforms areas.

7 The Boundary Operator

Let $(O, X_1, ..., X_n)$ be a cartesian coordinate system for a Grassmann algebra Ω. Every element ϕ of Ω may be written uniquely in the form

$$\phi = x + yO + x_1X_1 + ... + x_nX_n + ... + y_1OX_1 + ... + y_nOX_n$$
$$... + x_{12}X_1X_2 + ... + x_{1n}X_1X_n + ... + x_{n-1,n}X_{n-1}X_n$$
$$... + y_{12}OX_1X_2 + ... + x_{123}X_1X_2X_3 + ...$$

The *boundary operator* is the linear function $\partial : \Omega \to \Omega$ given by

$$\partial\phi = y + y_1X_1 + ... + y_nX_n... + y_{12}X_1X_2 + ... + y_{123}X_1X_2X_3 + ...;$$

∂ simply strips O from those basis elements $OX_i...X_k$ containing it and annihilates those that don't. Obviously $\partial^2 = 0$.

If $A = O + \Sigma a_iX_i$ and $B = O + \Sigma b_jX_j$ then

$$\partial(AB) = \partial(\Sigma b_jOX_j - \Sigma a_iOX_i + \Sigma a_ib_jX_iX_j)$$
$$= \Sigma b_jX_j - \Sigma a_iX_i$$
$$= B - A.$$

One easily shows that $\partial(ABC) = (B - A)(C - A) = BC - AC + AB$ and $\partial(ABCD) = (B - A)(C - A)(D - A) = BCD - ACD + ABD - ABC$ and, by induction,

$$\partial(A_1A_2...A_k) = (A_2 - A_1)...(A_k - A_1)$$
$$= A_2A_3...A_k - A_1A_3...A_k + ...$$

Since every element of a Grassmann algebra is a linear combination of products of points, we see that, despite the definition, ∂ is independent of the coordinate system.

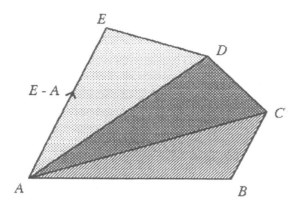

Fig. 7. Typical boundary calculations.

Typical boundary calculations (see Fig. 7) give:

$$\partial(AB + BC + CD + DE) = E - A,$$
$$\partial(AB + BC + CD + DE + EA) = 0,$$
$$\partial(ABC + ACD + ADE) = AB + BC + CD + DE + EA.$$

Observe that for any point A, $\partial(A) = 1$, while for any vector V, $\partial(V) = 0$; and, in general, $\partial(\Sigma a_i A_i) = \Sigma a_i$.

Recall that elements $\Sigma a_i A_i$ of the vector space Ω_0 spanned by all points are called *chains of dimension 0* or, briefly, *0-chains*; thus every 0-chain is either a point or a vector or a multiple of a point. Elements $\Sigma a_{ij} A_i A_j$ of the vector space Ω_1 spanned by all bipoints are called *1-chains*; these include bipoints and bivectors. In general, elements $\Sigma a_{i_0 \ldots i_k} A_{i_0} \ldots A_{i_k}$ of the vector space Ω_k spanned by all k-dimensional multipoints are called *k-chains*; for consistency, numbers are called -1-*chains*, and we write $\Omega_{-1} = \mathcal{K}$.

Some of our earlier proofs become simpler if we use the boundary operator. For example, the fact that

$$CD = AB \Rightarrow D - C = B - A$$

is obtained immediately by just applying ∂ to both sides of the hypothesis equation.

A k-chain ϕ is called a (*global*) *k-boundary* if it belongs to the range $ran(\partial)$ (that is to say, has the form $\partial(\psi)$ for some $k+1$-chain ψ) and a *k-cycle* if it belongs to the kernel $ker(\partial)$ (that is to say, satisfies $\partial(\phi) = 0$). From the fact that $\partial^2 = 0$ it follows that every k-boundary is a cycle. It is easy to see that

the converse also holds. For example, if the 1-chain $\phi = \Sigma y_i O X_i + \Sigma x_{i,j} X_i X_j$ is a cycle, then $0 = \partial(\phi) = \Sigma y_i X_i$, and so every $y_i = 0$ and ϕ has the form $\phi = \Sigma x_{i,j} X_i X_j = \partial \Sigma x_{i,j} O X_i X_j$, and this argument evidently works for k-chains in general. Thus the k-cycles (or k-boundaries) are precisely the pure vectorial k-chains.

In particular, areas of closed curves are 1-cycles, and for any P

$$m\langle A_0, A_1, \ldots, A_n \rangle = \partial(A_0 A_1 P + A_1 A_2 P + \ldots A_{n-1} A_n P + A_n A_0 P),$$

a formula that offers additional insight into our generalized notion of area.

Note that ∂ does not preserve products and so is not a ring homomorphism; however there is a simple interaction with multiplication:

Theorem 12. *For multipoints ϕ and ψ*

$$\partial(\phi\psi) = (\partial\phi)\psi + (-1)^{degree(\phi)} \phi \partial \psi.$$

This is a special case of Theorem 16 below.

8 Grassmann Polynomials

The following is a typical affine geometry theorem.

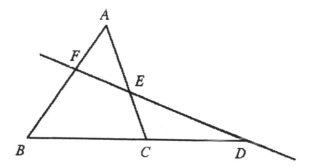

Fig. 8. The Menelaus theorem.

Geometry Theorem 6 (*The Menelaus Theorem*).

hyp$_1$ $D = aB + a'C$
hyp$_2$ $E = bC + b'A$
hyp$_3$ $F = cA + c'B$
hyp$_4$ $a'b'c' = -abc$
conc collinear(D, E, F)

The hypotheses are constraints on some points. We may think of drawing a diagram. Choose a point D on the line through B and C. Choose E on the line through C and A. Choose F on the line through A and B in such a way that the fourth condition is satisfied. The conclusion is another constraint that the three points must satisfy. The universe in which all this is interpreted is just a plane containing the points. What about the proof? Given our definition of collinearity, it is a simple algebraic calculation:

$$DEF = abcBCA + a'b'c'CAB = (abc + a'b'c')ABC = 0$$

Associated with the Menelaus theorem we have an algebra of Grassmann expressions, satisfying the usual rules for numbers and points. Generalising, we define the *Grassmann polynomial algebra* over the field $\mathcal{K}$, $\mathcal{K}[n, m]$, to be the (unique up to isomorphism) associative algebra over $\mathcal{K}$ freely generated by the n point indeterminates $P_1, P_2, \ldots, P_n$ and m number indeterminates $x_1, x_2, \ldots, x_m$, subject to the relations

1. $P_i P_j = -P_j P_i$, $1 \leq i, j \leq n$,
2. $x_i x_j = x_j x_i$, $1 \leq i, j \leq m$, and
3. $P_i x_j = x_j P_i$, $1 \leq i \leq n$, $1 \leq j \leq m$.

Elements of $\mathcal{K}[n, m]$ are called *(Grassmann) polynomials*. Sometimes we use $A, B, C, \ldots$ and $a, b, c, \ldots$ to signify point and number variables respectively. In $\mathcal{K}[n, m]$, two polynomials are equal if and only if their equality represents an identity involving up to n points and m numbers which holds in every Grassmann algebra $\Omega[\mathcal{K}, \mathcal{P}]$; that is, elements of $\mathcal{K}[n, m]$ are equal if and only if they define the same *polynomial function* on every Grassmann algebra over $\mathcal{K}$.

$\mathcal{K}[n, 0]$ is a copy of the Grassmann algebra $\Omega[\mathcal{K}, \mathcal{P}]$ with basis $\{P_1, P_2, \ldots, P_n\}$, while $\mathcal{K}[0, m]$ is just the familiar commutative ring of polynomials in $x_1, x_2, \ldots, x_m$ over $\mathcal{K}$. For this latter ring, the Gröbner basis algorithm of Buchberger provides a powerful algorithmic tool: see [2]. But much work has appeared in the literature generalising this algorithm to other sorts of structure, including to a broad class which includes Grassmann algebras: see Apel [1]. The Grassmann case is treated in isolation in [26], and we present below a streamlined version of that treatment which uses results from [1] and which incorporates the boundary map.

The Hilbert Basis theorem extends easily to $\mathcal{K}[n, m]$, since it is a finite-dimensional extension of a standard polynomial ring. Hence every left ideal, right ideal and two-sided ideal is finitely generated.

Throughout the remainder of this paper, f, g, h, k will stand for elements of $\mathcal{K}[n, m]$, F, G for finite subsets of $\mathcal{K}[n, m]$, r, s, t, u for elements of $T_{n,m}$, and $a, b, c, \alpha, \beta, \gamma$ for numbers in $\mathcal{K}$. Additionally, we will often use $A, B, C, \ldots$ for point indeterminates.

We define the *standard order* on the indeterminates of $\mathcal{K}[n, m]$ as follows:

$$P_1 < P_2 < \ldots < P_n < x_1 < x_2 < \ldots < x_m.$$

A non-zero product of the indeterminates in which they appear in ascending order is called a *term*. For example, $x_2^3 x_4 P_3 P_7$ is a term, but $x_4 P_7 P_3$ is not. Also 1 is a term. We denote by $T_{n,m}$ the set of all terms in $\mathcal{K}[n, m]$. Each polynomial in $\mathcal{K}[n, m]$ is uniquely expressible (modulo re-ordering of the summand terms) as a linear combination of terms $T_{n,m}$, obtained by expanding in the usual way. We call this a *canonical form* of the element. The set of terms occurring in this linear combination for a given $p \in \mathcal{K}[n, m]$ is denoted by $T(p)$. If $|T(p)| = 1$, then we denote the single element of $T(p)$ by $T(p)^*$.

For $t \in T_{n,m}$, let $P(t)$ denote the set of point variables appearing in t. For a set S of point variables, $T(S)$ denotes the unique $t \in T_{n,0}$ such that $P(t) = S$.

9 Ideals and Geometry Theorems

Let $\Omega = \Omega(\mathcal{K}, \mathcal{P})$ be a Grassmann algebra. As we have seen, affine geometry theorems may be expressed in terms of Grassmann algebra. The hypotheses of the theorem correspond to certain Grassmann expressions in the points and numbers mentioned in the theorem being zero, and so does the one (or possibly more) conclusion, and we can say the theorem is true for Ω if and only if all substitutions of points from $\mathcal{P}$ for the point variables mentioned in the Grassmann geometry theorem statement which satisfy the hypotheses also satisfy the conclusion. So, if the hypotheses are expressed in the form $f_1 = 0, f_2 = 0, \ldots, f_k = 0$ and the conclusion in the form $g = 0$ where all $f_i, g \in \mathcal{K}[n, m]$, then we say the f_i are the *hypothesis polynomials* and g is the *conclusion polynomial* of the *possible theorem*

$$f_1 = 0, f_2 = 0, \ldots, f_k = 0 \Rightarrow g = 0.$$

The *consequence space* associated with $F \subseteq \mathcal{K}[n, m]$ is defined to be $\mathcal{C}(F) = \{f : f \in \mathcal{K}[n, m], f(a) = 0$ for all $a \in \Omega^n \times \mathcal{K}^m$ that satisfy $g(a) = 0$ for all $g \in F\}$. Thus $\mathcal{C}(F)$ is the set of polynomials which vanish whenever all polynomials in F do. Thus the possible theorem $f_1 = 0, f_2 = 0, \ldots, f_k = 0 \Rightarrow g = 0$ is true for Ω if and only if $g \in \mathcal{C}(\{f_1, f_2, \ldots, f_k\})$.

For $F \subseteq \mathcal{K}[n, m]$, we denote the *ideal generated by* F by (F), so that

$$(F) = \{\sum_i p_i f_i q_i : p_i, q_i \in \mathcal{K}[n, m], f_i \in F\}.$$

Similarly, we denote the *left ideal generated by* F by $(F)_L$, so that $(F)_L = \{\sum_i p_i f_i | p_i \in \mathcal{K}[n, m], f_i \in F\}$. It is easy to see that $\mathcal{C}(F)$ is actually an ideal of $\mathcal{K}[n, m]$ for all $F \subseteq \mathcal{K}[n, m]$.

In earlier examples we used the fact that for any $F \subseteq \mathcal{K}[n, m]$, $(F) \subseteq \mathcal{C}(F)$. Our method often boiled down to doing some *equational reasoning*. For example, consider the simple theorem:

Geometry Theorem 7.

hyp parallelogram(A, B, C, D)
conc parallelogram(B, C, D, A).

Equivalently, we have the hypothesis $B - A - C + D = 0$ and conclusion $C - B - D + A = 0$, so the conclusion polynomial is just -1 times the hypothesis

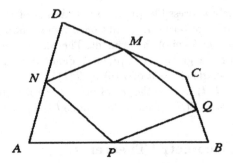

Fig. 9. A property of quadrilaterals.

polynomial, so $C - B - D + A \in (B - A - C + D)$ (in the appropriate Grassmann polynomial algebra).

Another example

Geometry Theorem 8.

hyp$_1$ $P = \mathrm{midpoint}(A, B)$
hyp$_2$ $Q = \mathrm{midpoint}(B, C)$
hyp$_3$ $M = \mathrm{midpoint}(C, D)$
hyp$_4$ $N = \mathrm{midpoint}(D, A)$
conc $\mathrm{parallelogram}(P, Q, M, N)$

(see Fig. 9) may be re-written as

hyp$_1$ $f_1 = 0$
hyp$_2$ $f_2 = 0$
hyp$_3$ $f_3 = 0$
hyp$_4$ $f_4 = 0$
conc $g = 0$

where $f_1 \equiv 2P - A - B$, $f_2 \equiv 2Q - B - C$, $f_3 \equiv 2M - C - D$, $f_4 \equiv 2N - D - A$ and $g \equiv Q - P - M + N$, It is easy to see that

$$g = \frac{1}{2}f_2 - \frac{1}{2}f_1 - \frac{1}{2}f_3 + \frac{1}{2}f_4$$

and hence $g \in (f_1, f_2, f_3, f_4)$.

Geometry Theorem 4, expressing a familiar fact about parallelograms, follows from the fact that

$$ABC - ACD = (-AC)(B - A - C + D)$$

so again the ideal generated by the hypothesis polynomial contains the conclusion polynomial. The same happens in the Menelaus theorem, the first example we see in which number variables occur in an essential way. The reader should

be convinced that any algebraic proof which makes use of substitution and simplification is really just showing that the conclusion polynomial is in the ideal
generated by the hypothesis polynomials.

We would like to know exactly when a theorem of Grassmann geometry is
true; that is, we would like a method of checking whether $f \in \mathcal{C}(F)$. In the case
where there are no point variables, it is clear that if some *power* of f is in (F),
then $f \in \mathcal{C}(F)$. Conversely, we have the famous

Theorem 13 (*The Hilbert Nullstellensatz*). *Suppose $\mathcal{K}$ is an algebraically closed
field. For any $F \subseteq \mathcal{K}[0, m]$, $f \in \mathcal{C}(F)$ if and only if there exists some integer
$\rho > 0$ such that $f^\rho \in (F)$.*

The Nullstellensatz implies the following very useful

Corollary 14. *Suppose $\mathcal{K}$ is an algebraically closed field. For any $F \subseteq \mathcal{K}[0, m]$
and $f \in \mathcal{K}[0, m]$, $f \in \mathcal{C}(F)$ if and only if $1 \in (F \cup \{fx_{m+1} - 1\})$ in $\mathcal{K}[0, m+1]$.*

This provides a necessary and sufficient condition for determining whether
a geometry theorem is true over the complex numbers if $n = 0$, and has been
exploited with great success: see [21] and [22] where this *refutational* approach
is discussed in detail, as well as [4] where the useful notion of *genericity* is given
a full and elegant treatment.

Unfortunately there is no known analog of the Nullstellensatz for Grassmann
polynomials, over any kinds of field. Moreover, the notion of genericity seems
unable to be captured in terms of ideals only. The sufficient condition $f \in (F)$
is quite useful as we have seen, but is not necessary, even if $m = 0$. A simple
counterexample is

$$ABC = 0 \Rightarrow AB + BC + CA = 0.$$

Clearly $AB + BC + CA \notin (ABC)$. Nevertheless, the implication holds, since
$\partial(ABC) = AB + BC + CA$.

We may define the boundary map on $\mathcal{K}[n, m]$ in the obvious manner. Linear-
ity of ∂ implies that for all $F \subseteq \mathcal{K}[n, m]$, $\mathcal{C}(F)$ is closed under taking boundaries:
$\partial(f) \in \mathcal{C}(F)$ for all $f \in \mathcal{C}(F)$. We call an ideal of $\mathcal{K}[n, m]$ closed under taking
boundaries a ∂-*ideal*.

It makes sense to talk about the smallest ∂-ideal containing a subset F of
$\mathcal{K}[n, m]$ — the ∂-ideal *generated by* F — essentially because $\mathcal{K}[n, m]$ together
with the unary operation ∂ is a *multi-operator group*: see [23]. Notation: $(F)_\partial$.
Thus for $F \subseteq \mathcal{K}[n, m]$, $(F) \subseteq (F)_\partial \subseteq \mathcal{C}(F)$.

Ideals of the form $\mathcal{C}(F)$ have another important algebraic property. We say
that $f \in \mathcal{K}[n, m]$ is *point homogeneous of degree* $p \geq 0$ (usually shortened to
just *homogeneous*) if $T(f) \subseteq \{t \in T_{n,m} | |P(t)| = p\}$; thus f is homogeneous
of degree p if every term in f contains a product of p point variables. The set
of homogeneous polynomials of degree p in $\mathcal{K}[n, m]$, $p < n$, will be denoted by
$\mathcal{K}[n, m]^{(p)}$. Clearly $\mathcal{K}[n, m] = \sum_{p=1}^n \mathcal{K}[n, m]^{(p)}$, so that every $\phi \in \mathcal{K}[n, m]$ may
be uniquely expressed as the sum of its *homogeneous components*.

A *homogeneous left ideal* I of $\mathcal{K}[n,m]$ is a left ideal for which $I = \sum_{p=1}^{n}(I \cap \mathcal{K}[n,m]^{(p)})$. A *homogeneous subset* of $\mathcal{K}[n,m]$ is a subset all elements of which are homogeneous.

We note that $\mathcal{K}[n,m]$ is a graded ring with respect to the grading determined by the degree of homogeneity. The above definition of 'homogeneous' is consistent with that used for graded rings generally.

The following results will be useful when we come to consider algorithms. The next one was proved in [26].

Theorem 15. *In $\mathcal{K}[n,m]$, a homogeneous left ideal is an ideal.*

We can now easily prove the following useful product rule.

Theorem 16. *For $\phi \in \mathcal{K}[n,m]^{(p)}$ and $\psi \in \mathcal{K}[n,m]^{(q)}$,*

$$\partial(\phi\psi) = (\partial\phi)\psi + (-1)^p \phi \partial\psi.$$

Proof. Let $\phi = \sum_i \alpha_i s_i$, $\psi = \sum_j \beta_j t_j$, where for all i,j, α_i and β_j are numbers, s_i is a term of degree p and t_j a term of degree q. Then

$$\phi\psi = \sum_i \alpha_i s_i \sum_j \beta_j t_j = \sum_{i,j} \alpha_i \beta_j s_i t_j,$$

so

$$\partial(\phi\psi) = \partial(\sum_{i,j} \alpha_i \beta_j s_i t_j)$$

$$= \sum_{i,j} \alpha_i \beta_j \partial(s_i t_j)$$

$$= \sum_{i,j} \alpha_i \beta_j (\partial(s_i)t_j + (-1)^p s_i \partial(t_j))$$

$$= \sum_i \alpha_i \partial(s_i) \sum_j \beta_j t_j + (-1)^p \sum_i \alpha_i s_i \sum_j \beta_j \partial(t_j)$$

$$= (\partial\phi)\psi + (-1)^p \phi \partial\psi.$$

$\square$

Theorem 17. *For every homogeneous subset F of $\mathcal{K}[n,m]$, $(F)_\partial = (F \cup \partial F)$.*

Proof. Since $F \cup \partial F \subseteq (F)_\partial$, we have $(F \cup \partial F) \subseteq (F)_\partial$.

Conversely, let $\theta \in (F \cup \partial F) = (F \cup \partial F)_L$ by Theorem 15. Then by using distributivity, it is possible to express θ in the form $\theta = \sum p_i f_i + \sum q_j \partial g_j$, where the f_i and g_j are not necessarily distinct elements of F, and the p_i and q_j are homogeneous elements of $\mathcal{K}[n,m]$. Then by Theorem 16,

$$\partial\theta = \sum \partial(p_i f_i) + \sum \partial(q_j \partial g_j)$$

$$= \sum ((\partial p_i)f_i + p_i' \partial f_i) + \sum (\partial q_j \partial g_j + q_j' \partial^2 g_j)$$

$$= \sum (\partial p_i)f_i + \sum p_i' \partial f_i + \sum \partial q_j \partial g_j$$

$$\in (F \cup \partial F),$$

where $p'_i = \pm p_i$ and $q'_j = \pm q_j$. Hence $(F \cup \partial F)$ is a ∂-ideal. It contains F, so $(F)_\partial \subseteq (F \cup \partial F)$, and the proof is complete. □

Because no non-trivial linear combination of m-points can equal a sum of non-trivial linear combinations of n-points for various $n \neq m$, we have the following

Theorem 18. *For $F \subseteq \mathcal{K}[n, m]$, $\mathcal{C}(F)$ is homogeneous.*

There are many homogeneous ideals not of the form $\mathcal{C}(F)$ however. For instance, (P) is a homogeneous ideal in $\mathcal{K}[1, 0]$, yet $\partial P = 1 \notin (P)$. So not every homogeneous ideal is a ∂-ideal. Similarly, for

$$f = P_1 - P_2 + (P_1 - P_3)(P_1 - P_4)(P_1 - P_5),$$

it is easily shown that $(f)_\partial = (f)_L$ and hence not every ∂-ideal is homogeneous. We shall later see an example of a homogeneous ∂-ideal not of the form $\mathcal{C}(F)$.

In the Grassmann algebraic formulation of a geometry theorem, it is homogeneous polynomials which feature, as they are the important polynomials in geometry.

Corollary 19. *If F is a set of homogeneous polynomials, then $(F)_\partial = (F \cup \partial F)_L$.*

Proof. If F is homogeneous, so is $F \cup \partial F$ so $(F \cup \partial F)_L$ is a homogeneous left ideal, as follows from a basic fact concerning graded rings. Thus $(F \cup \partial F)_L$ is an ideal by Theorem 15, and so $(F \cup \partial F)_L = (F \cup \partial F) = (F)_\partial$ by Theorem 17. □

10 Gröbner Bases and Theorem Proving

We next introduce the relevant variant of Gröbner bases and the Buchberger algorithm. There are some basic notions we must define first, such as admissible orders, reduction and so on; they are mostly straight-forward variants of the usual commutative notions, so we will be fairly brief. In any case, many generalisations of Buchberger's algorithm and the Gröbner basis idea to various kinds of generalised polynomial have already been considered in the literature. Indeed the current one may be viewed as being a special case of one considered by [1]. However, some of the special features of the Grassmann case, such as homogeneity and the boundary map, have not been considered elsewhere apart from in [26].

Let $\leq$ be a total ordering of $T_{n,m}$. For $p \in \mathcal{K}[n, m]$, we denote the highest term in $T_{n,m}$ occurring in f with respect to $\leq$ by $Hterm_\leq(f)$, or, if there is no ambiguity, by $Hterm(f)$.

The ordering $\leq$ is *admissible* if, for all s, t, u

1. $1 \leq t$,
2. if $s < t$ and $P(u) \cap P(s) = P(u) \cap P(t) = \emptyset$, then $T(us)^* < T(ut)^*$.

So 1 is the smallest term, and pre-multiplying by a term preserves the ordering providing neither term becomes zero. In fact, because of the commutativity and anti-commutativity relations in the Grassmann polynomial algebra, any such order also satisfies the condition that if $s < t$ and $P(u) \cap P(s) = P(u) \cap P(t) = \emptyset$, then $T(su)^* < T(tu)^*$, so post-multiplication is order preserving providing neither term becomes zero.

We note that if $P(u) \backslash P(t) = \emptyset$, then ut and tu are multiples of $T(ut)^* = T(tu)^* = Hterm(ut)^* = Hterm(tu)^*$. Examples are variants of the commutative case: for instance, we have the *total degree order* on $T_{n,m}$. With $n = 2$ and $m = 1$, this is as follows:

$$1 < P_1 < P_2 < x_1 < P_1 P_2 < P_1 x_1 < P_2 x_1 < x_1^2$$
$$< P_1 P_2 x_1 < P_1 x_1 < P_2 x_1 < x_1^3 < \cdots,$$

Here is the *lexicographic order* for the same n, m:

$$1 < P_1 < P_1 P_2 < P_2 < x_1 < P_1 x_1 < P_2 x_1 < P_1 P_2 x_1 < x_1^2 < P_1 x_1^2 < \cdots.$$

It should be clear how these generalise.

Let $\leq$ be a fixed admissible order in what follows (the total degree order in all computed examples). We use the following abbreviations throughout:

$coef(f, t)$ is the coefficient of the term t in f, where $f \in \mathcal{K}[n, m]$.
$Hcoef(f)$ is $coef(f, Hterm(f))$.

For example, with the total degree order on the terms of $\mathcal{K}[4, 0]$,

$coef(P_1 + 2P_3 P_1, P_1 P_3) = -2$,
$Hterm(P_1 P_2 + P_2 P_3 + P_4 P_3) = P_3 P_4$, and
$Hcoef(P_1 P_2 + P_2 P_3 + P_4 P_3) = coef(P_1 P_2 + P_2 P_3 + P_4 P_3, P_3 P_4) = -1$.

Every admissible order on $T_{n,m}$ is a well order (the proof of which is very similar to the commutative, $n = 0$ case), so the terms can be listed in order from the smallest element 1 up as far as one wishes. If $m = 0$, the list is finite.

We say $g \to_F h$ (g *left reduces to* h *modulo* F) if there are b, u, and $f \in F$ with $P(u) \cap P(Hterm(f)) = \emptyset$ and such that

$h = g - buf$,
$coef(g, T(u \cdot Hterm(f))) \neq 0$, and
$b = coef(g, T(u \cdot Hterm(f)))/Hcoef(uf) \neq 0$.

In such circumstances we say that $g \to_{f,b,u}$ and that $g \to_{t,f} h$ where $t = T(u \cdot Hterm(f))$. $\to_F$ is the *reduction relation for* F.

These are just the analogs of the commutative polynomial notions. To say that $g \to_F h$ is to say that there is a polynomial f in F whose largest term can be used to replace a term in g by a linear combination of smaller terms to give a polynomial h. Thus reduction is just a formalisation of the process of substitution and simplification which is the basis of our equational approach to geometric reasoning. The reduction relation $\to_F$ is a Noetherian relation (so that reductions cannot go on forever), a straightforward generalisation of the corresponding result for standard polynomials appearing in [2].

For example, in $\mathcal{K}[3,0]$, let $F = \{f\}$, $f = P_1P_3 - 2P_1P_2$, $g = P_1P_2P_3$. Then $g \to_F P_1P_2P_3 - (-1)P_2(P_1P_3 - 2P_1P_2) = 0$. Thus $g \to_{t,f} 0$ where $t = P_1P_2P_3$, and $g \to_{f,-1,P_2}$.

Let $\mathcal{K}$ be a computable field (meaning one which can be implemented on a computer, such as the rationals). The following algorithm yields a normal form $N(g)$ of a given polynomial g, modulo F.

begin
    $N(g) := g$
    while exist $f \in F, b, u$ such that $N(g) \to f, b, u$, do
        choose $f \in F$, $u \in T_{n,m}$, $b \in \mathcal{K}$ such that $N(g) \to_{f,b,u}$
    $N(g) := N(g) - b \cdot u \cdot f$.
end

Correctness is clear. Termination follows from the Noetherian property of $\to_F$.

For $u, v \in T_{n,m}$, with $P(u) = P(v) = \emptyset$, let $lcm(u,v)$ denote the usual least common multiple of u and v as polynomial terms. Then for arbitrary $s = fp, t = gq \in T_{n,m}$, with f, g containing only number variables and p, q only point variables, $lcm(s,t)$ is defined to be $lcm(f,g)T(P(s) \cup P(t))$. So $lcm(s,t)$ is the term of lowest degree (both number and point degree) divisible by both s and t.

The S-polynomial, $SP(f_1, f_2)$, corresponding to f_1 and f_2 is defined to be

$$SP(f_1, f_2) = a_1 \cdot Hcoef(u_2f_2) \cdot u_1f_1 - a_2 \cdot Hcoef(u_1f_1) \cdot u_2f_2,$$

with a_1, a_2, u_1, u_2 such that

$$P(u_1) \cap P(Hterm(f_1)) = P(u_2) \cap P(Hterm(f_2)) = \emptyset$$

and

$$\begin{aligned}
a_1u_1 \cdot Hterm(f_1) &= T(u_1 \cdot Hterm(f_1))^* \\
&= Hterm(u_1f_1) \\
&= lcm(Hterm(f_1), Hterm(f_2)) \\
&= a_2u_2 \cdot Hterm(f_2) \\
&= T(u_2Hterm(f_2))^* \\
&= Hterm(u_2f_2).
\end{aligned}$$

The existence and uniqueness of $SP(f_1, f_2)$ for given $f_1, f_2 \in P$ are easily shown - let $u_1 = T((P(Hterm(f_1)) \cup P(Hterm(f_2)))\backslash P(Hterm(f_1)))$, and likewise with u_2. The S-polynomial is just the difference between the results of reducing the lcm of the highest terms of two polynomials using each of them in turn.

To illustrate these definitions, letting $f = P_3P_5x_2^2 - 2P_2P_4$ and $g = P_3P_4x_2 + 3P_1P_2$, we have that $lcm(f,g) = lcm(P_3P_5x_2^2, P_3P_4x_2) = x_2^2P_3P_4P_5$, and so

$$\begin{aligned}
SP(f,g) &= -P_4(P_3P_5x_2^2 - 2P_2P_4) - P_5x_2(P_3P_4x_2 + 3P_1P_2) \\
&= P_3P_4P_5x_2^2 + 0 - P_3P_4P_5x_2^2 - 3P_1P_2P_5x_2 \\
&= -3P_1P_2P_5x_2.
\end{aligned}$$

F is a *Gröbner basis* if every $f \in (F)_L$ can be reduced to zero by a sequence of reductions involving $\to_F$.

Theorem 20. *The following statements are equivalent:*

1. F *is a Gröbner basis.*
2. *If* $f_1, f_2 \in F$, *then for any* $P_i \in P(Hterm(f_1))$, $P_i \cdot f_1 \to_F^* 0$ *and* $SP(f_1, f_2)$ $\to_F^* 0$.

The proof follows from a more general result of Apel [1].

Theorem 21. *Given a finite subset* F *of* $\mathcal{K}[n, m]$, *the following algorithm constructs a Gröbner basis* G *such that* $(F)_L = (G)_L$.

```
begin
  V := {point variables in P}
  G := F
  H := F
  B := {{f₁, f₂}|f₁, f₂ ∈ F, f₁ ≠ f₂}
  comment: H plays two roles in what follows
  while H ≠ ∅ do
  comment: in the next procedure H is the subset of G which supplies
  polynomials which, when multiplied by appropriate point variables, are to
  be included in G if they are not of normal form zero modulo G
  begin
    while H ≠ ∅ do
    begin
      f := an element of H
      H := H\{f}
      W := P(Hterm(f))
      comment: W is the set of all point variables that do not occur in
      Hterm(f)
      while W ≠ ∅ do
      begin
        P := an element of W
        k := P · f
        k' := N(G, k)
        if k' ≠ 0 then
            H := H ∪ {k'}
            G := G ∪ {k'}
            B := B ∪ {{g, k'}|g ∈ G}
            comment: if the normal form of P · f is not zero, then it is added to
            both G and H, thereby enlarging the basis and providing more
            polynomials for multiplication by appropriate terms as above
      end
    end
  end
```

comment: H is now empty; in the next procedure, H will comprise the
additions to G arising in the course of enlarging G by means of
S-polynomials

while $B \neq \emptyset$ do
begin
    $\{f_1, f_2\} :=$ an element of B
    $B := B \backslash \{\{f_1, f_2\}\}$
    $h := N(G, Spoly(f_1, f_2))$
    if $h \neq 0$ then
        $B := B \cup \{\{g, h'\} | g \in G\}$
        $G := G \cup \{h'\}$
        $H := H \cup \{h'\}$
  end
end

Proof. Termination occurs since the only polynomials that are used to extend G or H are in normal form modulo G.

The algorithm will not terminate unless G has the two properties as in (3) of Theorem 20, so that G is a Gröbner basis. Furthermore, every polynomial used to extend the original set F in the initial run through the routine is either a left multiple of an element of F or a linear combination of such left multiples (a left multiple of an S-polynomial). Hence it is in $(F)_L$, so by induction $G \subset (F)_L$, and hence $(G)_L = (F)_L$, since $F \subseteq G$. $\qquad\square$

Thus the left ideal membership problem has an algorithmic solution. In fact, if F is a set of homogeneous polynomials, $(F) = (F)_L$ by 15, and we have the obvious

Corollary 22. *Given a finite subset F of $\mathcal{K}[n, m]$ consisting of homogeneous polynomials, the above algorithm constructs a Gröbner basis G such that $(F) = (G)$.*

From Corollary 19, we also have the following

Corollary 23. *Given a finite subset F of $\mathcal{K}[n, m]$ consisting of homogeneous Grassmann polynomials, the above algorithm constructs a Gröbner basis G from $F' = F \cup \partial F$ such that $(G) = (F)_\partial$.*

Hence the ∂-ideal membership problem for homogeneous polynomials has an algorithmic solution; note that the boundary map is only needed initially in order to compute the boundaries of all elements of F. Thus any equational consequence of a set of hypotheses in which the boundary map is freely used will be reduced to zero by the Gröbner basis of the hypothesis polynomials together with their boundaries, and conversely.

In geometry, it is more natural to encode collinearity information (and the higher-dimensional analogs) using a single product: thus collinear(A, B, C) becomes $ABC = 0$; similarly $M = \text{midpoint}(A, C)$ can be rendered as $AM = MC$.

Then $ABCD = 0$ says that the four points A, B, C, D are coplanar, from which one can deduce via the boundary map that $ABC = ABD + BCD + CAD$, which has an interpretation in terms of oriented areas.

Consider the problem of proving Proposition 3. Encoding the hypotheses as $ADP = BCP = 0$, $AM = MC$ and $BN = ND$, it is impossible to prove the conclusion that $AB + BC + CD + DA = 4(MN + NP + PM)$ using only substitutions and ideal-theoretic manipulations generally: the conclusion is not in the ideal of the hypotheses, nor even is the stronger conclusion polynomial $ABC + ACD - 4MNP$ from which the original conclusion may be derived by applying the boundary map. Instead one must first apply the boundary map to the hypotheses, and then reduction of both conclusions to zero is possible.

A more striking example arises by considering the one-dimensional analog of oriented area. Thus if $AB = CD$, it follows that $AC + CD = AD$, although clearly $AC + CD - AD$ is not a multiple of $AB - CD$. However, taking the boundary of $AB - CD$ gives $A - B - C + D$, and then one can easily check that $AC + CD - AD$ is in the ideal $(AB - CD, A - B - C + D)$.

A final observation. For vectors U, V, W of a Grassmann algebra, it follows from the Exchange Theorem that if $UV = 0$ and $UW = 0$ then either $U = 0$ or $VW = 0$. Hence in $\mathcal{K}[5, 0]$, letting $F = \{(P_1 - P_2)(P_1 - P_3), (P_1 - P_2)(P_1 - P_4), (P_1 - P_2)(P_1 - P_5) - (P_1 - P_3)(P_1 - P_4)\}$, we have that $(P_1 - P_3)(P_1 - P_4) \in \mathcal{C}(F)$. Note that $\partial F = \{0\}$, so by Theorem 17, $(F) = (F)_\partial$, a homogeneous ∂-ideal. Now one can apply the above algorithm to show that $(P_1 - P_3)(P_1 - P_4) \notin (F) = (F)_\partial$. This shows that a theorem prover based solely upon the computation of Gröbner bases of ∂-ideals is not complete. We conjecture that there is an algorithm which incorporates use of the Exchange Theorem to produce two or more new subcases of the hypotheses every time it is used and which is complete; such an algorithm should also be able to deal with genericity issues.

11 Conclusion

The origins of the ideas in this paper lie in the great works *Die Lineale Ausdehnungslehre*, of 1844, and *Die Ausdehnungslehre*, of 1862, of Hermann Grassmann, in which he founded a core discipline of modern mathematics: linear algebra, including exterior algebra. Although the second of these books was published as a new edition of the first, it really offered a very different approach to the same material and both are included in full in the *Collected Works* [15]. An historical analysis of Grassmann's work appears in [10]; also, in [12] it is argued that Grassmann anticipated some aspects of modern universal algebra. The approach to area advocated in the present paper has been used by one of us in geometry courses for twenty years or so.

For an overview of automated geometry theorem proving and of algebraic methods, [29] and [18] are recommended. The algorithmic method for proving Euclidean geometry theorems, beautifully presented by Shang-Ching Chou in [4], has its origin in the pioneering work of Wu Wen-tsün; see also [3], [30] and [31].

A somewhat different approach is to use Buchberger's Gröbner base algorithm; see, for example, [20], [21], [22] and [24]. The algorithms presented in this paper (and in [26]) extend Buchberger's algorithm to take in exterior products.

Acknowledgements. We thank the anonymous referees of the first draft of this paper for their helpful suggestions. The work reported here was partially supported by Australian Research Council Large Grants A49132001 and A49331346.

References

1. Apel, J., *A relationship between Gröbner bases of ideals and vector modules of G-algebras*, Contemporary Mathematics **131** (1992), 195–204.
2. Buchberger, B., *Gröbner bases: An algorithmic method in polynomial ideal theory*, in Multidimensional Systems Theory, ed. Bose, N. K., Reidel, 1985, 184–232.
3. Chou, S., *Proving elementary geometry theorems using Wu's algorithm*, in Automated Theorem Proving: After 25 Years, ed. Bledsoe, W. W., and Loveland, D. W., Am. Math. Soc., 1984, 243–286.
4. Chou, S., *Mechanical Geometry Theorem Proving*, Reidel, 1988.
5. Chou, S., Gao, X., and Zhang, J., *Automated geometry theorem proving using vector calculation*, in Proc. ISSAC (Kiev, 1993), 284–291.
6. Chou, S., Gao, X., and Zhang, J., *Automated production of traditional proofs for constructive geometry theorems*, in Proc. 8th IEEE Symbolic Logic in Computer Science (1993), 48–56.
7. Chou, S., Gao, X., and Zhang, J., *Automated production of traditional proofs in solid geometry*, J. Aut. Reas. **14**, 257–291.
8. Coxeter, H. S. M., and Greitzer, G., *Geometry Revisited*, Math. Ass. Am., 1967.
9. Doubilet, P., Rota, G., and Stein, J., *On the foundations of combinatorial theory IX: Combinatorial methods in invariant theory*, Studies in Applied Math. **57** (1974), 185–216.
10. Fearnley-Sander, D., *Hermann Grassmann and the creation of linear algebra*, Am. Math. Monthly **86** (1979), 809–817.
11. Fearnley-Sander, D., *Affine geometry and exterior algebra*, Houston J. Math. **6** (1980), 53–58.
12. Fearnley-Sander, D., *Hermann Grassmann and the prehistory of universal algebra*, Am. Math. Monthly **89** (1982), 161–166.
13. Fearnley-Sander, D., *The Idea of a Diagram*, in Resolution of Equations in Algebraic Structures, ed. Ait-Kaçi, H., and Nivat, M., Academic Press, 1989, 127–150.
14. Feynman, R. P., Leighton, R. B., and Sands, M., *Lectures on Physics*, Addison-Wesley, 1963.
15. Grassmann, H. G., *Gesammelte Mathematische und Physikalische Werke*, ed. Engel, F., 3 vols. in 6 parts, Leipzig, 1894-1911.
16. Hestenes, D., *New Foundations for Classical Mechanics*, Kluwer, 1986.
17. Hestenes, D., and Ziegler, R., *Projective geometry with Clifford algebra*, Acta. Appl. Math. **23** (1991).
18. Hong, H., Wang, D., and Winkler, F. (eds.), *Algebraic approaches to geometric reasoning*, Ann. Math. and AI **13** (1, 2) (1995).
19. Hungerford, T. W., *Algebra*, Holt, Rinehart and Winston, 1974.

20. Kapur, D., *Using Gröbner bases to reason about geometry problems*, J. Symb. Comp. **2** (1986), 199–208.
21. Kapur, D., *Geometry theorem proving using Hilbert's Nullstellensatz*, in Proc. SSAC (1986), 202–208.
22. Kapur, D., *A refutational approach to theorem proving in geometry*, Artificial Intelligence **37** (1988), 61–93.
23. Kurosh, A. G., *Lectures on general algebra*, Chelsea Publishing Company, New York, 1965.
24. Kutzler, B., and Stifter, S., *On the application of Buchberger's algorithm to automated geometry theorem proving*, J. Symb. Comp. **2** (1986), 389–397.
25. Pilz, G., *Near rings*, North-Holland, 1983.
26. Stokes, T., *Gröbner bases in exterior algebra*, J. Aut. Reas. **6** (1990), 233–250.
27. Sturmfels, B., and Whiteley, W., *On the synthetic factorization of projectively invariant polynomials*, J. Symb. Comp. **11** (1991), 439–453.
28. White, N.L., and McMillan T., *Cayley Factorization*, IMA Preprint Series **371** (1987).
29. Wang, D., *Geometry machines: From AI to SMC*, in Proc. AISMC-3, LNCS **1138** (1996), 213–239.
30. Wu, W.-t., *Basic principles of mechanical theorem proving in geometries*, J. Aut. Reas. **2** (1986), 221–252.
31. Wu, W.-t., *Mechanical theorem proving in geometries: Basic principles*, Springer, New York, 1994.

Automated Production of Readable Proofs for Theorems in Non-Euclidean Geometries

Lu Yang[1], Xiao-Shan Gao[2], Shang-Ching Chou[3], and Jing-Zhong Zhang[1]

[1] Chengdu Institute of Computer Applications, Academia Sinica
[2] Institute of Systems Science, Academia Sinica
[3] Department of Computer Science, The Wichita State University

Abstract. We present a complete method which can be used to produce short and human readable proofs for a class of constructive geometry statements in non-Euclidean geometries. The method is a substantial extension of the area method for Euclidean geometry. The method is an *elimination algorithm* which is similar to the *variable elimination method* of Wu used for proving geometry theorems. The difference is that instead of eliminating coordinates of points from general algebraic expressions, our method eliminates points from high level geometry invariants. As a result the proofs produced by our method are generally short and each step of the elimination has clear geometric meaning. A computer program based on this method has been used to prove more than 90 theorems from non-Euclidean geometries including many new ones. The proofs produced by the program are generally very short and readable.

1 Introduction

Two of the main approaches to automated geometry reasoning are the approach based on *synthetic deduction* or heuristic methods [9, 14, 13, 6] and the approach based on *algebraic computation* which first transforms a geometry statement into an algebraic statement via coordinates and then deals with the algebraic statement using algebraic techniques such as Wu's method [23, 1, 20, 21, 26] and the Gröbner basis method [1, 15, 16, 17, 18]. Generally speaking, the algebraic approaches are decision procedures and are more powerful, while the synthetic approaches are not decision procedures but can be used to produce proofs in traditional style.

Automated theorem proving in non-Euclidean geometries with algebraic method has been reported in a few work. A method for non-Euclidean geometries was introduced by W. T. Wu [23] as a direct consequence of his algebraic method. Transformation theorems between hyperbolic and elliptic geometries were studied in [3, 8]. In [24], properties of trigonometric functions were used to derive new properties of non-Euclidean geometries. The polynomials involved in the proofs are generally larger than those in Euclidean geometry.

In [2], we introduced the *area method* for Euclidean geometry which is a combination of algebraic computation and synthetic deduction: it uses simple algebraic operations to achieve efficiency and uses synthetic deduction to preserve the geometric meaning of the proofs. The area method has been used to produce short and *human-readable proofs* automatically for hundreds of difficult geometry theorems from Euclidean geometry [2]. By *human-readable proofs*, we mean that the proofs are short enough for human to write down without difficulty and each step of the proofs has clear geometric meaning.

The method presented in this paper is a substantial extension of the area method for Euclidean geometry to two non-Euclidean geometries: the hyperbolic (or the Bolyai-Lobachevsky) geometry and the elliptic (or the Riemann-Cayley) geometry. The main difficulty of the extension is that the concept of area in these two non-Euclidean geometries does not have the properties that make the area method possible in Euclidean geometry. We thus need to find new types of geometry quantities. The basic quantities used by us in non-Euclidean geometries are the *argument* (roughly speaking, the sine of the area) and the *cosine of distance*. Furthermore, not all of the properties of these quantities needed in our method are in classical books of non-Euclidean geometry [7, 10]. We think some of the geometry results proved by us, such as Propositions 7 and 16, are new.

The method is based on an *elimination approach* which is similar to the *variable elimination method* of Wu used for proving geometry theorems [23]. The difference is that instead of eliminating point coordinates from algebraic expressions, our method eliminates points from high level geometry invariants or quantities. In other words, it is an elimination method at a higher level and as a result the proofs produced are generally human-readable. Our experiments confirm this: the proofs of the 90 theorems produced by our program are short. Actually, the average term of the maximal polynomials in these proofs is 2.2 (see Section 4), while using the coordinate approach, proofs involving polynomials with more than one hundred terms are not uncommon. The geometric meaning of each step of the proofs for the 90 theorems is also clear.

The format of proofs produced by our method is similar to that produced by rewrite rule methods. But there is a major difference. A typical rewrite method will first generate a canonical set of rules and then use these rules to reduce the conclusion. In our case, the rules are generated dynamically, i.e., they are generated only when they are needed during the process of proofs. Generating all the rules for a geometry statement is much time consuming and not needed.

We have implemented the algorithm using Common Lisp on a NeXT workstation. The prover is available via ftp at emcity.cs.twsu. edu: pub/geometry/software/euc.tar.Z. (The default method in the prover is for Euclidean geometry. To use the method for non-Euclidean geometries, you need to do (setq non-Euclidean t).) The prover has been used to prove 90 theorems totally automatically [5]. Of the 90 theorems, about 40 belong to the projective geometry and thus are true in non-Euclidean geometries. The validity of the others are generally non-trivial. Notably, some extensions of the Ceva and Menelaus theorems discovered recently [11] were extended to hyperbolic and elliptic geometries using the method developed in this paper.

The rest of the paper is organized as follows. Section 2 presents an outline of the method. Section 3 presents the detailed elimination method. Section 4 presents the experimental results and examples. Section 5 presents the conclusion remarks.

2 Outline of the Method for Hyperbolic Geometry

In this section, we give an outline of the method, leaving the detailed elimination method to Section 3. For the basic concepts of non-Euclidean geometry, the reader may consult [7].

2.1. An Example. We first use an example to show how the method works.

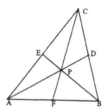

Figure 1

Example 1 (Ceva's Theorem). In Figure 1, ABC is a triangle; P is any point. Points D, E, F are the intersection points of lines AP, BP, CP with lines BC, CA, AB respectively. Show that

$$\frac{\sinh(AF)}{\sinh(FB)}\frac{\sinh(BD)}{\sinh(DC)}\frac{\sinh(CE)}{\sinh(EA)} = 1$$

where $\sinh(AF)$ is the hyperbolic sine of segment AF.

To prove this theorem, we use a geometry invariant or quantity: *the argument.* The absolute value of the argument of a triangle ABC is defined to be

$$|S_{ABC}| = |\sinh(AB) \cdot \sinh(BC) \cdot \sin \angle(AB, BC)|.$$

We also assume that the argument of a clock-wised triangle is positive and

$$S_{ABC} = S_{CAB} = S_{BCA} = -S_{ACB} = -S_{BAC} = -S_{CBA}.$$

It is clear that $S_{ABC} = 0$ iff A, B, and C are on the same line. Another quantity is the *ratio of hyperbolic sines* of directed line segments $\frac{\sinh(AB)}{\sinh(PQ)}$ where A, B, P, Q are collinear. Then it is clear that

$$\frac{\sinh(AB)}{\sinh(PQ)} = -\frac{\sinh(BA)}{\sinh(PQ)} = -\frac{\sinh(AB)}{\sinh(QP)}.$$

The following result is a direct consequence of the definition of the argument.

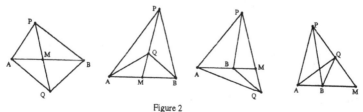

Figure 2

Proposition 1 (The Co-side Theorem). *Let M be the intersection of two lines AB and PQ and $Q \neq M$. Then $\frac{\sinh(PM)}{\sinh(QM)} = \frac{S_{PAB}}{S_{QAB}}$. Note that the result is true in all the four cases in Figure 2.*

Proof of Example 1. In our method, we need to give an order for the points in the geometry statement such that each point can be constructed from the previous points. In this example, a construction order of points is A, B, C, P, D, E, F: first taking four free points A, B, C, P, then taking intersection points of lines AP, BP, CP with lines BC, CA, AB respectively.

Our proof method is to *eliminate points* from the conclusion. By the co-side theorem, we can eliminate points F, E and D from the conclusion respectively:

$$\frac{\sinh(AF)}{\sinh(FB)} = \frac{S_{ACP}}{S_{BPC}}, \frac{\sinh(CE)}{\sinh(EA)} = \frac{S_{BCP}}{S_{BPA}}, \frac{\sinh(BD)}{\sinh(DC)} = \frac{S_{ABP}}{S_{APC}}.$$

Then it is clear that

$$\frac{\sinh(AF)}{\sinh(FB)} \frac{\sinh(BD)}{\sinh(DC)} \frac{\sinh(CE)}{\sinh(EA)} = \frac{S_{ACP}S_{BCP}S_{ABP}}{S_{BPC}S_{BPA}S_{APC}} = 1.$$

2.2. Constructive Geometry Statements. We have mentioned that the proving method is for constructive geometry statements, in which points are introduced by some constructions. A *construction* is one of the following ways of introducing new points. For each construction, we also give its *non-degenerate* (ndg) conditions which guarantee the construction introduces a point properly.

C1 (POINT[S], $Y_1, \cdots, Y_l$). Take arbitrary points $Y_1, \cdots, Y_l$ in the plane.

C2 (ON, Y, (LINE, U, V)). Take a point Y on the line passing through points U and V. The ndg condition of C2 is $U \neq V$.

C3 (MRATIO, Y, U, V, r). Take a point Y on a line UV such that $r = \frac{\sinh(UY)}{\sinh(YV)}$. The quantity r could be a number, an indeterminate, or an expression in geometry quantities. The ndg condition is $U \neq V$.

C4 (LRATIO, Y, U, V, r_1, r_2). Take a point Y on a line UV such that $r_1 = \frac{\sinh(UY)}{\sinh(UV)}, r_2 = \frac{\sinh(YV)}{\sinh(UV)}$. The quantities r_1 and r_2 could be numbers or indeterminates. The ndg condition is $U \neq V$.

C5 (INTER, Y, (LINE, U, V), (LINE, P, Q)). Point Y is the intersection point of line PQ and line UV. The ndg condition is that line UV and line PQ have a proper intersection point. The algebraic form is $S_{UPVQ} \neq 0$. The argument of a quadrilateral is defined in Section 3.1.

C6 (FOOT, Y, P, U, V). Y is the foot of the perpendicular line drawn from P to line UV. The ndg condition is $U \neq V$.

C7 (INTER, Y, (LINE, U, V), (TLINE, P, Q)). Point Y is the intersection of line UV and the line passing through P and perpendicular to PQ. The ndg condition is that the two lines have a proper intersection point.

C8 (INTER, Y, (LINE, P, Q), (CIR, O, P)). Point Y is the intersection point of line PQ and circle (CIR, O, P) (i.e., the circle with center O and passing through point P). The ndg conditions are $Y \neq P$, $O \neq P$, and $P \neq Q$.

C9 $(\text{INTER}, Y, (\text{CIR}, O_1, P), (\text{CIR}, O_2, P))$. Point Y is the intersection point of circle (CIR, O_1, P) and circle (CIR, O_2, P) other than point P. The ndg condition is that O_1, O_2, and P are not collinear.

Now class **C**, the class of *constructive geometry statements*, is defined as follows. A statement in class **C** is a list $S = (C_1, \cdots, C_k, G)$ where C_i, $i = 1, \cdots, k$, are constructions such that each C_i introduces a new point from the points introduced before; and G, which is either a geometry predicate like collinear and perpendicular or an algebraic equation in geometric quantities, is the conclusion of the statement.

Let $S = (C_1, \cdots, C_k, G)$ be a statement in **C**. The *ndg condition* of S is the set of ndg conditions of the C_i plus the condition that the denominators of the geometry quantities in G are not equal to zero.

The constructive description for Ceva's theorem (Example 1) is as follows.

$((\text{POINTS}, A, B, C, P)$

$(\text{INTER}, D, (\text{LINE}, B, C), (\text{LINE}, P, A))$

$(\text{INTER}, E, (\text{LINE}, A, C), (\text{LINE}, P, B))$

$(\text{INTER}, F, (\text{LINE}, A, B), (\text{LINE}, P, C))$

$(\frac{\sinh(AF)}{\sinh(FB)} \frac{\sinh(BD)}{\sinh(DC)} \frac{\sinh(CE)}{\sinh(EA)}, =, 1))$

The ndg conditions for Ceva's theorem are: AP intersects BC properly; BP intersects AC properly; CP intersects AB properly; $F \neq B$; $D \neq C$; and $E \neq A$.

It is obvious that each construction can be reduced to one or two *geometry predicates*. For example, construction C6 is equivalent to two predicates: collinear(Y,U,V) and perpendicular(Y,P,P,Q). For a geometry statement S in class C, let Pr be the conjunction of the predicates derived from the constructions and Nd be the set of the non-degenerate conditions. Then the *predicate form* of S is

$$\forall P_i[(Pr \wedge Nd) \Rightarrow G]$$

where P_i are the points in S and G is the conclusion of S.

2.3. The Algorithm. We have mentioned in Example 1 that the proving method is to eliminate points from the basic geometry quantities. Formally, we can describe the method as follows.

INPUT: $S = (C_1, C_2, \cdots, C_k, E)$ is a statement in **C**, where E is an equation in geometric quantities.

OUTPUT: The algorithm tells whether S is true or not, and if it is true, produces a proof for S.

S1. For $i = k, \cdots, 1$, do S2, S3, S4, and finally S5.

S2. Check whether the ndg conditions of C_i are satisfied. For instance, if the ndg condition is $A \neq B$, we will check whether $\cosh(AB) = 1$ with this algorithm. In other words, we need to prove a theorem whose conclusion is the negation of the ndg. If an ndg condition of a geometry statement is not satisfied, the statement is *trivially true*. The algorithm terminates.

S3. Let $G_1, \cdots, G_s$ be the geometric quantities occurring in E. For $j = 1, \cdots, s$ do S4.

S4. Let $H_j =$ ELIM(G_j, C_i) where ELIM is an algorithm which eliminates the point introduced by construction C_i from G_j using methods given in Section 3 to obtain a formula H_j. Replace G_j by H_j in E to obtain a new E.

S5. Now E is an equation in hyperbolic trigonometric functions of independent variables. We can use the method described in the last paragraph of Section 3.4 to check weather E is a hyperbolic trigonometric identity. If E is an identity then S is valid. Otherwise S is false.

Proof of the correctness. Only the last step needs explanation. If E is an identity, the statement is obviously true. Note that the ndg conditions ensure that the denominators of all the expressions occurring in the proof do not vanish. If E is not an identity. We divide it into two cases. If all the elimination is for linear quantities then the geometric quantities left in E are free parameters, i.e., in the geometric configuration of S they can take arbitrary values. Since E is not an identity, we can take some integer values for these quantities such that when replacing these quantities by the corresponding values in E, we obtain a non-vanishing number. In other words, we obtain a counter example in real geometry for S. In this case S is false in the real hyperbolic geometry. If some quadratic equations are used in the elimination, by a theorem in [22] the statement is false in the complex geometry. We do not know whether the statement is valid over the field of real numbers. ∎

It is clear that the key step of the algorithm is algorithm ELIM in Step S4, i.e., *how to eliminate points from geometry quantities*. This will be presented in detail in Section 3.

3 Elimination Methods for Hyperbolic Geometry

In Section 3.1, we give the geometric facts which will serve as the deduction basis or axioms of our method. Some of these properties are known or easy to prove, but some of them such as Propositions 7 and 16 are new. In the rest of this section, we present the elimination algorithm ELIM. We will try to give the proofs for some of the frequently used elimination steps. Proofs of others can be found in our full report [4]. Readers who are not concerned with the technical details may skip this section.

3.1. Geometry Preliminaries. We use capital English letters to denote points in the hyperbolic plane. The absolute value of *argument of a quadrilateral ABCD* is defined to be

$$|S_{ABCD}| = |\sinh(AC) \cdot \sinh(BD) \cdot \sin \angle(AC, BD)|.$$

We also assume that $S_{ABCD} = -S_{CBAD} = -S_{BADC}$. A more general form of the co-side theorem is as follows.

Proposition 2 (The Co-side Theorem). *Let M be the intersection of two lines AB and PQ and $Q \neq M$. Then*

$$\frac{\sinh(PM)}{\sinh(QM)} = \frac{S_{PAB}}{S_{QAB}}; \frac{\sinh(PM)}{\sinh(PQ)} = \frac{S_{PAB}}{S_{PAQB}}; \frac{\sinh(QM)}{\sinh(PQ)} = \frac{S_{QAB}}{S_{PAQB}}.$$

Proposition 3. *Let R be a point on line PQ. Then for two points A and B,*

$$S_{RAB} = \frac{\sinh(PR)}{\sinh(PQ)} S_{QAB} + \frac{\sinh(RQ)}{\sinh(PQ)} S_{PAB}.$$

Another basic geometry quantity is the hyperbolic cosine of a line segment.

Proposition 4. *Let R be a point on line PQ. Then for any point A, we have:*

$$\cosh(RA) = \frac{\sinh(PR)}{\sinh(PQ)} \cosh(QA) + \frac{\sinh(RQ)}{\sinh(PQ)} \cosh(PA).$$

Similar to the area method in Euclidean geometry, we introduce another geometry quantity: the *Pythagorean difference* for quadrilateral $ABCD$ is defined to be

$$P_{ABCD} = \cosh(AD) \cdot \cosh(BC) - \cosh(AB) \cdot \cosh(CD).$$

Then the *Pythagorean difference* for a triangle ABC is

$$P_{ABC} = P_{ABBC} = \cosh(AC) - \cosh(AB)\cosh(CB).$$

Proposition 5. *$AB \perp PQ$ iff $P_{APBQ} = 0$.*

Proposition 6. *Let Y be the foot of the perpendicular from point P to UV. Then*

$$\frac{\sinh(UY)}{\sinh(YV)} = \frac{P_{PUV}}{P_{PVU}}, \frac{\sinh(UY)}{\sinh(UV)} = \frac{P_{PUV}}{f_1(P,U,V)}, \frac{\sinh(YV)}{\sinh(UV)} = \frac{P_{PVU}}{f_1(P,U,V)}$$

where $f_1(P,U,V)^2 = \sinh(UV)^2(2\cosh(PU)\cosh(PV)\cosh(UV) - \cosh(PU)^2 - \cosh(PV)^2)$.

Proposition 7. *Let R be the intersection of line UV and the line passing through point P and perpendicular to PQ. Then*

$$\frac{\sinh(UR)}{\sinh(VR)} = \frac{P_{UPQ}}{P_{VPQ}}, \frac{\sinh(UR)}{\sinh(UV)} = \frac{P_{UPQ}}{f_2(P,Q,U,V)}, \frac{\sinh(VR)}{\sinh(UV)} = \frac{P_{VPQ}}{f_2(P,Q,U,V)}$$

where $f_2(P,Q,U,V)^2 = P_{UPQ}^2 + P_{VPQ}^2 - 2P_{UPQ}P_{VPQ}\cosh(UV)$.

The following proposition gives the relations among the parameters r, r_1, and r_2 in constructions C3 and C4.

Proposition 8. *Let R be a point on line UV, $r = \frac{\sinh(UR)}{\sinh(RV)}, r_1 = \frac{\sinh(UR)}{\sinh(UV)}, r_2 = \frac{\sinh(RV)}{\sinh(UV)}$. Then*

1. $r_1^2 + 2r_1r_2 \cosh(UV) + r_2^2 = 1$.
2. $r_1 = rr_2$, $r_2^2 = \frac{1}{1+2r\cosh(UV)+r^2}$.

We have defined four *geometry quantities*: the argument, the Pythagorean difference, the hyperbolic cosine of line segments, and the ratio of hyperbolic sines of directed line segments.

In the rest of this section, we will present algorithm ELIM. We need to show how to eliminate points introduced by constructions C1-C8 from the three basic geometry quantities: ratios, arguments, and hyperbolic cosines. We first show that the nine constructions are not independent. For instance, C2 can be reduced to C3: taking an arbitrary point on a line UV is equivalent to taking a point on UV such that $\frac{\sinh(UY)}{\sinh(UV)}$ is an indeterminate. In fact, constructions C3, C8 and C9 can also be reduced to other constructions [4]. Therefore, we need only consider five constructions C1, C4, C5, C6, and C7. Algorithm ELIM will be presented as eight elimination lemmas.

3.2. Eliminating Points from Ratios. Now we consider how to eliminate points from the ratios.

Lemma 9. $ELIM(f,c)$: $f = \frac{\sinh(AY)}{\sinh(BC)}$ and c is construction (LRATIO, Y, U, V, r_1, r_2), where A, B, C are points on line UV. We have:

$$f = \frac{\sinh(AU)}{\sinh(BC)}(r_1\cosh(UV) + r_2) + r_1\frac{\sinh(UV)}{\sinh(BC)}\cosh(AU).$$

Proof. By Proposition 4,

$$\frac{\sinh(AY)}{\sinh(BC)} = \frac{\sinh(AU + UY)}{\sinh(BC)}$$
$$= \frac{\sinh(AU)}{\sinh(BC)}\cosh(UY) + \frac{\sinh(UY)}{\sinh(BC)}\cosh(AU)$$
$$= \frac{\sinh(AU)}{\sinh(BC)}(r_1\cosh(UV) + r_2) + r_1\frac{\sinh(UV)}{\sinh(BC)}\cosh(AU).$$

Lemma 10. $ELIM(f,c)$: $f = \frac{\sinh(AY)}{\sinh(CD)}$ and c is (INTER, Y, (LINE, U, V), (LINE, P, Q)), where points A and Y are on line CD. We have:

$$f = \begin{cases} \frac{S_{AUV}}{S_{GUDV}} & \text{if A is not on UV;} \\ \frac{S_{APQ}}{S_{CPDQ}} & \text{otherwise.} \end{cases}$$

Proof. This proposition can be similarly proved as the co-side theorem.

Lemma 11. $ELIM(f,c)$: $f = \frac{\sinh(AY)}{\sinh(CD)}$ and c is (FOOT, Y, P, U, V), where points A and Y are on line CD. We have:

$$f = \begin{cases} \frac{\sinh(AV)}{\sinh(CD)}\frac{P_{PAV}}{f_1(P,A,V)} & \text{if } A \in UV \text{ and } A \neq V; \\ \frac{S_{AUV}}{S_{GUDV}} & \text{if } A \notin UV. \end{cases}$$

Proof. The first and second cases can be proved similarly as Proposition 6 and the co-side theorem respectively. ∎

Lemma 12. *ELIM(f,c): $f = \frac{\sinh(AY)}{\sinh(CD)}$ and c is (INTER, Y, (LINE, U, V), (TLINE, P, Q)), where points A and Y are on line CD. We have:*

$$\frac{\sinh(AY)}{\sinh(CD)} = \begin{cases} \frac{\sinh(AV)}{\sinh(CD)}\frac{P_{APQ}}{f_2(P,Q,A,V)} & \text{if } A \in UV \text{ and } A \neq V; \\ \frac{S_{AUV}}{S_{CUDV}} & \text{if } A \notin UV. \end{cases}$$

Proof. The first and second cases can be similarly proved as Proposition 7 and the co-side theorem respectively. ∎

3.3. Eliminating Points from Linear Quantities.

Let $G(Y)$ be S_{ABY}, $\cosh(AY)$, P_{ABY}, or P_{ABCY} for different points A, B, C, and Y. Then for three collinear points Y, U, and V, by Propositions 3 and 4 we have:

$$(I) \qquad G(Y) = \frac{\sinh(UY)}{\sinh(UV)}G(V) + \frac{\sinh(YV)}{\sinh(UV)}G(U).$$

We call $G(Y)$ a *linear geometry quantity* of point Y. Elimination procedures for linear geometry quantities are similar for constructions C4-C6.

Lemma 13. *ELIM($G(Y)$,c): Let $G(Y)$ be a linear geometry quantity. Then $G(Y)$ equals to*

$$\begin{cases} r_1 G(V) + r_2 G(U) & \text{if } c = (LRATIO, Y, U, V, r_1, r_2); \\ \frac{S_{UPQ}G(V)-S_{VPQ}G(U)}{S_{UPVQ}} & \text{if } c = (INTER, Y, (LINE, U, V), (LINE, P, Q)); \\ \frac{P_{PUV}G(V)+P_{PVU}G(U)}{f_1(PUV)} & \text{if } c = (FOOT, Y, P, U, V); \\ \frac{P_{UPQ}G(V)-P_{VPQ}G(U)}{f_2(P,Q,U,V)} & \text{if } c = (INTER, Y, (LINE, U, V), (TLINE, P, Q)). \end{cases}$$

Proof. Since point Y is on line UV in each case, equation (I) is true. The first case is just (I). For the second case, by the co-side theorem

$$\frac{\sinh(UY)}{\sinh(UV)} = \frac{S_{UPQ}}{S_{UPVQ}}, \frac{\sinh(YV)}{\sinh(UV)} = -\frac{S_{VPQ}}{S_{UPVQ}}.$$

Substituting these into (I), we prove the result. The third and fourth cases are consequences of (I) and Propositions 6 and 7. Note that the non-degenerate conditions of each construction guarantee that the denominator of the new expression does not vanish. ∎

3.4. Eliminating Points from Quadratic Quantities.

The functions f_1 and f_2 in Lemmas 13, 11, and 12 satisfy quadratic equations. We call such quantities quadratic quantities. Now, we present a mechanical method of eliminating points from them. First, another quadratic quantity is S_{ABCD}.

Proposition 14. *We have:*

$$S^2_{ABCD} = S_{ABD}S_{BCA}\cosh(CD) + S_{BCA}S_{CDB}$$
$$\cosh(AD) + S_{CDB}S_{DAC}\cosh(AB) + S_{ABD}S_{DAC}\cosh(BC).$$

To eliminate geometric quantities satisfying quadratic equations completely, we need more algebraic tools which can be found in our full report [4]. In what follows, we only describe briefly how this method works by showing how to prove hyperbolic identities.

Let P be a polynomial in hyperbolic trigonometric functions of variables $\pm x_1$, $\cdots$, $\pm x_n$, $\pm\frac{1}{2}x_1$, $\cdots$, $\pm\frac{1}{2}x_n$, $2x_1$, $\cdots$, $2x_n$, $\cdots$. We want to know weather $P = 0$ is an identity. We set $y_i = \frac{1}{m}x_i, i = 1,\cdots,n$ such that each hyperbolic trigonometric function in P has the form $\sinh(\sum k_i y_i)$ or $\cosh(\sum s_i y_i)$ for integers k_i and s_i. First, we can easily represent P as a polynomial R in $z_i = \sinh(y_i)$ and $w_i = \cosh(y_i)$, $i = 1,\cdots,n$. Then $R = 0$ is an identity iff R will become zero when replacing w_i^2 by $1 + z_i^2$ (for a proof see [8]).

3.5. Eliminating Free Points.

We now show how to eliminate the free points (points introduced by construction C1). To do that, we need the concept of *argument coordinates*. Let $A, O, U,$ and V be four points such that $O, U,$ and V are not collinear. The argument coordinates of A with respect to OUV are

$$x_A = \frac{S_{AUV}}{S_{OUV}}, \quad y_A = \frac{S_{OAV}}{S_{OUV}}, \quad z_A = \frac{S_{OUA}}{S_{OUV}}.$$

It is clear that the points in the plane are in a one to one correspondence with their argument coordinates. The following lemma reduces a linear quantity of free points to their argument coordinates.

Lemma 15. *ELIM($G(Y),c$): $G(Y)$ is a linear geometry quantity $G(Y)$; c is one of the constructions C4-C7. Let $O, U,$ and V be three non-collinear points. Then we have:*

$$G(Y) = \frac{S_{YUV}}{S_{OUV}}G(O) + \frac{S_{OYV}}{S_{OUV}}G(U) + \frac{S_{OUY}}{S_{OUV}}G(V).$$

Proof. Without loss of generality, let OY intersect UV at T. If OY does not intersect UV, we may consider the intersection of UY and OV or that of VY and OU since one of them must exist. By (I),

$$G(Y)=\tfrac{\sinh(OY)}{\sinh(OT)}G(T)+\tfrac{\sinh(YT)}{\sinh(OT)}G(O)=\tfrac{\sinh(OY)}{\sinh(OT)}(\tfrac{\sinh(UT)}{\sinh(UV)}G(V)+\tfrac{\sinh(TV)}{\sinh(UV)}G(U))+\tfrac{\sinh(YT)}{\sinh(OT)}G(O).$$

By the co-side theorem,

$$\frac{\sinh(YT)}{\sinh(OT)} = \frac{S_{YUV}}{S_{OUV}}; \quad \frac{\sinh(OY)}{\sinh(OT)} = \frac{S_{OUYV}}{S_{OUV}};$$
$$\frac{\sinh(UT)}{\sinh(UV)} = \frac{S_{OUY}}{S_{OUYV}}; \quad \frac{\sinh(TV)}{\sinh(UV)} = \frac{S_{OYV}}{S_{OUYV}}.$$

Substituting these into the above formula, we obtain the desired result. ∎

Using Lemma 15, any expression in geometric quantities can be written as an expression in $\cosh(OU)$, $\cosh(OV)$, $\cosh(UV)$, S_{OUV}, and the argument coordinates of the free points. These quantities are still not independent. First, it is known that [7]

$$S_{OUV}^2 = 1 - \cosh(OU)^2 - \cosh(OV)^2 - \cosh(UV)^2$$

$$+2\cosh(OU)\cosh(OV)\cosh(UV).$$

Second, the three argument coordinates of a point satisfy the following property.

Proposition 16. *Let x_A, y_A, z_A be the argument coordinates of A with respect to OUV. Then*

$$x_A^2 + y_A^2 + z_A^2 + 2y_A z_A \cosh(UV) + 2x_A z_A \cosh(OV) + 2x_A y_A \cosh(OU) = 1.$$

Proof. We will prove this result with the aide of our program. We introduce five points with the following constructions:

$$(\text{POINTS}, O, U, V)$$
$$(\text{LRATIO}, D, U, V, r_1, r_2)$$
$$(\text{LRATIO}, A, O, D, s_1, s_2).$$

Using Lemma 13, we can eliminate points D and A from

$$f = x_A^2 + y_A^2 + z_A^2 + 2y_A z_A \cosh UV + 2x_A z_A \cosh OV + 2x_A y_A \cosh OU - 1$$

where $x_A = \frac{S_{AUV}}{S_{OUV}}$, $\quad y_A = \frac{S_{OAV}}{S_{OUV}}$, $\quad z_A = \frac{S_{OUA}}{S_{OUV}}$. Denote the output by g. Let

$$h_1 = r_1^2 + r_2^2 + 2r_1 r_2 \cosh UV - 1,$$
$$h_2 = s_1^2 + s_2^2 + 2s_1 s_2 \cosh OD - 1,$$
$$h_3 = r_1 \cosh OV + r_2 \cosh OU - \cosh OD.$$

According Propositions 4 and 8, $h_1 = 0, h_2 = 0, h_3 = 0$. Doing successive pseudo-division of g with h_3, h_2, h_1, we know that g is a linear combination of h_1, h_2, h_3. This proves the result. ∎

Since the argument coordinates satisfy quadratic equations of triangular form, we can use the techniques presented in Section 3 to eliminate one of the three coordinates for each point.

3.6. Eliminating Cyclic Points. To deal with theorems involving cyclic points efficiently, we introduce a new construction.

C9 (CIRCLE, $A_1, \cdots, A_s$), $(s \geq 3)$. Points $A_1 \cdots A_s$ are on the same circle. There is no ndg condition for this construction.

Let $A_1, \cdots, A_s$ be points on a circle with center O. We choose a point, say A_1, as the reference point. Then each point A_i is uniquely determined by the oriented angle $\frac{\angle A_1 O A_i}{2}$ (we assume that all angles have values from $-\pi$ to π).

Lemma 17. *Let A, B, C, D be points on a circle with center O and radius δ, and A the reference point. We denote $\frac{\angle AOB}{2}$ by $\angle B$. Then*

1. $S_{BCD} = \frac{4\sinh(\angle \frac{BQC}{2}) \cdot \sinh(\angle \frac{CD}{2}) \cdot \sinh(\angle \frac{BD}{2})}{\tanh(\delta)}$.
2. $\sinh(\angle \frac{BC}{2}) = \sinh(\delta) \sin(\angle C - \angle B)$.

Using Lemma 17, an expression in arguments and Pythagorean differences of points on a circle can be reduced to an expression in the radius of the circle and hyperbolic trigonometric functions of independent angles. By Example 3, we can check whether such an expression is an identity. We thus have a complete proving method for this construction. Note that this construction can be used only as the first construction in a geometry statement.

4 Experimental Results and Extensions

4.1. Experimental Results. We have incorporated the method into the Common Lisp prover for the area method [2]. [5] is a collection of 90 geometry theorems proved by our prover. Of the 90 theorems, about 40 belong to the projective geometry and thus are true in non-Euclidean geometries. The validity of the other 50 theorems are not trivial. Many of the theorems such as Example 3 are new.

The following table contains the timing and proof length statistics for the 90 geometry problems solved by our computer program. Maxterm means the number of terms of the maximal polynomial occurring in a proof. Lemmano is the number of elimination lemmas used to eliminate points from geometry quantities. In other words, lemmano is the number of deduction steps in the proof.

Proving Time		Proof Length		Deduction Step	
Time (secs)	% of Thms	Maxterm	% of Thms	Lemmano	% of Thms
$t \leq 0.05$	46%	$m = 1$	62%	$l \leq 5$	14%
$t \leq 0.1$	62%	$m \leq 2$	82%	$l \leq 10$	46%
$t \leq 1$	94%	$m \leq 5$	91%	$l \leq 20$	77%
$t \leq 5$	97%	$m \leq 10$	96%	$l \leq 30$	90%
$t \leq 6$	100%	$m \leq 15$	100%	$l \leq 69$	100%
0.27	average	2.2	average	15.12	average

Table 1. Statistics for the 90 geometry problems

From Table 1, we can see that our program is very fast. More importantly, the average length of the maximal polynomials in the proofs for the 90 theorems is only 2.2. Considering the fact that the 90 theorems are quite non-trivial, for instance, Examples 2 and 4 are among the moderately difficult ones in the 90 theorems, we can say that the proofs produced by our prover are much shorter than the coordinate based methods.

4.2. Two Examples. In [11], B. Grünbaum and G.C. Shephard discovered many new results in Euclidean geometry about polygons using numerical search. We will prove the corresponding results in non-Euclidean geometries using the method developed in this paper. Furthermore, for any polygon with concrete number of sides, our prover may be used to prove the corresponding theorems automatically.

Example 2. Let $ABCDE$ be a pentagon, $P = AD \cap BE$, $Q = AC \cap BE$, $R = BD \cap AC$, $S = CE \cap BD$, and $T = AD \cap CE$. Then

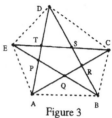

Figure 3

$$\frac{\sinh{(AP)}}{\sinh{(TD)}}\frac{\sinh{(DS)}}{\sinh{(RB)}}\frac{\sinh{(BQ)}}{\sinh{(PE)}}\frac{\sinh{(ET)}}{\sinh{(SC)}}\frac{\sinh{(CR)}}{\sinh{(QA)}} = 1;$$

$$\frac{\sinh{(AT)}}{\sinh{(PD)}}\frac{\sinh{(DR)}}{\sinh{(SB)}}\frac{\sinh{(BP)}}{\sinh{(QE)}}\frac{\sinh{(ES)}}{\sinh{(TC)}}\frac{\sinh{(CQ)}}{\sinh{(RA)}} = 1.$$

In general case, let $V_1, \cdots, V_m$ be a polygon and $1 \le d \le \frac{m}{2}$, $1 \le j \le \frac{m}{2}$ integers. We denote by $P_{d,j,i}$ the intersection of lines $V_i V_{i+d}$ and lines $V_{i+j} V_{i+j+d}$, $i = 1, \cdots, m$. Then $P_{d,j,i-j}$ is the intersection of line $V_{i-j} V_{i-j+d}$ and line $V_i V_{i+d}$. Let

$$T(m, d, j) = \prod_{i=1}^{m} \frac{\sinh{(V_i P_{d,j,i})}}{\sinh{(P_{d,j,i-j} V_{i+d})}}; \quad S(m, d, j) = \prod_{i=1}^{m} \frac{\sinh{(V_i P_{d,j,i-j})}}{\sinh{(P_{d,j,i} V_{i+d})}}.$$

Then Example 2 is a special case of the following theorem.

Example 3. (1) $T(m, d, j) = 1$ iff $d + 2j = m$ or $2d + j = m$. (2) $S(m, d, j) = 1$ iff $d + 2j = m$, $2d = j$, or $2j = d$.

All specific cases of Example 3 can be proved with our prover. For instance, the proof for the first case of Example 2 is as follows.

The machine proof:

$$-\frac{\frac{\sinh{(ET)}}{\sinh{(CE)}} \cdot \frac{\sinh{(DS)}}{\sinh{(BD)}} \cdot \frac{\sinh{(CR)}}{\sinh{(AC)}} \cdot \frac{\sinh{(BQ)}}{\sinh{(BE)}} \cdot \frac{\sinh{(AP)}}{\sinh{(AD)}}}{\frac{\sinh{(EP)}}{\sinh{(BE)}} \cdot \frac{\sinh{(DT)}}{\sinh{(AD)}} \cdot \frac{\sinh{(CS)}}{\sinh{(CE)}} \cdot \frac{\sinh{(BR)}}{\sinh{(BD)}} \cdot \frac{\sinh{(AQ)}}{\sinh{(AC)}}}$$

$$\overset{T}{=} \frac{S_{ADE} \cdot S_{ACDE} \cdot (-\frac{\sinh{(DS)}}{\sinh{(BD)}}) \cdot \frac{\sinh{(CR)}}{\sinh{(AC)}} \cdot \frac{\sinh{(BQ)}}{\sinh{(BE)}} \cdot \frac{\sinh{(AP)}}{\sinh{(AD)}}}{(-\frac{\sinh{(EP)}}{\sinh{(BE)}}) \cdot \frac{\sinh{(CS)}}{\sinh{(CE)}} \cdot \frac{\sinh{(BR)}}{\sinh{(BD)}} \cdot \frac{\sinh{(AQ)}}{\sinh{(AC)}} \cdot S_{CDE}) \cdot (-S_{ACDE})}$$

$$\overset{simplify}{=} \frac{-S_{ADE} \cdot \frac{\sinh{(AP)}}{\sinh{(AD)}} \cdot \frac{\sinh{(BQ)}}{\sinh{(BE)}} \cdot \frac{\sinh{(CR)}}{\sinh{(AC)}} \cdot \frac{\sinh{(DS)}}{\sinh{(BD)}}}{S_{CDE} \cdot \frac{\sinh{(AQ)}}{\sinh{(AC)}} \cdot \frac{\sinh{(BR)}}{\sinh{(BD)}} \cdot \frac{\sinh{(CS)}}{\sinh{(CE)}} \cdot \frac{\sinh{(EP)}}{\sinh{(BE)}}}$$

$$\overset{S}{=} \frac{-S_{ADE} \cdot (-S_{CDE}) \cdot (-S_{BCDE})}{S_{CDE} \cdot \frac{\sinh{(AQ)}}{\sinh{(AC)}} \cdot \frac{\sinh{(BR)}}{\sinh{(BD)}} \cdot (-S_{BCD}) \cdot \frac{\sinh{(EP)}}{\sinh{(BE)}} \cdot S_{BCDE}} \cdot \frac{\sinh{(AP)}}{\sinh{(AD)}} \cdot \frac{\sinh{(BQ)}}{\sinh{(BE)}} \cdot \frac{\sinh{(CR)}}{\sinh{(AC)}}$$

$$\overset{simplify}{=} \frac{S_{ADE}}{\frac{\sinh{(EP)}}{\sinh{(BE)}} \cdot S_{BCD} \cdot \frac{\sinh{(BR)}}{\sinh{(BD)}} \cdot \frac{\sinh{(AQ)}}{\sinh{(AC)}}} \cdot \frac{\sinh{(CR)}}{\sinh{(AC)}} \cdot \frac{\sinh{(BQ)}}{\sinh{(BE)}} \cdot \frac{\sinh{(AP)}}{\sinh{(AD)}}$$

$$\overset{R}{=} \frac{(-S_{BCD}) \cdot S_{ADE} \cdot (-S_{ABCD})}{\frac{\sinh{(EP)}}{\sinh{(BE)}} \cdot S_{BCD} \cdot (-S_{ABC}) \cdot \frac{\sinh{(AQ)}}{\sinh{(AC)}} \cdot S_{ABCD}} \cdot \frac{\sinh{(BQ)}}{\sinh{(BE)}} \cdot \frac{\sinh{(AP)}}{\sinh{(AD)}}$$

$$\overset{simplify}{=} \frac{-S_{ADE}}{\frac{\sinh{(AQ)}}{\sinh{(AC)}} \cdot S_{ABC} \cdot \frac{\sinh{(EP)}}{\sinh{(BE)}}} \cdot \frac{\sinh{(AP)}}{\sinh{(AD)}} \cdot \frac{\sinh{(BQ)}}{\sinh{(BE)}}$$

$$\overset{Q}{=} \frac{-S_{ADE} \cdot (-S_{ABC}) \cdot S_{ABCE}}{S_{ABE} \cdot S_{ABC} \cdot \frac{\sinh{(EP)}}{\sinh{(BE)}} \cdot (-S_{ABCE})} \cdot \frac{\sinh{(AP)}}{\sinh{(AD)}}$$

$$simplify \quad \frac{-S_{ADE}}{\frac{\sinh(EP)}{\sinh(BE)} \cdot S_{ABE}} \cdot \frac{\sinh(AP)}{\sinh(AD)}$$

$$\overset{P}{=} \frac{-S_{ABE} \cdot S_{ADE} \cdot (-S_{ABDE})}{S_{ADE} \cdot S_{ABE} \cdot S_{ABDE}} \overset{simplify}{=} 1$$

The elimination lemmas used in the proof:

$$\frac{\sinh(DT)}{\sinh(AD)} \overset{T}{=} \frac{-S_{CDE}}{S_{ACDE}}, \quad \frac{\sinh(ET)}{\sinh(CE)} \overset{T}{=} \frac{S_{ADE}}{-S_{ACDE}},$$

$$\frac{\sinh(CS)}{\sinh(CE)} \overset{S}{=} \frac{S_{BCD}}{S_{BCDE}}, \quad \frac{\sinh(DS)}{\sinh(BD)} \overset{S}{=} \frac{-S_{CDE}}{S_{BCDE}},$$

$$\frac{\sinh(BR)}{\sinh(BD)} \overset{R}{=} \frac{S_{ABC}}{S_{ABCD}}, \quad \frac{\sinh(CR)}{\sinh(AC)} \overset{R}{=} \frac{-S_{BCD}}{S_{ABCD}},$$

$$\frac{\sinh(AQ)}{\sinh(AC)} \overset{Q}{=} \frac{S_{ABE}}{S_{ABCE}}, \quad \frac{\sinh(BQ)}{\sinh(BE)} \overset{Q}{=} \frac{S_{ABC}}{S_{ABCE}},$$

$$\frac{\sinh(EP)}{\sinh(BE)} \overset{P}{=} \frac{S_{ADE}}{-S_{ABDE}}, \quad \frac{\sinh(AP)}{\sinh(AD)} \overset{P}{=} \frac{S_{ABE}}{S_{ABDE}}.$$

In the above *machine produced proof*, $a \overset{T}{=} b$ means that b is the result obtained by eliminating point T from a; $a \overset{simplify}{=} b$ means that b is obtained by canceling some common factors from the denominator and numerator of a; "eliminants" are the results obtained by eliminating points from separate geometry quantities.

Example 4 (Pascal's Theorem). Let $A, B, C, D, E,$ and F be six points on a circle. Let $P = AB \cap DF$, $Q = BC \cap EF$, and $S = CD \cap EA$. Show that $P, Q,$ and S are collinear.

Constructive description:

$$((CIRCLE, A, B, C, A_1, B_1, C_1)$$
$$(INTER, P, (LINE, A_1, B), (LINE, A, B_1))$$
$$(INTER, Q, (LINE, A, C_1), (LINE, A_1, C))$$
$$(INTER, S, (LINE, B_1, C), (LINE, B, C_1))$$
$$(INTER, Z_S, (LINE, Q, P), (LINE, B_1, C))$$
$$(\frac{\sinh(B_1 S)}{\sinh(CS)} = \frac{\sinh(B_1 Z_S)}{\sinh(CZ_S)}))$$

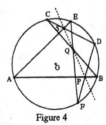

Figure 4

In the above description, by introducing a new Z_S, the fact that $P, Q,$ and S are collinear is equivalent to the fact $S = Z_S$ or $\frac{\sinh(B_1 S)}{\sinh(CS)} = \frac{\sinh(B_1 Z_S)}{\sinh(CZ_S)}$. We always use this trick if the conclusion of a geometry statement is collinear.

The machine proof:

$$\left(\frac{\sinh(B_1 S)}{\sinh(CS)}\right) / \left(\frac{\sinh(B_1 Z_S)}{\sinh(CZ_S)}\right)$$

$$\overset{Z_S}{=} \frac{-S_{CPQ}}{-S_{B_1 PQ}} \cdot \frac{\sinh(B_1 S)}{\sinh(CS)}$$

$$\overset{S}{=} \frac{(-S_{BB_1 C_1}) \cdot S_{CPQ}}{S_{B_1 PQ} \cdot (-S_{BCC_1})}$$

$$\overset{Q}{=} \frac{S_{CA_1 P} \cdot S_{ACC_1} \cdot S_{BB_1 C_1} \cdot S_{ACC_1 A_1}}{(-S_{B_1 C_1 P} \cdot S_{ACA_1}) \cdot S_{BCC_1} \cdot (-S_{ACC_1 A_1})}$$

$$\overset{simplify}{=} \frac{S_{BB_1 C_1} \cdot S_{ACC_1} \cdot S_{CA_1 P}}{S_{BCC_1} \cdot S_{ACA_1} \cdot S_{B_1 C_1 P}}$$

$$\stackrel{P}{=} \frac{S_{BB_1C_1} \cdot S_{ACC_1} \cdot S_{BCA_1} \cdot S_{AA_1B_1} \cdot S_{ABB_1A_1}}{S_{BCC_1} \cdot S_{ACA_1} \cdot (-S_{BA_1B_1} \cdot S_{AB_1C_1}) \cdot (-S_{ABB_1A_1})}$$

$$\stackrel{simplify}{=} \frac{S_{AA_1B_1} \cdot S_{BCA_1} \cdot S_{ACC_1} \cdot S_{BB_1C_1}}{S_{AB_1C_1} \cdot S_{BA_1B_1} \cdot S_{ACA_1} \cdot S_{BCC_1}}$$

$$= \frac{(\mathrm{sh}(\frac{A_1B_1}{2}) \cdot \mathrm{sh}(\frac{AB_1}{2}) \cdot \mathrm{sh}(\frac{AA_1}{2})) \cdot (\mathrm{sh}(\frac{CA_1}{2}) \cdot \mathrm{sh}(\frac{BA_1}{2}) \cdot \mathrm{sh}(\frac{BC}{2}))}{(\mathrm{sh}(\frac{B_1C_1}{2}) \cdot \mathrm{sh}(\frac{AC_1}{2}) \cdot \mathrm{sh}(\frac{AB_1}{2})) \cdot (\mathrm{sh}(\frac{A_1B_1}{2}) \cdot \mathrm{sh}(\frac{BB_1}{2}) \cdot \mathrm{sh}(\frac{BA_1}{2}))}$$

$$\frac{(\mathrm{sh}(\frac{CC_1}{2}) \cdot \mathrm{sh}(\frac{AC_1}{2}) \cdot \mathrm{sh}(\frac{AC}{2})) \cdot (\mathrm{sh}(\frac{B_1C_1}{2}) \cdot \mathrm{sh}(\frac{BC}{2}) \cdot \mathrm{sh}(\frac{BB_1}{2})) \cdot (\tan(\delta))^4}{(\mathrm{sh}(\frac{CA_1}{2}) \cdot \mathrm{sh}(\frac{AA_1}{2}) \cdot \mathrm{sh}(\frac{AC}{2})) \cdot (\mathrm{sh}(\frac{CC_1}{2}) \cdot \mathrm{sh}(\frac{BC_1}{2}) \cdot \mathrm{sh}(\frac{BC}{2})) \cdot (\tan(\delta))^4}$$

$$\stackrel{simplify}{=} 1$$

The elimination lemmas used in the proof:

$$\frac{\sinh(B_1 Z_S)}{\sinh(C Z_S)} \stackrel{Z}{=} \frac{S_{B_1 PQ}}{S_{CPQ}}, \quad \frac{\sinh(B_1 S)}{\sinh(CS)} \stackrel{S}{=} \frac{S_{BB_1C_1}}{S_{BCC_1}}$$

$$S_{B_1 PQ} \stackrel{Q}{=} \frac{-S_{B_1C_1P} \cdot S_{ACA_1}}{S_{ACC_1A_1}}, \quad S_{CPQ} \stackrel{Q}{=} \frac{S_{CA_1P} \cdot S_{ACC_1}}{-S_{ACC_1A_1}}$$

$$S_{B_1C_1P} \stackrel{P}{=} \frac{-S_{BA_1B_1} \cdot S_{AB_1C_1}}{S_{ABB_1A_1}}, \quad S_{CA_1P} \stackrel{P}{=} \frac{S_{BCA_1} \cdot S_{AA_1B_1}}{-S_{ABB_1A_1}}$$

$$S_{BCC_1} = \frac{4 \cdot \mathrm{sh}(\frac{BC_1}{2}) \cdot \mathrm{sh}(\frac{CC_1}{2}) \cdot \mathrm{sh}(\frac{BC}{2})}{-\tan(\delta)}, \quad S_{ACA_1} = \frac{4 \cdot \mathrm{sh}(\frac{AA_1}{2}) \cdot \mathrm{sh}(\frac{CA_1}{2}) \cdot \mathrm{sh}(\frac{AC}{2})}{-\tan(\delta)}$$

$$S_{BA_1B_1} = \frac{4 \cdot \mathrm{sh}(\frac{BB_1}{2}) \cdot \mathrm{sh}(\frac{A_1B_1}{2}) \cdot \mathrm{sh}(\frac{BA_1}{2})}{-\tan(\delta)}, \quad S_{AB_1C_1} = \frac{4 \cdot \mathrm{sh}(\frac{AC_1}{2}) \cdot \mathrm{sh}(\frac{B_1C_1}{2}) \cdot \mathrm{sh}(\frac{AB_1}{2})}{-\tan(\delta)}$$

$$S_{BB_1C_1} = \frac{4 \cdot \mathrm{sh}(\frac{BC_1}{2}) \cdot \mathrm{sh}(\frac{B_1C_1}{2}) \cdot \mathrm{sh}(\frac{BB_1}{2})}{-\tan(\delta)}, \quad S_{ACC_1} = \frac{4 \cdot \mathrm{sh}(\frac{AC_1}{2}) \cdot \mathrm{sh}(\frac{CC_1}{2}) \cdot \mathrm{sh}(\frac{AC}{2})}{-\tan(\delta)}$$

$$S_{BCA_1} = \frac{4 \cdot \mathrm{sh}(\frac{BA_1}{2}) \cdot \mathrm{sh}(\frac{CA_1}{2}) \cdot \mathrm{sh}(\frac{BC}{2})}{-\tan(\delta)}, \quad S_{AA_1B_1} = \frac{4 \cdot \mathrm{sh}(\frac{AB_1}{2}) \cdot \mathrm{sh}(\frac{A_1B_1}{2}) \cdot \mathrm{sh}(\frac{AA_1}{2})}{-\tan(\delta)}$$

In the last step of the above proof, Lemma 17 is used. Note that we use sh to represent sinh in order to save printing space.

4.3. Proving Theorems in Elliptic Geometry.

The proving method presented by now is for hyperbolic geometry. The proving method for constructive geometry statements in elliptic geometry is quite similar to that of the hyperbolic geometry. First, we have the following result [8].

Theorem 18. *Let points* $A, B, P, Q, R \cdots$ *satisfy the same geometry conditions in hyperbolic and elliptic geometries. Then*

$$P(\sin(AB), \cos(AB), \sin(\angle PQR), \cos(\angle PQR), \cdots) = 0$$

is a valid algebraic equation in elliptic geometry if and only if

$$P(\sqrt{-1}\sinh(AB), \cosh(AB), \sin(\angle PQR), \cos(\angle PQR), \cdots) = 0$$

is a valid algebraic equation in hyperbolic geometry.

By the above theorem, all the basic propositions and elimination lemmas proved in this paper can be transformed to similar results in elliptic geometry. We thus have a similar algorithm of theorem proving in elliptic geometry.

Another consequence of Theorem 18 is that a constructive geometry statement with conclusion $P(\sin(AB), \cos(AB), \cdots) = 0$ is true in elliptic geometry iff the statement with same hypotheses and with conclusion $P(\sqrt{-1}\sinh(AB), \cosh(AB), \cdots) = 0$ is true in hyperbolic geometry. For instance, Ceva' theorem (Example 1 with conclusion $\frac{\sin(AF)}{\sin(FB)}\frac{\sin(BD)}{\sin(DC)}\frac{\sin(CE)}{\sin(EA)} = 1$) is true in elliptic geometry.

5 Conclusion Remarks

Conventionally, studies on non-Euclidean geometries are mainly focused on the axiom systems, the representations of these geometries in Euclidean geometry, etc. The topic of proving interesting and difficult geometry problems is generally overlooked. Despite of this, we believe that an efficient mechanical proving method for non-Euclidean geometries like the one given in this paper is of importance: besides the fact that the method/program is used to find many new theorems, it is also important for educational purpose. One of the authors was once asked by a professor during the 1993 International Mathematical Olympiad about how to prove the centroid theorem in hyperbolic geometry, which is not available in common textbooks of non-Euclidean geometries. We take this as an indication of lack of a general method of solving problems in non-Euclidean geometries. The method developed in this paper may provide such a tool, because the proofs produced by the method are generally short and in a shape that a student of mathematics could easily learn to design with pencil and paper.

There are still many problems not solved or solved unsatisfactorily for this approach. Though a large portion of the geometry theorems in textbooks can be proved by our prover, there are still ones which are not in class **C**, e.g., theorems which cannot be described constructively. A moderate goal is to extend the elimination method to more constructions, such as the ones considered in [2]. The main difficulty: for more complicated constructions, the elimination formulas needed may become large and obscure. In this case, the goal of producing short and readable proofs may be dampened. Also the current method is limited to theorems of equational type: geometry inequalities cannot be handled by the current method.

Acknowledgment. The work reported here was supported in part by the NSF Grant CCR-9420857 and the Chinese National Science Foundation.

References

1. S. C. Chou, *Mechanical Geometry Theorem Proving*, D. Reidel Publishing Company, Dordrecht, Netherlands, 1988.
2. S. C. Chou, X. S. Gao, & J. Z. Zhang, *Machine Proofs in Geometry*, World Scientific Pub., Singapore, 1994.
3. S. C. Chou & X. S. Gao, Mechanical Theorem Proving in Riemann Geometry, *Computer Mathematics*, W. T. Wu ed., World Scientific Pub., Singapore, pp. 136-157, 1993.
4. S. C. Chou, X. S. Gao, L. Yang, & J. Z. Zhang, Automated Production of Readable Proofs for Theorems in non-Euclidean Geometries, TR-WSU-94-9, 1994.

5. S. C. Chou, X. S. Gao, L. Yang, & J. Z. Zhang, A Collection of 90 Automatically Proved Geometry Theorems in non-Euclidean Geometries, TR-WSU-94-10, 1994.

6. H. Coelho & L. M. Pereira, Automated Reasoning in Geometry with Prolog, *J. of Automated Reasoning*, 2, 329-390, 1987.

7. H. S. M. Coxeter, *Non-Euclidean Geometry*, Univ. of Toronto Press, 1968.

8. X. S. Gao, Transcendental Functions and Mechanical Theorem Proving in Elementary Geometries, *J. of Automated Reasoning*, 6, 403-417, 1990.

9. H. Gelernter, J. R. Hanson, & D. W. Loveland, Empirical Explorations of the Geometry-theorem Proving Machine, *Proc. West. Joint Computer Conf.*, pp. 143-147, 1960.

10. M. J. Greenberg, *Euclidean Geometry and non-Euclidean Geometry, Development and History*, Freeman, 1980.

11. B. Grünbaum & G. C. Shephard, From Mennelaus to Computer Assisted Proofs in Geometry, Preprint, Washington State University, 1993.

12. T. Havel, Some Examples of the Use of Distances as Coordinates for Euclidean Geometry, *J. of Symbolic Computation*, 11, 579-594, 1991.

13. K. R. Koedinger & J. R. Anderson, Abstract Planning and Perceptual Chunks: Elements of Expertise in Geometry, *Cognitive Science*, 14, 511-550, 1990.

14. A. J. Nevins, Plane Geometry Theorem Proving Using Forward Chaining, *Artificial Intelligence*, 6, 1-23, 1976.

15. D. Kapur, Geometry Theorem Proving Using Hilbert's Nullstellensatz, *Proc. of SYMSAC'86*, Waterloo, pp. 202-208, 1986.

16. B. Kutzler & S. Stifter, A Geometry Theorem Prover Based on Buchberger's Algorithm, *Proc. CADE-7*, Oxford, ed. J. H. Siekmann, LNCS, no. 230, pp. 693-694, 1987.

17. S. Stifter, Geometry Theorem Proving in Vector Spaces by Means of Gröbner Bases, *Proc. of ISSAC-93*, Kiev, ACM Press, pp. 301-310,1993.

18. T. E. Stokes, *On the Algebraic and Algorithmic Properties of Some Generalized Algebras*, Ph.D thesis, University of Tasmania, 1990.

19. A. Tarski, *A Decision Method for Elementary Algebra and Geometry*, Univ. of California Press, Berkeley, 1951.

20. D. M. Wang, On Wu's Method for Proving Constructive Geometry Theorems, *Proc. IJCAI'89*, Detroit, pp. 419-424, 1989.

21. D. M. Wang, Elimination Procedures for Mechanical Theorem Proving in Geometry, *Ann. of Math. and AI*, 13, 1-24, 1995.

22. W. T. Wu, On the Decision Problem and the Mechanization of Theorem Proving in Elementary Geometry, *Scientia Sinica* 21, 159-172, 1978; Also in *Automated Theorem Proving: After 25 years*, A.M.S., Contemporary Mathematics, 29, pp. 213-234, 1984.

23. W. T. Wu, *Basic Principles of Mechanical Theorem Proving in Geometries*, Volume I: Part of Elementary Geometries, Science Press, Beijing (in Chinese), 1984. English Edition, Springer-Verlag, 1994.

24. L. Yang, Computer-Aided Proving for New Theorems of non-Euclidean Geometry, Research Report, No.4, Mathematics Research Section, Australian National University, 1989.

25. J. Z. Zhang, L. Yang, & M. Deng, The Parallel Numerical Method of Mechanical Theorem Proving, *Theoretical Computer Science*, 74, 253-271, 1990.
26. J. Z. Zhang, L. Yang, & X. Hou, A Criterion for Dependency of Algebraic Equations, with Applications to Automated Theorem Proving, *Science in China* Ser.A, 37, 547-554, 1994.

Points on Algebraic Curves and the Parametrization Problem [*]

Erik Hillgarter and Franz Winkler

Institut für Mathematik and RISC-LINZ
Johannes Kepler Universität
A-4040 Linz, Austria

Abstract

A plane algebraic curve is given as the zeros of a bivariate polynomial. However, this implicit representation is badly suited for many applications, for instance in computer aided geometric design. What we want in many of these applications is a rational parametrization of an algebraic curve. There are several approaches to deciding whether an algebraic curve is rationally parametrizable and if so computing such a parametrization. In all these approaches we ultimately need some simple points on the curve. The field in which we can find such points crucially influences the coefficients in the resulting parametrization. We show how to find simple points over some practically interesting fields. Consequently, we are able to decide whether an algebraic curve defined over the rational numbers can be parametrized over the rationals or the reals. Some of these ideas also apply to parametrization of surfaces. If in the term geometric reasoning we do not only include the process of proving or disproving geometric statements, but also the analysis and manipulation of geometric objects, then algorithms for parametrization play an important role in this wider view of geometric reasoning.

I. The parametrization problem

An algebraic curve $\mathcal{C}$ in the affine plane $\mathbb{A}^2(\mathbb{K})$ over the algebraically closed field $\mathbb{K}$ is defined as

$$\mathcal{C} = \{(x,y) \in \mathbb{K}^2 \,|\, f(x,y) = 0\},$$

where $f(x,y)$ is a bivariate square–free polynomial with coefficients in $\mathbb{K}$, called the defining polynomial of $\mathcal{C}$. Observe that the defining polynomial of a plane algebraic curve is determined up to multiplication by non-zero constants in $\mathbb{K}$. The curve $\mathcal{C}$ is irreducible iff its defining polynomial is irreducible. The degree of $\mathcal{C}$, $\deg(\mathcal{C})$, is simply the degree of the defining polynomial f.

[*] Partially supported by *Österr. Fonds zur Förderung der wissenschaftlichen Forschung*, Proj. HySaX, Nr. P11160-TEC and ÖAD, Proj. Acción Integrada 30/97.

The singularity structure of an algebraic curve is not fully apparent in the affine plane, since the curve might have "singularities at infinity". So from time to time we will need to view an algebraic curve as an object in the projective plane $\mathbb{P}^2(\mathbb{K})$. The affine plane is embedded into the projective plane by identifying an affine point (a, b) with the projective point $(a : b : 1)$. In addition to these affine points, the projective plane contains *points at infinity*, with projective coordinates $(a : b : 0)$, where $(a, b) \neq (0, 0)$. A projective curve agrees with the corresponding affine curve, except that finitely many points at infinity are added. The points on the projective curve are the solutions of $F(x, y, z) = 0$, where $F(x, y, z)$ is the homogenization of $f(x, y)$ w.r.t. the homogenizing variable z.

Some algebraic curves can also be represented parametrically, i.e. their points can be generated by rational functions

$$x(t) = \frac{p_1(t)}{q_1(t)}, \quad y(t) = \frac{p_2(t)}{q_2(t)},$$

in $\mathbb{K}(t)$, i.e. $p_1(t), p_2(t), q_1(t), q_2(t) \in \mathbb{K}[t]$. More precisely, we have the following definition.

Definition I.1: If the irreducible affine plane curve $\mathcal{C}$ is defined by $f(x, y) \in \mathbb{K}[x, y]$, $\mathbb{K}$ an algebraically closed field of characteristic 0, then $\mathcal{P}(t) = (x(t), y(t)) \in \mathbb{K}(t) \times \mathbb{K}(t)$ is a *rational parametrization* of $\mathcal{C}$ iff, except for finitely many exceptions, every evaluation $(x(t_0), y(t_0))$ at $t_0 \in \mathbb{K}$ is a point on $\mathcal{C}$, and conversely almost every point on $\mathcal{C}$ is the result of evaluating the parametrization at some element of $\mathbb{K}$.

In this case $\mathcal{C}$ is called *parametrizable* or *rational*.

Equivalently, $\mathcal{P}(t) = (x(t), y(t))$ is a rational parametrization of $\mathcal{C}$ if $\mathcal{P} : \mathbb{K} \longrightarrow \mathcal{C}$ is rational and not both $x(t)$ and $y(t)$ are constant. Furthermore, if $\mathcal{P}$ is birational we say that $\mathcal{P}(t)$ is a *proper parametrization*. ∎

In computer aided geometric design (cagd) one usually requires that the algebraic curves are rational, see e.g. [3]. In fact, transformation methods between these two representations of algebraic curves of genus 0 are of great interest in cagd. This problem appears as one of the 10 most important problems in cagd in [8].

The problem of deciding whether an algebraic curve over an algebraically closed field of characteristic 0 is rational was solved by Hilbert and Hurwitz more than 100 years ago [1]. In fact, they prove that if the rational curve $\mathcal{C}$ is defined by a polynomial $f(x, y) \in \mathbb{K}[x, y]$, where $\mathbb{K}$ is not necessarily algebraically closed, then $\mathcal{C}$ has a parametrization with coefficients in an algebraic extension field $\mathbb{L}$ of $\mathbb{K}$ with $[\mathbb{L} : \mathbb{K}] \leq 2$ and $\mathbb{L} = \mathbb{K}$ for $\deg(f)$ odd. The construction of such an "optimal" parametrization requires $\mathcal{O}(\deg(f))$ birational transformations. Every one of these transformations decreases the degree of the curve by 2, and thus ultimately leads to either a line (in the odd degree case) or to an irreducible conic $\mathcal{C}_2$ defined over the same field as the original curve $\mathcal{C}$. This reduction process was abbreviated in [12] and actually the corresponding conic can simply be interpolated. Every point on the conic C_2 corresponds rationally to a point on C and vice versa (with finitely many exceptions). So, if we can find a point of C_2 in an extension field $\mathbb{L}$ of $\mathbb{K}$, then we can parametrize C_2 and C with coefficients in $\mathbb{L}$.

In a geometric approach to the parametrization problem, we consider a linear system $\mathcal{L}$ of curves having prescribed multiplicities at the singular points of C. In particular, it is reasonable (but not necessary) to consider a system $\mathcal{L}$ of curves of degree $\deg(C) - 2$ and require that every singular point P of C be a point on every curve in $\mathcal{L}$ with multiplicity at least $\text{mult}_P(C) - 1$. $\mathcal{L}$ is called the system of adjoint curves of C. The dimension of $\mathcal{L}$ is $\deg(C) - 2$. So by fixing $\deg(C) - 3$ simple points on C to also be simple points on every curve in $\mathcal{L}$, we reduce the dimension of $\mathcal{L}$ to 1, i.e. every curve in $\mathcal{L}$ will intersect C in the fixed points (the singularities and some fixed simple points of C) and exactly 1 additional "free" simple intersection point on C, which depends rationally on the free parameter of the system $\mathcal{L}$. Thus, the rational expression of the "free" intersection point immediately yields the desired parametrization of C. This geometric approach was theoretically described in [14] and was investigated from the computer algebra point of view in [11].

Example I.1: The irreducible curve C in the affine plane over the complex numbers $\mathbb{C}$ is defined by

$$f(x, y) = x^5 + 3x^4 y - 2x^3 y^2 + x^2 y^3 + y^5 - 2x^4 - 4x^3 y + x^2 y^2$$
$$-xy^3 - 2y^4 + x^3 + x^2 y + xy^2 + y^3.$$

C has a triple point P_1 at the origin $(0,0)$, and double points $P_2 = (0,1), P_3 = (1,1), P_4 = (1,0)$. So the genus of C is 0, and it is parametrizable. In order to construct a subsystem of dimension 1 of the system of adjoints of C, we need 2 simple points on C. Intersecting C by the line $\overline{P_2 P_3}$ we get the simple point $P_5 = (-3, 1)$, and intersecting C by the line $\overline{P_2 P_4}$ we get the simple point $P_6 = (5/6, 1/6)$. The system $\mathcal{A}$ of adjoints of degree 3 having multiplicity 2 at the base point P_1 and 1 at the base points P_2, P_3, P_4, P_5, P_6 is defined by

$$h(x, y, t) = -tx^3 - 3tx^2 y + (3t - 1)xy^2 + (5 - 7t)y^3 + tx^2 + xy + (7t - 5)y^2.$$

Because of Bézout's theorem we expect the intersection multiplicity of curves in $\mathcal{A}$ with C to be $5 \cdot 3 = 15$. In the base points of $\mathcal{A}$ we get an intersection multiplicity of $6 + 2 + 2 + 2 + 1 + 1 = 14$. So a general element of $\mathcal{A}$ has one more intersection point with C, depending rationally on the parameter t. This point traverses the curve C as t traverses the affine line, yielding the following rational parametrization of C:

$$x(t) = \frac{-648t^5 + 2502t^4 - 3900t^3 + 3067t^2 - 1217t + 195}{702t^5 - 2673t^4 + 4113t^3 - 3196t^2 + 1254t - 199},$$

$$y(t) = \frac{-216t^5 + 810t^4 - 1194t^3 + 859t^2 - 298t + 39}{702t^5 - 2673t^4 + 4113t^3 - 3196t^2 + 1254t - 199}.$$

Since the simple points P_5, P_6 have coefficients in $\mathbb{Q}$, also the parametrization has coefficients in $\mathbb{Q}$, i.e. we need no algebraic extension of the field of coefficients. ∎

Alternative algebraic approaches to the problem of parametrization of algebraic curves are investigated in [13] and [6]. But also in these approaches the field of coefficients $\mathbb{L}$ of the parametrization is precisely the field in which we can construct simple points of the curve C.

So in any case, the determination of an optimal field of parametrization $\mathbb{L}$, i.e. a field achieving the bound on the extension degree in the paper of Hilbert and Hurwitz, hinges on the ability to determine points on irreducible conics. Let us demonstrate this fact by a simple example.

Example I.2: We consider the tacnode curve C in the affine plane over $\mathbb{C}$ defined by

$$f(x,y) = 2x^4 - 3x^2 y + y^2 - 2y^3 + y^4.$$

The tacnode curve has double points at $(0,0)$ and $(0,1)$, and one more double point in the first neighborhood of $(0,0)$. So we need one more point on C for determining a system of adjoints of degree 2 and dimension 1.

(a) Intersection of C with the line $y = -1$ yields the simple point $P_1 = (\alpha, -1)$ on C, where α is any one of the 4 roots of the irreducible polynomial $2z^4 + 3z^2 + 4$. This leads to the following parametrization over $\mathbb{Q}(\alpha)$:

$$x(t) = \frac{-36\alpha t^4 + (48\alpha^2 + 144)t^3 + (108\alpha^3 + 18\alpha)t^2 - (40\alpha^2 - 84)t + 24\alpha^3 + 20\alpha}{4 \cdot (9t^4 + 24\alpha t^3 - (16\alpha^2 + 60)t^2 - (24\alpha^3 + 20\alpha)t + 12\alpha^2 + 6)},$$

$$y(t) = \frac{-9t^4 - (18\alpha^3 + 3\alpha)t^3 + (2\alpha^2 - 33)t^2 - (12\alpha^3 + 2\alpha)t - 4}{(9t^4 + 24\alpha t^3 - (16\alpha^2 + 60)t^2 - (24\alpha^3 + 20\alpha)t + 12\alpha^2 + 6)}.$$

The coefficients of this parametrization are in $\mathbb{C}\backslash\mathbb{R}$.

(b) Intersection of C with the line $y = 1$ yields the simple point $P_2 = (\beta, 1)$ on C, where β is any one of the 2 roots of the irreducible polynomial $2z^2 - 3$. This leads to the following parametrization over $\mathbb{Q}(\beta)$:

$$x(t) = \frac{2\beta t^4 + 9t^3 - 27t - 18\beta}{11t^4 + 24\beta t^3 + 12t^2 + 18},$$

$$y(t) = \frac{2t^4 + 12\beta t^3 + 39t^2 + 36\beta t + 18}{11t^4 + 24\beta t^3 + 12t^2 + 18}.$$

The coefficients of this parametrization are in $\mathbb{R}$.

(c) Intersection of C with the line $x = 1$ yields the simple point $P_3 = (1,2)$ on C. This leads to the following parametrization over $\mathbb{Q}$:

$$x(t) = \frac{2t^4 + 7t^3 - 21t - 18}{9t^4 + 40t^3 + 64t^2 + 48t + 18},$$

$$y(t) = \frac{4t^4 + 28t^3 + 73t^2 + 84t + 36}{9t^4 + 40t^3 + 64t^2 + 48t + 18}.$$

Thus, the field in which we can find a simple point on $\mathcal{C}$ determines the coefficient field of the resulting parametrization.

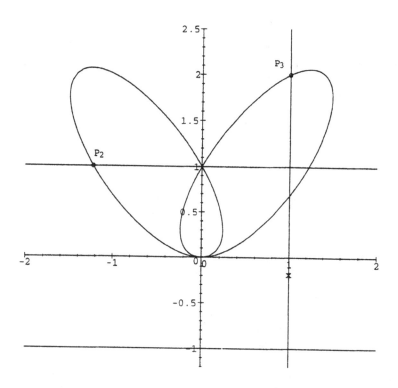

Fig. 1

∎

In fact, because every rational curve can be birationally transformed into a conic over the same coefficient field, it suffices to find simple points on irreducible conics.

Relation of parametrization to geometric reasoning

If in the term geometric reasoning we do not only include the process of proving or disproving geometric statements, but also the analysis and manipulation of geometric objects, then algorithms for parametrization play an important role in this wider view of geometric reasoning. Implicitization and parametrization are operations for changing the algebraic representation of geometric objects. For some operations we want implicit representations, e.g. for deciding whether a point actually belongs to a curve or surface. For other operations we want (some of) the geometric objects in parametric representation, e.g. for intersection of objects or for visualization.

II. Rational points on conics

In this section we consider irreducible conics defined over $\mathbb{Q}$, i.e. curves of degree two with rational coefficients. Such an irreducible conic in the projective plane over $\overline{\mathbb{Q}}$, the field of algebraic numbers, is defined by an irreducible homogeneous polynomial $G \in \mathbb{Q}[x, y, z]$ of degree two as the set $\{(\overline{x} : \overline{y} : \overline{z}) \in \mathbb{P}^2(\overline{\mathbb{Q}}) \mid G(\overline{x}, \overline{y}, \overline{z}) = 0\}$. In the sequel we refer to

$$G(x, y, z) = ax^2 + bxy + cy^2 + dxz + eyz + fz^2 = 0, \text{ or}$$

$$\tag{1}$$

$$g(x, y) = G(x, y, 1) = ax^2 + bxy + cy^2 + dx + ey + f = 0,$$

as the General Conic Equation. (1) defines the projective and the corresponding affine conic, respectively. We denote the projective conic by C^* and the affine conic by C.

Definition II.1: We call $P = (\overline{x} : \overline{y} : \overline{z}) \in C^*$ a *rational point* on C^* iff $P \in \mathbb{P}^2(\mathbb{Q})$. Analogously for the corresponding affine curve. ∎

Our goal is to decide whether there is a rational point on the conic C^*, and if so, compute one. We follow the presentation of [2], where any missing details can be found. [2] in turn is based on [4], [5] and [10].

The following theorem shows that the existence of one rational point on an irreducible conic implies that there are infinitely many rational points on it. In particular, if the projective conic C^* has a rational point, then the affine conic C has a rational point in $\mathbb{A}^2(\mathbb{Q})$. Indeed, we will basically not distinguish between C^* and C and treat them quite interchangeably.

Theorem II.1: *An irreducible conic defined over $\mathbb{Q}$ has no or infinitely many rational points.*

Proof: We give only a sketch of the proof. Suppose there is a rational point P on the conic. Then we intersect the conic with a line through this rational point having a rational direction vector. We will usually get two intersection points – the original rational point P and an additional rational point. Varying the slope of the line leads to infinitely many other rational points on the conic. ∎

It makes sense to distinguish between parabolas on the one hand and ellipses and hyperbolas on the other hand, since on a parabola we are guaranteed to find one (and therefore infinitely many) rational point(s). Indeed, in the parabolic case we can give a formula for a rational point on the conic (namely a rational function in the coefficients of (1)). On the other hand, on an ellipse or hyperbola we are not guaranteed that such a rational point even exists. In case it does (we will show how to decide that) we can compute such a point by an algorithm that is based on a constructive proof of the so called *Legendre Theorem*. First we deal with the parabolic case.

The parabolic case

C is a parabola if and only if the coefficients of (1) satisfy one of the following relations :

$$b^2 = 4ac, \text{ or } d^2 = 4af, \text{ or } e^2 = 4cf.$$

W.l.o.g. we assume now the case $b^2 = 4ac$, i.e. we consider a parabola with respect to x and y, whereas z is the homogenizing variable.

First, let us assume $c \neq 0$. By simple expansion we have

$$4cg(x, y) = (bx + 2cy + e)^2 + d'x + f',$$

where $d' = 4cd - 2be$, $f' = 4cf - e^2$. Because C is irreducible, we have $d' \neq 0$. Thus, a rational solution is given by

$$\bar{x} = -\frac{f'}{d'}, \quad \bar{y} = -\frac{e + b\bar{x}}{2c}, \quad (\bar{z} = 1).$$

Now the remaining case to treat is $c = 0$. Again by irreducibility, we have $a \neq 0$. A rational solution is then given by

$$\bar{x} = -\frac{d + b\bar{y}}{2a}, \quad \bar{y} = -\frac{f'}{d'}, \quad (\bar{z} = 1),$$

where $d' = 4ae - 2bd$ and $f' = 4af - d^2$.

Example II.1: Consider the parabola defined by

$$g(x, y) = x^2 + 2xy + y^2 + x + 2y - 2,$$

i.e. $(a, b, c, d, e, f) = (1, 2, 1, 1, 2, -2)$. Since $a \neq 0$ and $c \neq 0$, we might use both formulae. Let us first use the formula for the case $c \neq 0$:

$$d' = 4cd - 2be = -4, \text{ and } f' = 4cf - e^2 = -12.$$

So we get the rational point

$$\bar{x} = -\frac{f'}{d'} = -3, \quad \bar{y} = -\frac{e + b\bar{x}}{2c} = 2$$

on the parabola.

Now we use the formula for the case $a \neq 0$:

$$d' = 4ae - 2bd = 4, \text{ and } f' = 4af - d^2 = -9.$$

So we get the rational point

$$\bar{x} = -\frac{d + b\bar{y}}{2a} = -\frac{11}{4}, \quad \bar{y} = -\frac{f'}{d'} = \frac{9}{4}$$

on the parabola. ■

The hyperbolic/elliptic case

We consider (1), but we impose other conditions on the coefficients. We use ideas from [5]. The hyperbolic/elliptic case is characterized by

$$b^2 \neq 4ac \text{ and } d^2 \neq 4af \text{ and } e^2 \neq 4cf.$$

We consider the dehomogenization with respect to z (i.e. in what follows, we will only make use of $b^2 \neq 4ac$). Let us define

$$N = 4de - 4bf, \ D = 4ac - b^2,$$
$$M_1 = 4c^2d^2 - 4bcde + 4ace^2 + 4b^2cf - 16ac^2f,$$
$$M_2 = 4a^2e^2 - 4bade + 4acd^2 + 4b^2af - 16ca^2f.$$

We consider two cases.
(CASE $a = c = 0$) In this case we have $b \neq 0$ and $N \neq 0$ (by irreducibility). In the new coordinates

$$x' = b(x + y) + d + e,$$
$$y' = b(x - y) - d + e$$

the equation $4bg(x, y) = 0$ has the following form :

$$(x')^2 - (y')^2 = N.$$

(CASE $c \neq 0$) We have $M_1 \neq 0$ and $(D > 0 \Rightarrow M_1 > 0)$ (both conditions are consequences of irreducibility). Under the coordinate change

$$x' = Dx + 2dc - be,$$
$$y' = bx + 2cy + e$$

the equation $4cDg(x, y) = 0$ becomes

$$(x')^2 + D(y')^2 = M_1.$$

The case $a \neq 0$ is totally analogous to the case $c \neq 0$ (just interchange the roles of x and y and therefore also those of a and c and those of d and e; in addition use M_2 instead of M_1).

In both cases we arrive at an equation of the form

$$X^2 + KY^2 = L, \tag{2}$$

where $K, L \in \mathbb{Q}$, and in both cases we do not have $(K > 0 \wedge L < 0)$, which would exclude the existence of a real solution.

Hence we can restrict us to equations of this form . Switching to homogeneous coordinates we set

$$X = \frac{x}{z}, \ Y = \frac{y}{z}, \ K = \frac{b'}{a'}, \ L = -\frac{c'}{a'}.$$

Note that if $K = k_1/k_2$, $L = l_1/l_2$ we may choose $a' = \text{lcm}(k_2, l_2)$, $b' = k_1 l_2 / \gcd(k_2, l_2)$, and $c' = -l_1 k_2 / \gcd(k_2, l_2)$. Then (2) becomes the diophantine equation

$$a'x^2 + b'y^2 + c'z^2 = 0. \tag{3}$$

Clearly a', b', and c' are nonzero and do not all have the same sign (look at their definitions and use $\neg(K > 0 \wedge L < 0)$). But we want to achieve more, namely the reduction of (3) to an equation of similar form whose coefficients are squarefree and pairwise relatively prime. We use ideas from [10]. Let us assume that

$$a' = a_1' \, r_1^2, \; b' = b_1' \, r_2^2, \; c' = c_1' \, r_3^2,$$

where a_1', b_1', and c_1' are squarefree[2]. Consider

$$a_1' x^2 + b_1' y^2 + c_1' z^2 = 0. \tag{4}$$

(4) has an integral solution iff (3) has one.
Now, we divide (4) by $\gcd(a_1', b_1', c_1')$, getting

$$a'' x^2 + b'' y^2 + c'' z^2 = 0. \tag{5}$$

What remains is to make the coefficients pairwise relatively prime.
Let $g_1 = \gcd(a'', b'')$, $a''' = a''/g_1$, $b''' = b''/g_1$, and let $(\overline{x}, \overline{y}, \overline{z})$ be an integral solution of (5). Then $g_1 \mid c'' \overline{z}^2$, and hence, since $\gcd(a'', b'', c'') = 1$, we have $g_1 \mid \overline{z}^2$. Furthermore, since g_1 is squarefree (since a'', b'' are), we have $g_1 \mid \overline{z}$. So, letting $z = g_1 z'$ and cancelling (5) by g_1, we arrive at

$$a''' x^2 + b''' y^2 + \underbrace{c'' g_1}_{c'''} (z')^2 = 0. \tag{6}$$

We have $\gcd(a''', b''') = 1$ and c''' is squarefree since g_1 and c'' are relatively prime. Repeating this process with $g_2 = \gcd(a''', c''')$ and $g_3 = \gcd(b'''', c'''')$ we arrive at

$$a(x')^2 + b(y')^2 + c(z')^2 = 0, \tag{7}$$

the so called Legendre Equation. We note : a, b, and c are nonzero, do not all have the same sign, are squarefree, and pairwise relatively prime. We will now try to find an integral solution of this diophantine equation that can then be

---

[2] For actually determining r_1, r_2 and r_3 we are confronted with integer factorization. Although there are no polynomial-time algorithms known for the factorization of large integers (the most powerful general purpose factoring method leads to a factorization of an integer m in time $O[\exp(2L\{L[L(m)]\})]$, where $L(m)$ denotes the length of m) this does not lead to problems in practical computations. Usually the integers to be factored are small enough such that succesful and fast application of integer factorization commands as provided by computer algebra systems is guaranteed.

Things are trivial if a', b', and c' are polynomials (this will occur if we deal with conic equations over $\mathbb{Q}(t)$ as in the following section) since squarefree factorization of polynomials poses no problems at all.

transformed back to a rational solution of the original equation. Algorithmic formulations (in pseudocode) of the above steps (including the parabolic case) can be found in [2], where in the appendix one can also find a Maple implementation.

Hence the problem of finding a rational point on an ellipse/ hyperbola reduces to the problem of finding a nontrivial integral solution of the so called Legendre Equation

$$ax^2 + by^2 + cz^2 = 0, \qquad (8)$$

where a, b, and c are integers such that $abc \neq 0$. By a nontrivial integral solution we mean a solution $(\overline{x}, \overline{y}, \overline{z}) \in \mathbb{Z}^3$ with $(\overline{x}, \overline{y}, \overline{z}) \neq (0, 0, 0)$ and $\gcd(\overline{x}, \overline{y}, \overline{z}) = 1$. We also pointed out that we may assume, w. l. o. g.,

$$a > 0, \ b < 0 \text{ and } c < 0, \qquad (9)$$

$$a, \ b, \text{ and } c \text{ are squarefree}, \qquad (10)$$

$$\gcd(a, b) = \gcd(a, c) = \gcd(b, c) = 1. \qquad (11)$$

We now deal with necessary and sufficient conditions in order that (8) has nontrivial integral solutions. Such conditions are given by the Theorem of Legendre. For a formulation of Legendre's Theorem we need the notion of quadratical residues.

Definition II.1: Let m, n be nonzero integers. Then m is a *quadratic residue* modulo n (written $m \, R \, n$) iff $\exists x \in \mathbb{Z} : x^2 \equiv_n m$. ∎

Now we can state the theorem.

Theorem II.3: (Legendre, Version 1) *Suppose a, b, and c satisfy (9), (10) and (11). Then (8) has a nontrivial integral solution iff*

$$-ab \, R \, c, \ -bc \, R \, a, \text{ and } -ac \, R \, b. \qquad (12)$$

We prove only the necessity of (12) for this first version of the theorem and prove the sufficiency then for a second (equivalent) version.

Proof: (Legendre's Theorem, necessity of (12))
Let $(\overline{x}, \overline{y}, \overline{z})$ be a solution of (8); it follows that $\gcd(c, \overline{x}) = 1$. For if any prime p divides $\gcd(c, \overline{x})$, then p divides $b\overline{y}^2$ but p does not divide b (since $\gcd(b, c) = 1$ by (11)) and so p divides $\overline{y}$. Consequently we have p^2 divides $a\overline{x}^2 + b\overline{y}^2$ and hence p^2 divides $c\overline{z}^2$. But c is squarefree and so p divides $\overline{z}$. This contradicts the assumption $\gcd(\overline{x}, \overline{y}, \overline{z}) = 1$.

As $\gcd(c, \overline{x}) = 1$ we can find $\overline{x}'$ satisfying $\overline{x}\overline{x}' \equiv_c 1$. Also, clearly

$$a\overline{x}^2 + b\overline{y}^2 \equiv_c 0,$$

and so, by multiplying with $b(\overline{x}')^2$,

$$b^2(\overline{x}')^2\overline{y}^2 \equiv_c -ab(\overline{x}\overline{x}')^2 \equiv_c -ab.$$

Thus $-ab\,R\,c$ holds. The remaining conditions can be derived similarly. ∎

Theorem II.4: (Legendre, Version 2) *Let a and b be positive squarefree integers. Then*

$$ax^2 + by^2 = z^2 \qquad (13)$$

has a nontrivial solution if and only if the following three conditions are satisfied:

$$a\,R\,b, \qquad (14)$$
$$b\,R\,a, \qquad (15)$$
$$-\frac{ab}{\gcd(a,b)^2}\,R\,\gcd(a,b). \qquad (16)$$

The equivalence of these two versions is easily established (see [2]). The following constructive proof of the Legendre Equation can be found in [4], we give the presentation from [2] (where all missing details can be found).

Proof: (Theorem II.4) The necessity of (14) to (16) is established by the necessity of (12) for the solvability of (8) and the claimed equivalence of the two versions of Legendres Theorem. So we show sufficiency and hence assume now that (14) to (16) hold.

Let us first of all consider two special (simple) instances of (13)

(CASE $a = 1$) Obviously, $(\overline{x}, \overline{y}, \overline{z}) = (1, 0, 1)$ is a solution.

(CASE $a = b$) Condition (16) requires -1 to be a square modulo b. If this is the case, we can find integers r and s such that $b = r^2 + s^2$ (consider this as an easy lemma), leading to a solution $(\overline{x}, \overline{y}, \overline{z}) = (r, s, r^2 + s^2)$.

Now we proceed to the general case. We may assume $a > b$, for if $b > a$ just interchange the roles of x and y. The strategy will be the following : We construct a new form $Ax^2 + by^2 = z^2$ satisfying the same hypotheses as (13), $0 < A < a$, and having a nontrivial solution iff (13) does so (and a solution of (13) can be computed from a solution of the new form). After a finite number of steps, interchanging A and b in case A is less than b, we arrive at one of the cases $A = 1$ or $A = b$, each of which has been settled. Now for the details.

We assume now that (14) - (16) hold. By (15) there exist integers x and k such that

$$x^2 = b + ka. \qquad (17)$$

Let $k = Am^2$, where A is the squarefree part of k. Also note that we can choose x such that $|x| \leq a/2$ by choosing the absolute least residue of x modulo a (*"symmetric representation* of the integers modulo a"). Let us now restate (17) as

$$x^2 = b + Am^2a. \qquad (18)$$

First of all we show that $0 < A < a$. Since by (18) and $b < a$

$$0 \leq x^2 = b + Am^2a < a + Am^2a = a(1 + Am^2)$$

we have $0 < 1 + Am^2$, and hence $A \geq 0$. But if $A = 0$, then (18) gives $x^2 = b$, contradicting the fact that b is squarefree. So we established $A > 0$. On the other hand by (18) and $b > 0$ and since $|x| \leq a/2$ we have

$$Am^2 a < x^2 \leq \frac{a^2}{4},$$

and so we have $A \leq Am^2 < a/4 (< a)$. So we consider now

$$Ax^2 + bY^2 = Z^2. \tag{19}$$

Clearly A, b are positive and squarefree integers. So we want to show

$$A \, R \, b, \tag{20}$$

$$b \, R \, A, \tag{21}$$

$$-\frac{Ab}{\gcd(A,b)^2} \, R \, \gcd(A, b). \tag{22}$$

In addition, we need that (13) has a nontrivial solution if and only if (19) has one, which will be shown constructively.

(*Show* (21)) With $g = \gcd(a, b)$, let $b_1 = b/g$, $a_1 = a/g$. We show $A \, R \, g$ and $A \, R \, b_1$. Then, we have $A \, R \, b_1 g$ (consider this as a lemma), i.e. $A \, R \, b$. First of all, note that (18) may be written as

$$x^2 = b_1 g + Am^2 a_1 g. \tag{23}$$

Since g is squarefree we have that g divides x. Setting $x_1 = \frac{x}{g}$ and cancelling gives

$$g x_1^2 = b_1 + Am^2 a_1. \tag{24}$$

Thus $Am^2 a_1 \equiv_g -b_1$, and hence

$$Am^2 a_1^2 \equiv_g -a_1 b_1. \tag{25}$$

Also note that $\gcd(m, g) = 1$, since a common factor would divide b_1 (by (24)) and hence $b = b_1 g$ would not be squarefree. But also $\gcd(a_1, g) = 1$ since $a = a_1 g$ is squarefree. Let m' and a_1' be the inverses of m respectively a_1 modulo g. By (16) (i.e. by $-a_1 b_1 \, R \, g$) we may choose y such that $y^2 \equiv_g -a_1 b_1$. Now (25) becomes $A \equiv_g (m')^2 (a_1')^2 y^2$, i.e. $A \, R \, g$. So this part is done. It remains to show $A \, R \, b_1$.

By (23) we have

$$x^2 \equiv_{b_1} Am^2 a. \tag{26}$$

By (14) (i.e. by $a \, R \, b$) we have $a \, R \, b_1$. Note also that $\gcd(a, b_1) = 1$ since a common factor would divide b_1 and g, contradicting the fact that $b = b_1 g$ is squarefree. Similarly, $\gcd(m, b_1) = 1$ (use (23)). Let a^* and m^* be the inverses of a respectively m modulo b_1. Let z be such that $z^2 \equiv_{b_1} a$ and let z^* be its inverse modulo b_1. Now (26) becomes

$$A \equiv_{b_1} x^2 (m^*)^2 a^* \equiv_{b_1} x^2 (m^*)^2 (z^*)^2,$$

i.e. $A \, R \, b_1$.

(*Show* (22)) By (18), we have $b \, R \, A$ immediately.

(*Show* (23)) With $r = \gcd(A, b)$ let $A_1 = A/r$, $b_2 = b/r$. We have to show $-A_1 b_2 \, R \, r$.

From (18) we conclude

$$x^2 = b_2 r + A_1 r m^2 a.$$

Since r is squarefree we have r divides x. So

$$A_1 m^2 a \equiv -b_2 \pmod{r}, \text{ or}$$
$$-A_1 b_2 m^2 a \equiv b_2^2 \pmod{r}. \tag{27}$$

Since $\gcd(a, r) = \gcd(m, r) = 1$, we may choose a^+ and m^+ as the inverses of a respectively m modulo r. Furthermore, from (14) (i.e. from $a \, R \, b$) we obtain $a \, R \, r$. Choose w such that $w^2 \equiv_r a$. Denote by w^+ the inverse of w modulo r. Then (27) becomes

$$-A_1 b_2 \equiv_r b_2^2 (m^+)^2 a^+ \equiv_r b_2^2 (m^+)^2 (w^+)^2,$$

i.e. $-A_1 b_2 \, R \, r$.

So we established (20) - (22) for (19). Assume now that (19) has a nontrivial solution $(\overline{X}, \overline{Y}, \overline{Z})$. Then

$$A\overline{X}^2 = \overline{Z}^2 - b\overline{Y}^2. \tag{28}$$

Multiplying this by (18) (i.e. by $Am^2 a = x^2 - b$) gives

$$a(A\overline{X}m)^2 = (\overline{Z}^2 - b\overline{Y}^2)(x^2 - b) =$$
$$= (\overline{Z}x + b\overline{Y})^2 - b(x\overline{Y} + \overline{Z})^2.$$

Thus a solution of (13) is

$$\overline{x} = A\overline{X}m, \; \overline{y} = x\overline{Y} + \overline{Z}, \; \overline{z} = \overline{Z}x + b\overline{Y}.$$

Written in matrix notation we have

$$\begin{bmatrix} \overline{x} \\ \overline{y} \\ \overline{z} \end{bmatrix} = \begin{bmatrix} Am & 0 & 0 \\ 0 & x & 1 \\ 0 & b & x \end{bmatrix} \cdot \begin{bmatrix} \overline{X} \\ \overline{Y} \\ \overline{Z} \end{bmatrix}.$$

The matrix is invertible since its two blocks are : the second (2×2) block has determinant $x^2 - b \neq 0$ (since b is squarefree). The solution is nontrivial since we claim that $\overline{x} = Am\overline{X} \neq 0$. Suppose $Am = 0$. Then by (18) we have $x^2 = b$, contradicting the squarefreeness of b. Suppose $\overline{X} = 0$. Then by (28) we have $\overline{Z}^2 = b\overline{Y}^2$, contradicting the squarefreeness of b. ∎

Algorithm for solving the Legendre Equation

The constructive proof for the existence of a nontrivial integral solution of the Legendre Equation in the previous section leads to a recursive algorithm

for computing such a solution. We will give a formulation in a Pascal-like pseudocode. We do not consider an algorithmic formulation of the transformations
that lead from the General Conic Equation to the Legendre Equation, see [2] for
that purpose. We assume the procedure $msqrt$ ("modular squareroot"), that has
the following meaning : for integers a, b with $a\,R\,b$ we have

$$msqrt(a, b)^2 \equiv_b a.$$

Such a procedure exists for example in Maple^{TM}. We work in symmetric representation of the integers modulo any number. In addition, we use the knowledge
that for a natural number r with $-1\,R\,r$ there exist integers x, y such that
$r = x^2 + y^2$ (for a proof of that fact and a procedure $Circle$ that computes such
x and y for given r see again [2]). Finally we need a procedure $sqfrp$ ("squarefree
part") for computing the squarefree part of an integer, i.e. for $n = \prod_{p \; prime} p^{n_p}$
we have

$$sqfrp(n) = \prod_{p \; prime} p^{\mathrm{mod}(n_p, 2)}.$$

Now we can give the pseudocode.

PROC LegendreSolve($\downarrow a \downarrow b \uparrow x \uparrow y \uparrow z$)
IN :
$a, b \in \mathbb{Z}$:
    positive, squarefree with $a\,R\,b$, $b\,R\,a$, $-ab/gcd(a,b)^2\,R\,gcd(a,b)$.
OUT :
$x, y, z \in \mathbb{Z}$
    such that $ax^2 + by^2 = z^2$.
LOCAL
$r, s, T, A, B, X, Y, Z, m \in \mathbb{Z}$
BEGIN
    if $a == 1$ **then**
        $x := 1; y := 0; z := 1$
    elseif $a == b$ **then**
        Call $Circle(\downarrow b, \uparrow x, \uparrow y)$;
        $z := x^2 + y^2$
    elseif $a > b$ **then**
        $s := msqrt(b, a)$;
        $T := (s^2 - b)/a$;
        $A := sqfrp(T); m := sqrt(T/A)$;
        Call $LegendreSolve(\downarrow A, \downarrow b, \uparrow X, \uparrow Y, \uparrow Z)$;
        $x := AXm; y := sY + Z; z := sZ + bY$
    else
        $s := msqrt(a, b)$;
        $T := (s^2 - a)/b$;
        $B := sqfrp(T); m := sqrt(T/B)$;
        Call $LegendreSolve(\downarrow B, \downarrow a, \uparrow Y, \uparrow X, \uparrow Z)$;

$$y := BY\,m; \; x := sX + Z; \; z := sZ + aX$$
  end if
END LegendreSolve

Some words on the number of self-references in *LegendreSolve*. The worst thing that can happen is that we reduce both coefficients of

$$ax^2 + by^2 = z^2$$

to 1. The number of self-references of *LegendreSolve* needed to achieve this is bounded by $2\log_4(\max(a, b))$, since every time we reduce a coefficient, it is reduced by a factor of 4 at least. In the situation $a = b$ we call *Circle* (and no more call to *LegendreSolve* is needed), which calls itself not more than $log(a)$ times (compare [2]). So in all cases, the maximal number of any procedure calls is $O(\log(\max(a, b)))$. The (theoretical) time complexity of the main algorithm (input an irreducible conic over the rational numbers and output a rational point if existent) would be at least exponential in any way since we use integer factorization in the implementation of the procedure *sqfrp*. But this fact has not turned out to be an obstacle in practical computations.

Some words on the space complexity of the main algorithm: If we denote by l the maximum length of any numerator or denominator of the coefficients of the General Conic Equation, then we have for the (integer) coefficients a and b of the associated Special Legendre Equation

$$ax^2 + by^2 = z^2$$

that[3]

$$L(\max(a, b)) \leq 12l.$$

(This worst case bound may be reached if the numerators and denominators of the General Conic Equation are all of equal length). For the diophantine solution $(\overline{x}, \overline{y})$ of the Special Legendre Equation we can give the bound

$$L(\max(\overline{x}, \overline{y})) \leq 5L(\max(a, b))^2.$$

Concluding we obtain (the backward transformations do not influence the order) that the maximal length of any numerator or denominator for the (rational) solution of the General Conic Equation is $O(l^2)$.

Example II.2: Consider the conic defined by

$$g(x, y) = x^2 - 4xy - 3y^2 + 4x + 8y - 5.$$

Carrying out the above described transformations leads to the corresponding (Special) Legendre Equation

$$7x^2 + 21z^2 = y^2.$$

---

[3] We assume integers in decimal representation.

Now we call LegendreSolve($\downarrow 7, \downarrow 21, \uparrow x, \uparrow z, \uparrow y$). Here is a trace of the corresponding values of the local variables s, T, B, m and the recursive calls of LegendreSolve :

$$(s, T, B, m) = (7, 2, 2, 1); \textbf{CallLegendreSolve}(\downarrow 2, \downarrow 7, \uparrow Y, \uparrow X, \uparrow Z);$$
$$(s, T, B, m) = (3, 1, 1, 1); \textbf{CallLegendreSolve}(\downarrow 1, \downarrow 2, \uparrow Y, \uparrow X, \uparrow Z);$$

Now the equation $Y^2 + 2X^2 = Z^2$ has the integral solution $(Y, X, Z) = (1, 0, 1)$. So the procedure produces the following integral solutions of the stacked equations :

$$(X, Y, Z) = (1, 1, 3),$$
$$(z, x, y) = (2, 10, 28),$$

the latter being an integral solution of $7x^2 + 21z^2 = y^2$, the (Special) Legendre Equation. Inverting the above indicated transformations we arrive at a rational solution of $g(x, y) = 0$:

$$(x, y) = (-\frac{3}{7}, \frac{16}{7}).$$

$\blacksquare$

Real points on conics[4]

Let us assume that no rational point lies on the conic. In this case we ask whether there is at least a real point on the conic, i.e. whether there exists $(\overline{x}, \overline{y}) \in \mathbb{R}^2$ such that

$$g(\overline{x}, \overline{y}) = 0.$$

Since every parabola contains a rational point, we only have to consider the elliptic/hyperbolic case. Again we transform the General Conic Equation to an equation of the form

$$x^2 + Ky^2 = L, \tag{29}$$

where K, L are rational numbers. A real point on the conic exists if and only if $\neg(K \geq 0 \wedge L < 0)$. In this case, a real solution of (29) is given by

$$(\overline{x}, \overline{y}) = (\sqrt{L}, 0) \quad \text{if} \quad L > 0,$$
$$(\overline{x}, \overline{y}) = (0, \sqrt{\frac{L}{K}}) \quad \text{if} \quad L < 0.$$

By back transformation, we arrive at a real solution for the General Conic Equation.

---

[4] The question when a rational algebraic plane curve over $\mathbf{Q}$ is parametrizable over $\mathbf{R}$ is treated in section 3.3 ("Parametrizing over the reals") of [12]. We state here the main result.

Theorem 3.2 (in [12]) *A rational algebraic plane curve over* $\mathbf{Q}$ *is parametrizable over* $\mathbf{R}$ *if and only if it is not birationally equivalent over* $\mathbf{R}$ *to the conic* $x^2 + y^2 + z^2$.

III. Points on conics over the rational function field

As in the rational case, we only have to consider the reduced equation

$$X^2 + K(t)Y^2 = L(t), \tag{30}$$

where $K, L \in \mathbb{Q}(t)$. Our goal is to find rational functions $X(t)$, $Y(t)$ satisfying (30). This solves the problem of finding rational functions satisfying the General Conic Equation with coefficients in $\mathbb{Q}(t)$ completely. For solving (30), we try to exploit the method used for the rational case. In order to point out the analogy between these cases, we note that $\mathbb{Q}(t)$ is the quotient field of $\mathbb{Q}[t]$, a *Euclidean Domain* (ED for short), like $\mathbb{Q}$ is the quotient field of $\mathbb{Z}$ (*the* standard example of an ED). This means that we can make use of modular arithmetic, as we did in the rational case. Also those details of the rational case depending on factorization can be adapted, since every ED is a *Unique Factorization Domain* (UFD). So we can do all transformations that we did in the rational case and finally arrive at an equation of the form

$$a(t)x^2 + b(t)y^2 = z^2, \tag{31}$$

where a and b are nonzero and squarefree polynomials satisfying (at least if there exists a rational solution)

$$a \, R \, b, \tag{32}$$
$$b \, R \, a, \tag{33}$$
$$-\frac{ab}{\gcd(a,b)^2} \, R \, \gcd(a, b). \tag{34}$$

(The notion of quadratic residue for polynomials is analogous to the one for integers). W. l. o. g. let us assume $\deg(a) \geq \deg(b)$. From the proof of Legendre's Theorem for the rational case we know that in the new coordinates

$$x = AXm,$$
$$y = sY + Z,$$
$$z = sZ + bY,$$

where[5]

$$s(t) = pmsqrt(b(t), a(t)),$$
$$k(t) = \frac{s(t)^2 - b(t)}{a(t)},$$
$$A(t) = sqfrp(k(t)),$$
$$m(t) = \sqrt{\frac{k(t)}{A(t)}},$$

---

[5] *pmsqrt* : "polynomial modular squareroot", i. e.

$$pmsqrt(b(t), a(t))^2 = b(t) \bmod a(t).$$

(31) has the form

$$AX^2 + bY^2 = Z^2.$$

In analogy to the rational case A is smaller than a in some sense : in the rational case it was the absolute value of a that dropped; here it is the degree of the polynomial $a(t)$ that drops. The point now is that by iterated application of the above transformation (as in the rational case) we arrive at some simple instances of the (polynomial) Legendre Equation, where we can decide the existence of a rational (function) solution and - if one exists - give a solution. Technical details and algorithmic formulations can be found in [2].

The problem treated in this chapter arises in the context of parametrizing surfaces over $\mathbb{Q}$. In particular, the following two problems are closely related :

1. Parametrize a conic $f(x, y) = 0$ (where $f \in \mathbb{Q}(t)[x, y]$) with rational functions in s and coefficients in $\mathbb{Q}(t)$.
2. Parametrize a surface $F(x, y, t) = 0$ (where $F \in \mathbb{Q}[x, y, t]$ is of total degree 2 in x and y) with rational functions in s and t.

The exact relationship and the application of our results to the parametrization of such surfaces needs to be investigated further.

Conclusion

Given an algebraic curve defined over the rational number field, we can decide whether the curve has genus 0 and infinitely many points over the rational numbers and therefore can be parametrized over the rationals. Similarly we can decide whether a real curve can be parametrized over the reals. We are able to extend these decision methods from $\mathbb{Q}$ to $\mathbb{Q}(t)$, the field of rational functions over $\mathbb{Q}$. We conjecture that this extension should lead to parametrization algorithms for certain surfaces of interest in computer-aided-geometric-design, such as discussed in [9]. This needs further investigation.

References

1. Hilbert, D., Hurwitz, A.: Über die Diophantischen Gleichungen vom Geschlecht Null. Acta math. 14 (1890) 217–224
2. Hillgarter, E.: Rational Points on Conics. Diploma Thesis, RISC-Linz, J. Kepler Universität Linz, Austria (1996)
3. Hoschek, J., Lasser, D.: Fundamentals of Computer Aided Geometric Design, A.K. Peters, Wellesley MA (1993)
4. Ireland, K., Rosen, M.: A Classical Introduction to Modern Number Theory, Springer Verlag, New York Heidelberg Berlin (1982)
5. Krätzel, E.: Zahlentheorie, VEB Dt. Verlag der Wissenschaften, Berlin (1981)

6. Mňuk, M.: Algebraic and Geometric Approach to Parametrization of Rational Curves, Ph.D. Dissertation, RISC-Linz, J. Kepler Universität Linz, Austria (1995)
7. Mňuk, M., Sendra, J.R., Winkler, F.: On the Complexity of Parametrizing Curves. Beiträge zur Algebra und Geometrie 37/2 (1996) 309–328
8. Nielson, G.M.: Cagd's Top Ten: What to Watch. IEEE Computer Graphics & Applications (Jan. 1993)
9. Peternell, M., Pottmann, H.: Computing Rational Parametrizations of Canal Surfaces. J. Symbolic Comp. 23/2&3 (1996) 255–266
10. Rose, H.E.: A Course in Number Theory, Clarendon Press, Oxford (1988)
11. Sendra, J.R., Winkler, F.: Symbolic Parametrization of Curves. J. Symbolic Comp. 12 (1991) 607–631
12. Sendra, J.R., Winkler, F.: Parametrization of Algebraic Curves over Optimal Field Extensions. J. Symbolic Comp. 23/2&3 (1996) 191–207
13. van Hoeij, M.: Rational Parametrization of Algebraic Curves using a Canonical Divisor. J. Symbolic Comp. 23/2&3 (1996) 209–227
14. Walker, R.J.: Algebraic Curves, Princeton University Press (1950)

Flat Central Configurations of Four Planet Motions

He Shi and Fengmei Zou

Institute of Systems Science, Academia Sinica
Beijing 100080, P.R. China

Abstract. The flat central configurations of four planet motions are investigated with Wu's elimination method. We obtain 12 collinear central configurations and a necessary condition for determining flat but non-collinear central configurations. We also prove that the number of central configurations in planet motions of 4 bodies is finite under the condition that the masses and angular velocity of the planets do not satisfy any algebraic relations.

Key words. Central configuration, Wu elimination.

1   Introduction

Wintner [1] studied the configurations of planet motions and conjectured that there are only finitely many central configurations with any number of particles with arbitrary masses (see [2]). For the collinear case, Moulton [3] obtained all the $n!/2$ solutions for arbitrary n particles with distinct masses. This result was reverified by Smale and his followers using topological methods [4]. Waldvogel [5] showed that for any n ($n > 4$), there exists a group of n particles which possess a non-flat central configuration. On the other hand, Wu [2] reduced the determination of the central configurations to a problem of polynomial equations-solving, and verified the case of three particles using the characteristic set (char-set) method [6]. In this paper, we follow Wu's work to determine the possible flat central configurations of four planets using the characteristic method.

For the problem of determining flat central configurations of four bodies, it is known from [1] that for most sets of masses there exists a variety of (up to ten or more) central configurations: a square is a central configuration only in the case of the four equal masses; for 4 given masses $m_i, i = 1, 2, 3, 4$, there exists at least one non-collinear flat central configuration only under the condition that the m_i satisfy certain inequalities. In this paper, we give all the possible solutions of the collinear case; for the non-collinear case, we show that, the number of central configurations in planet motions of 4 bodies is finite under the condition that the masses and angular velocity of the planets do not satisfy any algebraic relations. In other words, the Wintner conjecture for four planets is true in the general case. The computation is done using Maple in an Alpha Station-600. Initial computation is done interactively. Then we use a package of computing characteristic sets to find the solutions of central configurations.

2 Notation and the Wintner Conjecture

Let $m_1, m_2, \ldots, m_n$ be masses of n particles moving under mutual Newtonian gravitational attractions, and $r_1, r_2, \ldots, r_n$ be the positions of these masses at a certain moment, with $r_i \neq r_j$ for $i \neq j$. A configuration formed of masses m_i at position $r_i, i = 1, 2, \ldots, n$, is denoted by

$$\begin{bmatrix} m_1 & m_2 & \cdots & m_n \\ r_1 & r_2 & \cdots & r_n \end{bmatrix}$$

Definition 1. The *center* of n particles $m_1, \ldots, m_n$ at the positions $r_1, \ldots, r_n$ is defined to be the position $\sum_{i=1}^{n} m_i r_i / \sum_{i=1}^{n} m_i$.

Definition 2. A configuration is called a *central configuration* with respect to the masses $m_1, m_2, \ldots, m_n$, if there are initial velocities of the masses m_i such that under the Newtonian gravitational attractions the configurations formed by the masses during the motion will remain similar to the initial one.

From this definition, we know that a central configuration defines a class of configurations. As a direct consequence of Newtonian mechanics, for a central configuration the center of mass of the masses $m_1, m_2, \ldots, m_n$ may be considered to be fixed during motion.

Definition 3. The *inertial coordinate system* associated to a central configuration is defined to be a Cartesian coordinate system for which the origin is at the fixed center of mass of the masses $m_1, m_2, \ldots, m_n$.

In the following, we assume that $r_1, \ldots, r_n$ are the initial coordinates of the particles $m_1, \ldots, m_n$. By the definition of the center and the inertial coordinate system, we have

$$\sum_{i=1}^{n} m_i r_i = 0.$$

We set

$q_3(n) = $ the number of central configurations with given masses $m_1, \ldots, m_n$;
$q_2(n) = $ the number of central configurations with given masses $m_1, \ldots, m_n$ for which the masses are in the same plane;
$q_1(n) = $ the number of central configurations with given masses $m_1, \ldots, m_n$ for which the masses are on the same line.

Wintner Conjecture. $q_3(n)$ and $q_2(n)$ are finite for all n and masses $m_1, \ldots, m_n$.

3 Reduction of the Wintner Conjecture to Equations-Solving

In this section, we explain briefly how to reduce the problem of determining central configurations of planets to a problem of solving polynomial equations. More details about this can be found in [2].

Let $p_i, i = 1, 2, \ldots, n$, be polynomials and $PS = \{p_1, p_2, \ldots, p_n\}$ be a polynomial set. We denote the zero set of PS by $Zero(PS)$ and denote the restricted zero set of PS by $Zero_{rc}(PS)$ for which the reality conditions are observed. We have to observe some reality conditions, for example, $r_{ij} > 0, m_i > 0$ $(i, j = 1, 2, \ldots, n)$.

Taking the planar inertial coordinate system (x, y) with the origin at the center of masses. Let

$$r_i = (x_i, y_i), \ x_{ij} = x_i - x_j, \ y_{ij} = y_i - y_j, \ r_{ij} = \sqrt{(x_{ij}^2 + y_{ij}^2)},$$

$$i, j = 1, 2, \ldots, n, i \neq j.$$

By Newtonian mechanics, the Newtonian attraction of m_i at position r_i is equal to its centrifugal force. Hence we have a general system of equations (see [2])

$$\omega^2 x_i = \sum_{j \neq i} m_j \frac{x_{ij}}{r_{ij}^3},$$

$$\omega^2 y_i = \sum_{j \neq i} m_j \frac{y_{ij}}{r_{ij}^3}, \ i = 1, 2, \ldots, n,$$

where ω is an angular velocity and is a constant.

Let $u_j = x_j + iy_j, v_j = x_j - iy_j$ $(j = 1, 2, \ldots, n)$, where $i = \sqrt{-1}$, and $m_0 = -\omega^2$. Then the above equations become the following system of equations

$$m_0 u_i + \sum_{j \neq i} \frac{m_j}{(v_i - v_j) r_{ij}} = 0, i = 1, 2, \ldots, n, \tag{I}$$

$$m_0 v_i + \sum_{j \neq i} \frac{m_j}{(u_i - u_j) r_{ij}} = 0, i = 1, 2, \ldots, n, \tag{II}$$

and

$$r_{ij} - (u_i - u_j)(v_i - v_j) = 0, \ i \neq j, \qquad i, j = 1, 2, \ldots, n,$$

where $r_{ij} = r_{ji}, i \neq j, i, j = 1, 2, \ldots, n$.

Thus, the determination of flat central configurations is reduced to equations-solving with variables u_i, v_j, r_{ij} as unknowns. The Wintner conjecture is true if the above system of polynomial equations has a finite number of solutions for any possible masses of planets.

To simplify the notations, we set

$$u_{ij} = u_i - u_j, \ v_{ij} = v_i - v_j, \ i < j,$$

$$u_{ij} = u_{1j} - u_{1i}, \ v_{ij} = v_{1j} - v_{1i}, \ i \neq j, \ i, j = 2, \ldots, n.$$

4 The Central Configurations of Four Planets

The problem of the central configurations of three bodies was studied in [2] with the char-set method. In this section we study the problem for four bodies. In this case, equations (I) become the following form

$$q_1 \equiv m_0 u_1 + \frac{m_2}{(v_1 - v_2)r_{12}} + \frac{m_3}{(v_1 - v_3)r_{13}} + \frac{m_4}{(v_1 - v_4)r_{14}} = 0,$$

$$q_2 \equiv m_0 u_2 - \frac{m_1}{(v_1 - v_2)r_{12}} + \frac{m_3}{(v_2 - v_3)r_{23}} + \frac{m_4}{(v_2 - v_4)r_{24}} = 0,$$

$$q_3 \equiv m_0 u_3 - \frac{m_1}{(v_1 - v_3)r_{13}} - \frac{m_2}{(v_2 - v_3)r_{23}} + \frac{m_4}{(v_3 - v_4)r_{34}} = 0,$$

$$q_4 \equiv m_0 u_4 - \frac{m_1}{(v_1 - v_4)r_{14}} - \frac{m_2}{(v_2 - v_4)r_{24}} - \frac{m_3}{(v_3 - v_4)r_{34}} = 0.$$

It is easy to see that this system is equivalent to the system consisting of $q_0 \equiv m_1 u_1 + m_2 u_2 + m_3 u_3 + m_4 u_4 = 0$ and any three from q_1, q_2, q_3, q_4.

Note that q_0 is identical with any one of the following:

$$q_{01} \equiv (m_1 + m_2 + m_3 + m_4)u_1 - m_2 u_{12} - m_3 u_{13} - m_4 u_{14} = 0,$$

$$q_{02} \equiv (m_1 + m_2 + m_3 + m_4)u_2 + m_1 u_{12} - m_3 u_{23} - m_4 u_{24} = 0,$$

$$q_{03} \equiv (m_1 + m_2 + m_3 + m_4)u_3 + m_1 u_{13} + m_2 u_{23} - m_4 u_{34} = 0,$$

$$q_{04} \equiv (m_1 + m_2 + m_3 + m_4)u_4 + m_1 u_{14} + m_2 u_{24} + m_4 u_{34} = 0.$$

Clearing the fractions of q_1, q_2, q_3, q_4, we obtain the following polynomial equations

$$p_1 \equiv v_{12}r_{12}v_{13}r_{13}v_{14}r_{14}m_0 u_1 + m_2 v_{13}r_{13}v_{14}r_{14}$$
$$+ m_3 v_{12}r_{12}v_{14}r_{14} + m_4 v_{12}r_{12}v_{13}r_{13} = 0,$$

$$p_2 \equiv v_{12}r_{12}v_{23}r_{23}v_{24}r_{24}m_0 u_2 - m_1 v_{23}r_{23}v_{24}r_{24}$$
$$+ m_3 v_{12}r_{12}v_{24}r_{24} + m_4 v_{12}r_{12}v_{23}r_{23} = 0,$$

$$p_3 \equiv v_{13}r_{13}v_{23}r_{23}v_{34}r_{34}m_0 u_3 - m_1 v_{23}r_{23}v_{34}r_{34}$$
$$- m_2 v_{13}r_{13}v_{34}r_{34} + m_4 v_{13}r_{13}v_{23}r_{23} = 0,$$

$$p_4 \equiv v_{14}r_{14}v_{24}r_{24}v_{34}r_{34}m_0 u_4 - m_1 v_{24}r_{24}v_{34}r_{34}$$
$$- m_2 v_{14}r_{14}v_{34}r_{34} - m_3 v_{14}r_{14}v_{24}r_{24} = 0.$$

Let

$$p_9 \equiv r_{12}^2 - u_{12}v_{12} = 0,$$

$$p_{10} \equiv r_{13}^2 - u_{13}v_{13} = 0,$$

$$p_{11} \equiv r_{14}^2 - u_{14}v_{14} = 0,$$

$$p_{12} \equiv r_{23}^2 - u_{23}v_{23} = 0,$$

$$p_{13} \equiv r_{24}^2 - u_{24}v_{24} = 0,$$

$$p_{14} \equiv r_{34}^2 - u_{34}v_{34} = 0.$$

For simplification, we introduce the notations $a = m_1 + m_2 + m_3 + m_4$ and

$$f_{ij} = a + m_0 r_{13}^3, \quad i < j, \ i, j = 1, 2, 3, 4.$$

Now consider the set of polynomials

$$\{q_{01}, p_1, p_2, p_4, p_9, p_{10}, p_{11}, p_{12}, p_{13}, p_{14}\}.$$

After eliminating the variables $u_1, u_{12}, u_{13}, u_{14}$ from p_1 step by step via pseudo-division using polynomial equations $[q_{01} = 0, p_9 = 0, p_{10} = 0, p_{11} = 0]$, we obtain a new polynomial

$$p_{15} \equiv m_2 r_{13} r_{14} v_{13} v_{14} f_{12} + m_3 r_{12} r_{14} v_{12} v_{14} f_{13} + m_4 r_{12} r_{13} v_{12} v_{13} f_{14} = 0.$$

For p_2 (or p_4, resp.), after eliminating the variables $u_2, u_{12}, u_{23}, u_{24}$ (or $u_4, u_{14}, u_{24}, u_{34}$, resp.) using polynomial equations $q_{02} = 0, p_9 = 0, p_{12} = 0, p_{13} = 0$ (or $q_{04} = 0, p_{11} = 0, p_{13} = 0, p_{14} = 0$, respectively) and replacing v_{ij} ($i, j \neq 1$) with $(v_{1j} - v_{1i})$ in both of the polynomials we obtain

$$q_{16} \equiv m_1 r_{23} r_{24} (v_{13} - v_{12})(v_{14} - v_{12}) f_{12} - m_3 r_{12} r_{24} v_{12} (v_{14} - v_{12}) f_{23}$$
$$-m_4 r_{12} r_{23} v_{12} (v_{13} - v_{12}) f_{24} = 0,$$
$$q_{18} \equiv m_1 r_{24} r_{34} (v_{14} - v_{12})(v_{14} - v_{13}) f_{14} + m_2 r_{14} r_{34} v_{14} (v_{14} - v_{13}) f_{24}$$
$$+m_3 r_{14} r_{24} v_{14} (v_{14} - v_{12}) f_{34} = 0.$$

For system (II), we have the similar results

$$qq_0 \equiv m_1 v_1 + m_2 v_2 + m_3 v_3 + m_4 v_4 = 0,$$

$$q_5 \equiv m_0 v_1 + \frac{m_2}{(u_1 - u_2) r_{12}} + \frac{m_3}{(u_1 - u_3) r_{13}} + \frac{m_4}{(u_1 - u_4) r_{14}} = 0,$$

$$q_6 \equiv m_0 v_2 - \frac{m_1}{(u_1 - u_2) r_{12}} + \frac{m_3}{(u_2 - u_3) r_{23}} + \frac{m_4}{(u_2 - u_4) r_{24}} = 0,$$

$$q_7 \equiv m_0 v_3 - \frac{m_1}{(u_1 - u_3) r_{13}} - \frac{m_2}{(u_2 - u_3) r_{23}} + \frac{m_4}{(u_3 - u_4) r_{34}} = 0,$$

$$q_8 \equiv m_0 v_4 - \frac{m_1}{(u_1 - u_4) r_{14}} - \frac{m_2}{(u_2 - u_4) r_{24}} - \frac{m_3}{(u_3 - u_4) r_{34}} = 0.$$

Clearing the fractions, we have the following polynomial equations

$$p_5 \equiv u_{12} r_{12} u_{13} r_{13} u_{14} r_{14} m_0 v_1 + m_2 u_{13} r_{13} u_{14} r_{14} + m_3 u_{12} r_{12} u_{14} r_{14}$$
$$+m_4 u_{12} r_{12} u_{13} r_{13} = 0,$$
$$p_6 \equiv u_{12} r_{12} u_{23} r_{23} u_{24} r_{24} m_0 v_2 - m_1 u_{23} r_{23} u_{24} r_{24} + m_3 u_{12} r_{12} u_{24} r_{24}$$
$$+m_4 u_{12} r_{12} u_{23} r_{23} = 0,$$
$$p_7 \equiv u_{13} r_{13} u_{23} r_{23} u_{34} r_{34} m_0 v_3 - m_1 u_{23} r_{23} u_{34} r_{34} - m_2 u_{13} r_{13} u_{34} r_{34}$$
$$+m_4 u_{13} r_{13} u_{23} r_{23} = 0,$$
$$p_8 \equiv u_{14} r_{14} u_{24} r_{24} u_{34} r_{34} m_0 v_4 - m_1 u_{24} r_{24} u_{34} r_{34} - m_2 u_{14} r_{14} u_{34} r_{34}$$
$$-m_3 u_{14} r_{14} u_{24} r_{24} = 0.$$

Similar to the polynomials p_1, p_2, p_3, p_4, we can obtain polynomials p_{19}, p_{20}, p_{21} from p_5, p_6, p_8 respectively, where

$$p_{19} \equiv m_2 r_{13} r_{14} u_{13} u_{14} f_{12} + m_3 r_{12} r_{14} u_{12} u_{14} f_{13}$$
$$+ m_4 r_{12} r_{13} u_{12} u_{13} f_{14} = 0,$$
$$p_{20} \equiv m_1 r_{23} r_{24} (u_{13} - u_{12})(u_{14} - u_{12}) f_{12} - m_3 r_{12} r_{24} u_{12} (u_{14} - u_{12}) f_{23}$$
$$- m_4 r_{12} r_{23} u_{12} (u_{13} - u_{12}) f_{24} = 0,$$
$$p_{21} \equiv m_1 r_{24} r_{34} (u_{14} - u_{12})(u_{14} - u_{13}) f_{14} + m_2 r_{14} r_{34} u_{14} (u_{14} - u_{13}) f_{24}$$
$$+ m_3 r_{14} r_{24} u_{14} (u_{14} - u_{12}) f_{34} = 0.$$

And we have $qq_{01} \equiv (m_1 + m_2 + m_3 + m_4) v_1 - m_2 v_{12} - m_3 v_{13} - m_4 v_{14} = 0$. The polynomials $p_{15}, p_{16}, p_{18}, p_{19}, p_{20}, p_{21}$ are obtained from $p_1, p_2, p_4, p_5, p_6, p_8$ by pseudo-division. Since all the initials of the middle polynomials are not zero, we have

$$Zero(\{p_1, p_2, \ldots, p_{14}\}) = Zero(\{q_{01}, p_1, p_2, p_4, p_9, p_{10}, p_{11}, p_{12}, p_{13}, p_{14}\})$$
$$\cap Zero(\{qq_{01}, p_5, p_6, p_8, p_9, p_{10}, p_{11}, p_{12}, p_{13}, p_{14}\}),$$

and

$$Zero_{rc}(\{q_{01}, p_1, p_2, p_4, p_9, p_{10}, p_{11}, p_{12}, p_{13}, p_{14}\})$$
$$= Zero_{rc}(\{q_{01}, p_{15}, p_{16}, p_{18}, p_9, p_{10}, p_{11}, p_{12}, p_{13}, p_{14}\}),$$
$$Zero_{rc}(\{qq_{01}, p_5, p_6, p_8, p_9, p_{10}, p_{11}, p_{12}, p_{13}, p_{14}\})$$
$$= Zero_{rc}(\{qq_{01}, p_{19}, p_{20}, p_{21}, p_9, p_{10}, p_{11}, p_{12}, p_{13}, p_{14}\}).$$

To find zeros of $\{q_{01}, p_{15}, p_{16}, p_{18}, p_9, p_{10}, p_{11}, p_{12}, p_{13}, p_{14}\}$, we set the following order

$$m_1 \prec m_2 \prec m_3 \prec m_4 \prec r_{12} \prec r_{13} \prec r_{14} \prec r_{23} \prec r_{24} \prec r_{34} \prec m_0 \prec f_{12} \prec f_{13}$$
$$\prec f_{14} \prec f_{23} \prec f_{24} \prec f_{34} \prec v_{14} \prec v_{13} \prec v_{12} \prec u_1.$$

Using Wu elimination, we decompose the set $\{q_{01}, q_{15}, q_{16}, q_{18}\}$ and obtain five systems of polynomial equations. They are

(1). $ZS1 = Zero(\{z_{11}, z_{12}, z_{13}, z_{14}, z_{15}, z_{16}\})$, in which

$$z_{11} \equiv a u_1 - m_2 u_{12} - m_3 u_{13} - m_4 u_{14} = 0,$$
$$z_{12} \equiv f_{12} = 0,$$
$$z_{13} \equiv m_3 r_{14} v_{14} f_{13} + m_4 r_{13} v_{13} f_{14} = 0,$$
$$z_{14} \equiv f_{24} = 0,$$
$$z_{15} \equiv f_{23} = 0,$$
$$z_{16} \equiv m_4 r_{13} m_1 r_{34} f_{14} + m_4 r_{13} m_3 r_{14} f_{34} + m_3 r_{14} f_{13} m_1 r_{34} = 0.$$

(2). $ZS2 = Zero(\{z_{21}, z_{22}, z_{23}, z_{24}, z_{25}\})$, in which $z_{21} = z_{11}, z_{22} = z_{12}, z_{23} = z_{13}$ and

$$z_{24} \equiv -v_{12} m_3 r_{24} f_{23} - v_{12} f_{24} r_{23} m_4 + m_3 r_{24} f_{23} v_{14} + v_{13} f_{24} r_{23} m_4 = 0,$$
$$z_{25} \equiv r_{13} r_{23} m_4 m_2 r_{14} r_{34} f_{24} + r_{13} r_{23} m_4 m_1 r_{24} r_{34} f_{14} + r_{13} m_3 r_{24} f_{23} m_2 r_{14} r_{34}$$
$$+ r_{13} r_{23} m_4 m_3 r_{14} r_{24} f_{34} + m_3 r_{14} f_{13} r_{23} m_1 r_{24} r_{34} = 0.$$

(3). $ZS3 = Zero(\{z_{31}, z_{32}, z_{33}, z_{34}, z_{35}, z_{36}\})$, in which $z_{31} = z_{11}, z_{32} = z_{14}$ and

$$z_{33} \equiv v_{12}m_1r_{23}f_{12} + v_{12}f_{23}r_{12}m_3 - m_1r_{23}f_{12}v_{13} = 0,$$

$$z_{34} \equiv -m_1r_{34}f_{14}v_{14} + m_1r_{34}f_{14}v_{13} - m_3r_{14}v_{14}f_{34} = 0,$$

$$z_{35} \equiv m_1r_{34}f_{14}m_3r_{12}r_{14}f_{13} + m_3r_{14}f_{34}m_4r_{12}r_{13}f_{14} + m_3r_{14}^2f_{34}r_{13}m_2f_{12}$$
$$+m_1r_{34}f_{14}^2m_4r_{12}r_{13} + m_1r_{34}f_{14}r_{13}m_2r_{14}f_{12} = 0,$$

$$z_{36} \equiv r_{34}m_2r_{13}r_{14}m_1r_{23}f_{12} + r_{34}m_2r_{13}r_{14}m_3r_{12}f_{23} + r_{34}r_{12}m_3f_{13}m_1r_{23}r_{14}$$
$$+m_4r_{12}r_{23}r_{13}m_3r_{14}f_{34} + m_4r_{12}r_{23}r_{13}m_1r_{34}f_{14} = 0.$$

(4). $ZS4 = Zero(\{z_{41}, z_{42}, z_{43}, z_{44}, z_{45}, z_{46}\})$, in which $z_{41} = z_{11}, z_{42} = z_{12}$ and

$$z_{43} \equiv f_{13} = 0,$$

$$z_{44} \equiv f_{14} = 0,$$

$$z_{45} \equiv m_2m_4r_{23}r_{34}f_{24} + m_2m_3r_{24}r_{34}f_{23} + m_3m_4r_{23}r_{24}f_{34} = 0,$$

$$z_{46} \equiv m_3r_{24}f_{34}v_{12} + m_2r_{34}f_{24}v_{13} - (m_2r_{34}f_{24} + m_3r_{24}f_{34})v_{14} = 0.$$

(5). $ZS5 = Zero(\{z_{51}, z_{52}, z_{53}, z_{54}\})$, in which $z_{51} = z_{11}$ and

$$z_{52} \equiv -m_2r_{13}r_{14}v_{13}v_{14}f_{12} - m_3r_{12}r_{14}v_{12}v_{14}f_{13} - m_4r_{12}r_{13}v_{12}v_{13}f_{14} = 0,$$

$$z_{53} \equiv -v_{14}^2m_3r_{12}r_{14}f_{13}m_1r_{24}r_{34}f_{14} - v_{14}^2m_3r_{12}r_{14}^2f_{13}m_2r_{34}f_{24}$$
$$-v_{14}^2m_3^2r_{12}r_{14}^2f_{13}r_{24}f_{34} + v_{14}m_3r_{12}r_{14}f_{13}m_1r_{24}r_{34}f_{14}v_{13}$$
$$-v_{14}m_2r_{13}r_{14}^2v_{13}f_{12}m_3r_{24}f_{34} - v_{14}m_4r_{12}r_{13}v_{13}f_{14}^2m_1r_{24}r_{34}$$
$$-v_{14}m_2r_{13}r_{14}v_{13}f_{12}m_1r_{24}r_{34}f_{14} + v_{14}m_3r_{12}r_{14}^2f_{13}m_2r_{34}f_{24}v_{13}$$
$$-v_{14}m_4r_{12}r_{13}v_{13}f_{14}m_3r_{14}r_{24}f_{34} - v_{14}m_4r_{12}r_{13}v_{13}f_{14}m_2r_{14}r_{34}f_{24}$$
$$+m_4r_{12}r_{13}v_{13}^2f_{14}m_2r_{14}r_{34}f_{24} + m_4r_{12}r_{13}v_{13}^2f_{14}^2m_1r_{24}r_{34}$$
$$+m_2r_{13}r_{14}v_{13}^2f_{12}m_1r_{24}r_{34}f_{14} = 0,$$

$$z_{54} \equiv m_2r_{13}r_{14}r_{34}m_1r_{23}r_{24}f_{12} + m_2r_{13}r_{14}r_{34}m_4r_{12}r_{23}f_{24}$$
$$+m_2r_{13}r_{14}r_{34}m_3r_{12}r_{24}f_{23} + r_{13}m_4r_{12}r_{23}m_3r_{24}f_{34}r_{14}$$
$$+r_{13}m_4r_{12}r_{23}m_1r_{24}r_{34}f_{14} + r_{34}m_3r_{12}f_{13}m_1r_{23}r_{24}r_{14} = 0.$$

Similarly decomposing the set $\{qq_{01}, p_{19}, p_{20}, p_{21}\}$ we also obtain five systems of polynomial equations. They may be got from the above (1)–(5) via replacing the v_1, u_{ij} with u_1, v_{ij}, respectively.

Combining these two groups of systems of polynomial equations and adding the polynomials $p_9, p_{10}, p_{11}, p_{12}, p_{13}, p_{14}$, we get 15 systems of polynomial equations. Solving these systems by using the char-set method, we find that all the possible solutions of each system are divided into three cases according to how many $f_{ij} = 0$ as follows:

Case 1. Only one f_{ij} equals to zero. It gives to the possible solutions of collinear case.

Case 2. Two f_{ij} equal to zero. It is easy to see that there are at least three $f_{ij} = 0$. When the reality conditions are considered, there is no solutions in this case.

Case 3. No f_{ij} equals to zero. It leads to the possible solutions of non-collinear case.

Most of the 15 systems are reduced to Case 2, in which either all $r_{ij} = 0$ or three of the four points are both collinear and co-circular. The two cases in which the systems have solutions are discussed in detail below.

4.1   Collinear case

In Case 1, we have the following set of polynomials

$$EQS1 = \{z_{21}, z_{22}, z_{23}, z_{24}, z_{25}, y_{11}, y_{12}, y_{13}, p_9, p_{10}, p_{11}, p_{12}, p_{13}, p_{14}\},$$

in which $z_{21}, z_{22}, z_{23}, z_{24}, z_{25}$ are given in the system $ZS2$, $p_9, p_{10}, \cdots, p_{14}$ are given in Section 3 and y_{11}, y_{12}, y_{13} are given by the z_{21}, z_{24}, z_{25} via replacing the v_1, u_{ij} with u_1, v_{ij}, $i, j = 1, 2, 3, 4$, respectively.

Note that the total degree of some polynomials in the set $EQS1$ is 7. So, it is very complex to get the zeros of $EQS1$. However, we can get the characteristic sets of $EQS1$ by using Wu elimination. Set the order as follows

$$m_1 \prec m_2 \prec m_3 \prec m_4 \prec r_{12} \prec r_{13} \prec r_{23} \prec r_{24} \prec r_{34} \prec m_0 \prec f_{12} \prec f_{13}$$
$$\prec f_{14} \prec f_{23} \prec f_{24} \prec f_{34} \prec u_{14} \prec u_{13} \prec u_{12} \prec v_{14} \prec v_{13} \prec v_{12} \prec u_1 \prec v_1.$$

We obtain 12 characteristic sets from $EQS1$. They correspond to the 12 possible collinear central configurations. For example, one of the 12 characteristic consists of:

$$cs_1 \equiv r_{13} + r_{12} - r_{23},$$
$$cs_2 \equiv r_{13} + r_{24} + r_{14} - r_{23},$$
$$cs_3 \equiv r_{13} + r_{14} - r_{34},$$
$$cs_4 \equiv f_{12},$$
$$cs_5 \equiv r_{14}^2 m_3 f_{13} - f_{14} r_{13}^2 m_4,$$
$$cs_6 \equiv f_{23} r_{14} r_{24} m_3 + m_3 r_{13} f_{23} r_{24} - m_4 r_{23}^2 f_{24} - r_{23} f_{23} r_{24} m_3,$$
$$cs_7 \equiv r_{13} r_{23} m_4 m_2 r_{14} r_{34} f_{24} + r_{13} r_{23} m_4 m_1 r_{24} r_{34} f_{14} + r_{13} m_3 r_{24} f_{23} m_2 r_{14} r_{34}$$
$$+ r_{13} r_{23} m_4 m_3 r_{14} r_{24} f_{34} + m_3 r_{14} f_{13} r_{23} m_1 r_{24} r_{34},$$
$$cs_8 \equiv r_{14} u_{13} + u_{14} r_{13},$$
$$cs_9 \equiv r_{23} u_{14} + r_{14} u_{13} - r_{14} u_{12} + r_{13} u_{13} - r_{13} u_{12} - u_{13} r_{23},$$
$$cs_{10} \equiv r_{14}^2 - u_{14} v_{14},$$
$$cs_{11} \equiv v_{13} r_{14} + r_{13} v_{14},$$
$$cs_{12} \equiv v_{14} r_{23} + v_{13} r_{14} - r_{14} v_{12} + v_{13} r_{13} - r_{13} v_{12} - v_{13} r_{23},$$
$$cs_{13} \equiv a u_1 - m_2 u_{12} - m_3 u_{13} - m_4 u_{14},$$
$$cs_{14} \equiv a v_1 - m_2 v_{12} - m_3 v_{13} - m_4 v_{14}.$$

Hence, we have the following

Theorem 1. There are 12 collinear central configurations for four planets.

4.2 Flat but non-collinear case

In Case 3, we have the following set of polynomials:

$$EQS2 = \{z_{51}, z_{52}, z_{53}, z_{54}, y_{21}, y_{22}, y_{23}, y_{24}, p_9, p_{10}, p_{11}, p_{12}, p_{13}, p_{14}\},$$

in which $z_{51}, z_{52}, z_{53}, z_{54}$ are given in the system $ZS5$, $p_9, p_{10}, \cdots, p_{14}$ are as in the above system $EQS1$ and $y_{21}, y_{22}, y_{23}, y_{24}$ are given by the $z_{51}, z_{52}, z_{53}, z_{54}$ via replacing v_1, u_{ij} with u_1, v_{ij} respectively.

We set the following variable order

$$m_1 \prec m_2 \prec m_3 \prec m_4 \prec r_{12} \prec r_{13} \prec r_{23} \prec r_{24} \prec r_{34} \prec m_0 \prec g_{12} \prec g_{13}$$
$$\prec g_{14} \prec g_{23} \prec g_{24} \prec g_{34} \prec u_{14} \prec u_{13} \prec u_{12} \prec v_{14} \prec v_{13} \prec v_{12} \prec u_1 \prec v_1.$$

Using Wu elimination we obtain the following characteristic set which has possible real solutions

$$\begin{aligned}
cs_{21} \equiv\ & r_{12}^4 r_{34}^2 - r_{23}^2 r_{12}^2 r_{34}^2 + r_{23}^2 r_{12}^2 r_{13}^2 - r_{12}^2 r_{24}^2 r_{34}^2 - r_{13}^2 r_{12}^2 r_{34}^2 + r_{14}^2 r_{12}^2 r_{24}^2 \\
& + r_{12}^2 r_{34}^4 - r_{12}^2 r_{34}^2 r_{14}^2 - r_{12}^2 r_{13}^2 r_{24}^2 - r_{14}^2 r_{23}^2 r_{12}^2 + r_{14}^4 r_{23}^2 + r_{14}^2 r_{23}^4 \\
& + r_{13}^2 r_{34}^2 r_{14}^2 - r_{14}^2 r_{34}^2 r_{23}^2 + r_{23}^2 r_{24}^2 r_{34}^2 - r_{14}^2 r_{23}^2 r_{24}^2 + r_{13}^2 r_{34}^4 + r_{13}^2 r_{24}^4 \\
& - r_{13}^2 r_{23}^2 r_{24}^2 - r_{13}^2 r_{14}^2 r_{24}^2 - r_{13}^2 r_{14}^2 r_{23}^2 - r_{13}^2 r_{24}^2 r_{34}^2,
\end{aligned}$$

$$\begin{aligned}
cs_{22} \equiv\ & 2r_{12}^2 r_{13}^2 r_{14}^2 g_{13} - 2g_{12} r_{13}^2 r_{14}^2 r_{23}^2 + r_{13}^2 g_{12} r_{14}^2 r_{12}^2 + r_{13}^2 g_{12} r_{14}^4 \\
& - r_{12}^2 g_{13} r_{34}^4 + 2r_{12}^2 r_{13}^2 g_{13} r_{34}^2 + 2r_{12}^2 g_{13} r_{34}^2 r_{14}^2 + g_{12} r_{13}^2 r_{12}^2 r_{34}^2 \\
& - g_{12} r_{13}^4 r_{12}^2 + r_{13}^2 g_{12} r_{14}^2 r_{34}^2 - r_{13}^2 g_{12} r_{14}^4 - g_{12} r_{13}^2 r_{24}^2 r_{34}^2 \\
& + r_{13}^4 g_{12} r_{24}^2 + r_{13}^2 g_{12} r_{24}^2 r_{14}^2 - r_{12}^2 r_{13}^2 g_{13} - r_{14}^4 r_{12}^2 g_{13},
\end{aligned}$$

$$\begin{aligned}
cs_{23} \equiv\ & 2r_{12}^2 g_{14} r_{13}^2 - r_{12}^2 g_{13} r_{34}^2 + r_{12}^2 r_{13}^2 g_{12} + r_{14}^2 g_{12} r_{13}^2 \\
& - g_{12} r_{13}^2 r_{24}^2 + r_{12}^2 g_{13} r_{13}^2 + r_{14}^2 r_{12}^2 g_{13},
\end{aligned}$$

$$\begin{aligned}
cs_{24} \equiv\ & r_{13}^2 r_{12}^2 g_{14}^2 - g_{13} r_{14}^2 g_{23} r_{12}^2 - g_{23} r_{12}^2 g_{14} r_{13}^2 - r_{14}^2 g_{23} g_{12} r_{13}^2 \\
& - r_{14}^2 r_{12}^2 g_{13}^2 - r_{14}^2 g_{13} g_{12} r_{13}^2 - r_{14}^2 r_{12}^2 g_{13} g_{12} - r_{14}^2 r_{13}^2 g_{12}^2,
\end{aligned}$$

$$cs_{25} \equiv r_{13}^2 g_{12} g_{14} + r_{13}^2 g_{14}^2 + r_{13}^2 g_{14} g_{24} - r_{14}^2 g_{13} g_{23} - r_{14}^2 g_{13}^2 - r_{14}^2 g_{13} g_{12},$$

$$cs_{26} \equiv g_{34} + g_{24} + g_{23} + g_{14} + g_{13} + g_{12},$$

$$\begin{aligned}
cs_{27} \equiv\ & r_{14}^2 r_{13}^2 g_{12}^2 u_{13} u_{14} - u_{14} r_{13}^2 r_{12}^2 g_{14}^2 u_{13} - u_{14}^2 r_{13}^2 r_{12}^2 g_{14} g_{13} \\
& - r_{14}^2 u_{13}^2 r_{12}^2 g_{13} g_{14} - r_{14}^2 u_{13} r_{12}^2 g_{13}^2 u_{14},
\end{aligned}$$

$$cs_{28} \equiv u_{14} r_{13}^2 r_{12}^2 g_{14} + r_{14}^2 u_{13} r_{12}^2 g_{13} + r_{14}^2 r_{13}^2 u_{12} g_{12},$$

$$cs_{29} \equiv r_{14}^2 - u_{14} v_{14},$$

$$cs_{30} \equiv r_{13}^2 - u_{13} v_{13},$$

$$cs_{31} \equiv r_{12}^2 - u_{12} v_{12},$$

$$cs_{32} \equiv a u_1 - m_2 u_{12} - m_3 u_{13} - m_4 u_{14},$$

$$cs_{33} \equiv a v_1 - m_2 v_{12} - m_3 v_{13} - m_4 v_{14},$$

in which the notations $g_{12}, g_{13}, g_{14}, g_{23}, g_{24}, g_{34}$ are introduced by the following relations

$$m_1 m_2 f_{12} - r_{12} g_{12} = 0,$$
$$m_1 m_3 f_{13} - r_{13} g_{13} = 0,$$
$$m_1 m_4 f_{14} - r_{14} g_{14} = 0,$$
$$m_2 m_3 f_{23} - r_{23} g_{23} = 0,$$
$$m_2 m_4 f_{24} - r_{24} g_{24} = 0,$$
$$m_3 m_4 f_{34} - r_{34} g_{34} = 0.$$

Hence, from the above characteristic set we know that all of the zeros are uniquely determined by m_i, m_0, or m_i, ω. So the set has finitely many zeros which are determined by m_i, m_0, or m_i, ω. Thus we have

Theorem 2. The number of central configurations in planet motions of 4 bodies is finite under the condition that the masses and angular velocity of the planets do not satisfy any algebraic relations. In other words, the Wintner conjecture for $n = 4$ is true in the general case.

The complete version of Wintner conjecture has not been solved yet.

References

1. A. Wintner, *The Analytical Foundations of Celestial Mechanics*, Second Printing, Princeton Univ. Press, Princeton, 1947.
2. Wu Wen-tsün, *Central Configurations in Planet Motions and Vortex Motions*, MM Research Preprints No. 13(1995) 1–14.
3. F. R. Moulton, *The Straight Line Solutions for the Problem of n-bodies*, Annals of Math. 12(1910) 1–17.
4. S. Smale, *Topology and Mechanics*, I, II. Invent. Math. 10(1970) 305–331; 11(1970) 45–64.
5. J. Waldvogel, *Note Concerning a Conjecture by A. Wintner*, Celestial Mechanics 5(1972) 37–40.
6. Wu Wen-tsün, *Basic Principles of Mechanical Theorem Proving in Geometries*, Volume I: Part on Elementary Geometries, Science Press, Beijing (in Chinese), 1984; English edition, Springer, Wien New York, 1994.

Integration of Reasoning and Algebraic Calculus in Geometry

Stéphane Fèvre

LEIBNIZ–IMAG
46, avenue Félix Viallet 38031 Grenoble Cedex, France
e-mail: Stephane.Fevre@imag.fr

Abstract. This paper presents a new framework for merging reasoning and algebraic calculus in elementary geometry. This approach is based on the use of algebraic constraints in clause-based calculi. These constraints are considered as contexts for reasoning. It allows to introduce new inference rules and to use the powerful algebraic methods which have been successful in geometry theorem proving. The semantics of classical first-order logic has been modified to correspond with this framework and classical results in proof theory such as Herbrand's theorem, lifting lemma and completeness are proved to remain true within it. Some examples give an idea of the possibilities provided by this framework. A few strategies are also discussed and a comparison with related techniques is done.

1 Introduction

This section presents the main motivations for this work and gives a proof-theoretic point of view on the problem of integration of algebraic techniques in a logic framework.

1.1 Main threads in geometry theorem proving

The following distinction for the main threads of Geometry Theorem Proving is generally admitted. The axiomatic approach consists in general purpose methods developed in automated deduction for a specific logic. Generally the considered logic is the first-order logic with equality. Some experiments using OTTER [10] have been reported by A.Quaife [14]. Other methods use deductive database mechanisms to infer lemmae for proving a given theorem [8].

Another approach consists in using an algorithm for deciding validity of formulae for a given theory. Historically it is the first approach [17] but its main interest is theoretic: one can prove that a geometry is decidable though proving theorems using such a method is not necessary tractable.

The logic programming approach uses mechanisms developed in this field to make logic interpreters efficient (backtracking techniques, Horn's clauses management ...) to provide an easy way of programming provers. In this approach logic programs are used to both program a method and formalize a theorem [9].

The algebraic approach consists in translating a geometrical problem into an algebraic one and uses techniques such as variable elimination (e.g. [21, 22, 19]) and Gröbner bases [2, 12] to solve the problem algebraically. This approach has been proved to be very fruitful and efficient. Last, some specialized methods have been developed to optimize theorem proving in a specific geometry. They are very efficient and are based on algebraic properties of the geometry such as invariants. For instance, several methods developed by Chou, Gao and Zhang [7, 8] and based on the area or the concept of full-angles have been used to prove and discover a lot of difficult theorems. Once the algebraic formalism specific to their method is well understood it is possible for the reader to get a precise idea of the algebraic reasoning using a trace of the successive calculi. Most of the proofs are short but they do not give the geometric insight that the mathematician is looking for.

The interest of researchers in some of these approaches seems to have dropped away while specialized methods such as Wu's method, the area method and those using Gröbner bases are actively developed. One can say to simplify that the main interest of the first approach is its efficiency while the second one uses general techniques interesting for several communities.

1.2 Motivations

Generally speaking these techniques do not answer entirely to the desire of most users. We try to analyze the main reasons in what follows. First of all the proofs produced by most of these methods require the user to learn a new formalism which is an important drawback for learning geometry: people tend to spend more time in learning the formalism than geometric properties. While these techniques are interesting by themselves they are of little help for learning geometry. Moreover they are very different from proofs written by people. That is why these methods are more interesting for checking the validity of a geometry conjecture than for giving a formal proof.

For instance, the area method gives an algebraic point of view of a geometric problem skipping the logic view. Some produced proofs are very short and hide both the difficulty and the interest of the theorem. However a mathematician could consider some of these theorems to be difficult and proving it this way is no more illumining than knowing that the conjecture is true. It is obvious that one can reproach the same thing to logic-based methods: it is well-known that formal and complete logic proofs are much longer than proofs written by people although the size factor is far from being constant.

That is why we argue that automatic production of interesting proofs relies on the following three criteria. First one has to skip from proofs what is *logically* obvious. For instance structural rules in the sequent calculus may be considered as logically trivial. In the calculi based on resolution, the factorization rule, although important for the mechanization of deduction, may be considered as an obvious step. Using commutativity of the equality is another example. Second one has to avoid *mathematically* obvious materials such as normalization of algebraic expressions, or explicit substitution of a term by another proved to be equal. Last, and this point is certainly the most difficult to reach, to exhibit simple (deep) reasons for a theorem to be true.

As it is shown by the experience of specialized methods in geometry theorem proving, the last point requires to use several formal systems to make a fundamental reason appear. One of the main feature of proofs by hand is the use of various formalisms both for reducing the size of proofs and for stressing the explanative insight of proofs. In geometry, the introduction of algebraic computations allows sometimes to give a simpler argument to convince the reader of the validity of a geometric fact. Moreover the logic part gives a sketch of reasoning while algebraic computations check uninteresting details of this reasoning. This endeavour for making interact algebraic and logic techniques is called *Hybrid Geometry Theorem Proving* in what follows.

1.3 A short presentation

The *ATINF* research team [1, 3] has developed the concept of inference laboratory and defined three levels of components. The first one is a library of automated deduction tools, the second a set of experimental provers (resolution, tableaux, proof generalization, simultaneous search for proofs and counter-examples [5, 1], reduction of proof complexity using symmetries, and geother [19, 20], a set of algebra-based provers for geometry). The last one consists of a graphical environment [3] for integrating these tools, verifying, editing and presenting proofs. Hybrid geometry theorem proving may be considered as a new research subject both to illustrate the capabilities of existing tools and developing automated deduction to be useful for mathematicians and teachers in logics and geometry.

Previous attempts had been made to show that algebra and logic could interact. For instance it has been showed that algebraic computations occurring in the Wu's method [21, 22] can be converted into traditional geometry proofs [11] based on an axiomatization. It consists in associating points and constructions to algebraic expressions, and in generating a geometric interpretation of algebraic terms. It is also possible to apply algebraic techniques for proving theorems in first-order logic [13].

The main goal of this paper consists in illustrating how to merge algebraic and logic techniques to design new proof methods for geometry theorem proving. This work is intended to provide a basis for studying an intelligent integration of various techniques. It is also a guide for designing applicated logical frameworks.

Several levels of integration may be distinguished depending on the various operations intended to work simultaneously. The first level is based on *competition* and can be easily built: it suffices to apply in parallel two methods to a same problem without allowing any of them to use information inferred by the other. The second level is called *cooperation* and consists in using methods which are able to use information coming from one prover for the other. The third level is *integration* and consists in using external information to guide the proof search.

As the reader can imagine, the most obvious manner consists in translating logic formulae into algebraic expressions. This corresponds to a cooperation method and is what we have chosen to present in this paper. The current state-of-the-art allows us to integrate a lot of different techniques and particularly most of the general algebraic techniques. It is possible to adapt this technique to various logic-based provers though we have chosen a resolution-based framework to illustrate the method. It should not be difficult to use another calculus such as natural deduction though it is not included in this paper for sake of simplicity.

The next section describes some background knowledge and notations necessary for the presentation of our method.

2 Preliminaries

The introduction mentions that our method could be adapted to several calculi and is not restricted to resolution. However in what follows the reader is assumed to have some basic knowledge about this automated deduction method and about at least one algebraic method such as Gröbner bases. Actually it is not necessary to know the technical details because a lot of different techniques could be used instead. But it is important for understanding the possibilities and the limits of an integration. Note that some necessary conditions for these techniques to be integrated are given.

The calculus is based on resolution and deals with first-order logic with equality. The language extends the language of clauses. Last, although clauses are often considered as multisets or sometimes lists, the usual logic connectives are used instead of the set notation.

2.1 Definitions and Notations

Consider a signature (a finite set of symbols) Σ with an infinite collection of variables X and terms built upon them. A *substitution* is a total function σ mapping variables to terms. Let P be a predicate symbol of arity n $(n > 0)$ and $t_1, \ldots, t_n$ be n terms. $P(t_1, \ldots, t_n)$ is called an *atomic formula* (or *atom*) and any atomic formula is of the previous form. A *literal* is either an atomic formula or its negation. A *clause* is a disjunction of literals and is denoted by $A_1 \vee \cdots \vee A_n$. The result of the application of a substitution σ to a literal L is the literal obtained from L by replacing every occurrence of every variable x with $\sigma(x)$ and is denoted by $L\sigma$.

A *contextual clause* is a clause equipped with an equational presentation of an algebraic domain (ideal, radical, variety, ...) e.g. a system of polynomial equations $\{E_1 = 0, \ldots, E_n = 0\}$ where E_i $(i = 1, \ldots, n)$ can be considered as a definition for a finitely presented polynomial algebra. The kind of expressions depends on the chosen algebra. Note that here algebra must be understood in its mathematical sense; it does not refer to abstract algebra as it does usually in automated reasoning. The existence of a computable intersection operation between these algebraic domains is also assumed and denoted by $\cap$. A contextual clause is denoted by $C : S$ where C is the clausal part and S the algebraic part. The existence of a computable function called *translation* mapping some atomic formulas to algebraic systems is also assumed; translations are partial functions, but their source domain must be non-empty. Substitutions are extended to contextual clauses in the following way: if $C : S$ is a contextual clause, the substituted one is $C\sigma : S\cap\Sigma$ where $C\sigma$ is the natural application of σ to each literal in C and Σ is the translation of the equational system $\{x = \sigma(x)^1, \ldots\}$ where x is any variable in the domain of σ.

Example 1. Let K be a field of characteristic 0 (say the field of complex numbers), and $A = K[X_1, Y_1, \ldots, X_n, Y_n]$ be the considered algebra. A system of polynomial equations represents an ideal I of A with which a Gröbner basis is associated. Thus I is represented by such a basis B and the elements of I are represented by polynomials of A reduced by B. Inn this example the intersection operation is the intersection of two ideals and consists essentially in computing Gröbner bases. Let T be an axiomatization in the first-order logic with sorts of the elementary euclidian geometry for the K-plane in which one of the sort is intended to represent points. A translation τ may be defined by the usual cartesian coordinates of points. For instance, if p_1 and p_2 stand for points, $\tau(p_1 = p_2) = \{X_1 = X_2, Y_1 = Y_2\}$. Collinearity and parallelism may be also translated, but not betweeness. It is sufficient to consider an infinite number of variables of sort point and their corresponding coordinates for defining rigorously a translation. No variable is shared by clauses, and to every resolution step corresponds an embedding of two algebras into another one.

Informally, contexts are the domains of algebraic unknowns. That is, a contextual clause is satisfied if and only if for every value of the variables either their translation do not lie into the defined domain either the clausal part is satisfied. The notion of satisfaction is more precisely defined in the next section.

2.2 Semantics

An interpretation I is a triple $I = (D, i, \tau)$ where D is the domain of interpretation, i is a map from the signature to functions and predicates of corresponding arities, and τ is a partial function from atomic formulae and terms to algebraic domains. Instead of $i(f)$ where f is a functional or predicate symbol, f^I is used when there is no possible confusion. Thus if f is a functional symbol of arity n then f^I denotes a mapping from D^n to D, and if P is a predicate symbol of arity p then P^I denotes a map from D^p to $\{\top, \bot\}$ the minimal boolean algebra. An assignment of a set of variables X in the domain D of an interpretation I is a map from X to D. This gives rise to an evaluation function. The domain of τ must contain every equation of terms. τ may be seen as a translation from the interpretation I to the algebra. To avoid any confusion, τ may also be denoted τ_I.

An assignment of variables is a couple $\sigma = (\sigma_1, \sigma_2)$ of applications from the set of variables X in C to D (σ_1) and the algebra of terms built upon X and the signature (σ_2) such that if a is a ground term satisfying $\sigma_1(x) = a^I$ and

---

[1] Here $\sigma(x)$ denotes the term substituted to x and not a term whose root symbol is σ.

$\tau(x), \tau(a)$ are defined then $\tau(\sigma_2(x))$ and $\tau(a)$ are equal (in the algebra). Note that assignments depend on translations. A contextual clause $C : S$ is said to be satisfied by an interpretation I if and only if for any assignment $\sigma = (\sigma_1, \sigma_2)$, either the translations of these elements do not lie into the algebraic context or at least one literal in the substituted clause $C^I \sigma_1$ is evaluated to $\top$ if it is positive, to $\bot$ if it is negative. Note that every interpretation I is associated with an algebraic translation τ_I from I to the framework algebra A such that if $J = (E, j, \tau_J)$ is another interpretation and $h : I \to J$ is a morphism (i.e. if f is a functional symbol of arity n then $j(f)(h(x_1), \ldots, h(x_n)) = h(i(f)(x_1, \ldots, x_n)))$ then there exists τ_h such that $\tau_J = \tau_h \circ \tau_I$. This means that whatever the considered interpretation is, the translation of an interpreted expression must be consistent with the other interpretations. These interpretations are said to be **compatible**.

Example 2. It is possible to consider translations in several domains using the notion of compatible interpretation, for instance both in the cartesian plane (J) and the vectorial plane based on reals (I). If $\tau_I(P) = v_P$ and $\tau_J(P) = (x_P, y_P)$ then $\tau_h(v_P) = (x_P, y_P)$.

$$perpendicular(P, Q, R) : (v_P - v_Q) \cdot (x_R - x_Q, y_R - y_Q) = 0$$

is satisfied by interpretations in which the literal is satisfied or in which the algebraic relations are satisfied provided that (according to τ_h) v_P has the cartesian coordinates (x_P, y_P) etc.

The codomain of a translation is not related to the domain of the interpretation. In some sense algebraic domains replace the term algebra as a contextual clause is satisfied by an interpretation only when the latter satisfies the algebraic contexts. However a translation is a partial function and interpretations of non-translated elements are freely definable by sets. Interpretations are only required to match on translated expressions.

Remark. The notion of compatible interpretation is central because it ensures some consistency through translations. But these translations do not always exist; for instance, it is not possible to translate first-order logic into elementary arithmetic. However it is important to notice that it is possible to use less expressive algebras and to use them to reason schematically. It is similar to the so-called *abstractions* used to guide the proof search. Moreover translations are partial functions and they need only to be defined on terms. This allows to consider a large class of translations but restricts the inference rules, notably the rule (A) defined in the next section. The current definition is a compromise.

Translation of expressions appearing in the clausal part provides algebraic domains which intersect the context of a clause. From a more practical point of view, these domains are often presented by a set of equations (though it is not required in our method). That is why in this case for any variable in a clause there are some associated algebraic unknowns in the constraints such that there exists a computable function that associates to any value of a variable corresponding values for the corresponding algebraic unknowns (see the example in the previous section). Actually, it is enough to say that one maps the domain of each interpretation on a sub-algebra of the algebraic framework A such that their associated translations coincide on ground (without any variable) expressions.

As opposed to the interpretation used in semantic resolution, the algebraic context is part of the computation itself and is not used at a meta-level for designing a strategy: algebraic contexts are at the same level as clauses and not a representation of a partial interpretation.

3 A Hybrid Deduction System

In this section an inference system extending resolution to deal with algebraic contexts. Other automated theorem proving methods based on extended resolution calculi use similar or more elaborated rules [5, 1]. While they are very clever and provide sometimes interesting results, they do not correspond to our main goal. For a comparison see Section 6. It is possible that some of these methods be extended to treat algebraic contexts but our purpose consists more in describing how to build an extension of the resolution method rather than to propose another method to the reader. What is interesting is how algebraic computations can be used and which conditions the algebraic methods should satisfy to be used with logic ones according to our proposition.

Literals will be denoted by indexed capitals, rests of clauses by capitals and terms by small letters.

3.1 An extension of resolution

In the following inference system, denoted by $\mathcal{A}$, rules are preceded by their abbreviated name, Σ is the algebraic translation of σ, p is a position in a tree representing a term, and *mgu* stands for *most general unifier*:

resolution (R)
$$\frac{A_1 \vee D_1 {:} S_1 \quad \neg A_2 \vee D_2 {:} S_2}{D_1 \sigma \vee D_2 \sigma {:} S_1 \cap S_2} \qquad \sigma = mgu(A_1, A_2),$$

factorization (F)
$$\frac{A_1 \vee A_2 \vee D {:} S}{A_1 \sigma \vee D \sigma {:} S} \qquad \sigma = mgu(A_1, A_2),$$

paramodulation (P)
$$\frac{l_1 = r_1 \vee D_1 {:} S_1 \quad A_2 \vee D_2 {:} S_2}{A_2[p/r_1] \sigma \vee D_1 \sigma \vee D_2 \sigma {:} S_1 \cap S_2} \qquad \sigma = mgu(l_1, A_{2|p}),$$

algebraization (A)
$$\frac{\neg(l = r) \vee D {:} S}{\neg(l = r) \vee D {:} S \cap \{l = r\}}$$

Let this set of inference rules be denoted by (A). The following additional rules are also considered as they can be of interest for achieving some tasks presented in the section 5:

naming (N)	$$\dfrac{A_1[p/t]\vee D{:}S}{A[p/x]\vee D{:}S\cap\{x=t\}}$$	x a variable not occurring in the formula,
literal elimination (LE)	$$\dfrac{\neg A_1\vee D{:}S}{D{:}S}$$	when the formula $A_1 : S$ is valid in every model of the theory,
disequation elimination (DE)	$$\dfrac{\neg(l=r)\vee D{:}S}{D{:}S\cap\{l=r\}}$$	

It is easy to see that the deletion rule is a combination of algebraization and simplification. Some precisions are necessary for a full understanding of these rules. First let us recall that *algebraic translation* of σ is the process of associating an algebraic domain Σ with σ according to a (syntactic) translation. The rule (LE) may be used when one is able to translate predicates into algebraic relations. It means that a negative literal may be suppressed from the clausal part when its translation does not restrict the algebraic context (i.e. when the translation of the atom underlying to the literal is in the domain defined in the algebraic part of the clause).

The rule (DE) is an important case of (LE): for any equation may be translated into a domain, disequations may be systematically removed by applying successively the rules (A) and (LE). It is also possible to delete the initial formulae while preserving unsatisfiability (see Section 5).

The soundness of this system is straightforward: it suffices to use the previously given semantics. The case of (DE) is presented below as an example. Let I be an interpretation, s be an assignment. If $D :$ is not satisfied by I, s but $\neg(l = r) \vee D : S$ is, then $\neg(l = r)$ is satisfied by I, s too i.e. $l^I s \neq r^I s$ thus $: l = r$ is satisfied by I in the assignment s. This proves that $D : S \cap \{l = r\}$ holds.

It is well-known (see [6]) that the clausal part of $(\mathcal{A})$ defines a refutationally complete system in the following sense: for any denumerable set of unsatisfiable clauses with empty algebraic parts it is possible to finitely derive the empty clause using rules in $(\mathcal{A})$. But this does not prove that if a set of contextual clauses is unsatisfiable then it is possible to derive the empty clause.

That is what is proposed in the following section.

3.2 Completeness

First an adequate notion of completeness for the studied formal system is presented and a sketch of the proof that the previous deduction system called $\mathcal{A}$ is complete is presented. It is not possible to use the usual completeness notion because it is not adequate when considering a specific algebra; the truth value of a formula in first-order logic relies only on its syntax while in this case the class of interpretations is restricted to the previously defined compatible interpretations for contextual clauses: it suffices to define the notion of validity in restricting the class of interpretations to these compatible interpretations.

Let us call a *compatible model* for a set of contextual clauses, a compatible interpretation which satisfies each contextual clause in this set. As a consequence, a calculus is said to be complete if and only if for any denumerable set of contextual clauses for which there does not exist any compatible model, one can derive the empty contextual clause in a finite number of steps.

Starting from this definition, it is clear that the previous system (A) is complete because one can reproduce the standard proof of completeness of the resolution calculus with the paramodulation rule for first-order clause logic with equality. This is due to the following principles in the design of the inference system:

- rules of the resolution method for standard clauses are embedded in ours;
- no literal is discarded in $(\mathcal{A})$. Note that the algebraization rule preserves the disequation while it is not the case of the rule (DE);
- contexts are strengthened by application of rules (that is they are reduced).

The completeness theorem follows from these arguments. To convince the reader, a sketch of proof is given below.

First prove the completeness for the ground case, which is straightforward as ground literals and terms either have a defined translation either may be treated as in the standard case. Then use a lifting lemma standardly used for completeness proofs (see e.g. [15]) in order to lift the proof to the predicate level, the only difficulty being due to the definition of an assignment. This requires the use of an equivalent of the Herbrand's theorem, which is easy to establish, as one can restricts contexts to finitely presented ones containing translations of ground expressions: it is then possible to define it using a new predicate that one defines on ground terms; the Herbrand's theorem is a simple consequence. Last, the case of the rule (P) is similar to the classic one with only more cases to consider, due to the presence of contexts and the partiality of translations. As in the standard case, it is necessary to add the reflexive axiom of equality and all the reflexive axioms for functions. The proof is based on a case study starting from the existence of a refutation for the initial set of contextual clauses augmented by the denumerable set of equality axioms based on symbols used.

As a consequence, a fair use of the inference rules to an unsatisfiable set of contextual clauses infer the empty clause in a finite number of steps. However the inference rules which are not in $\mathcal{A}$ are important and are of interest for designing efficient strategies. This aspect is considered in the section 5 though the section 4 gives an illustration of some possible use of $\mathcal{A}$.

4 Examples

In this part some simple examples are exhibited to illustrate the interest of our approach.

Example 3. Let A, B, C be three points lying on a circle C with center O. Let H be the orthocenter of $\triangle ABC$. Let A' a point lying on C such that O, A and A'

are collinear. Then the lines BH and $A'C$ are parallel. A human-style reasoning could be the following:

- ACA' is an inscribed triangle and AA' is the diameter of the circle. Thus, AC is perpendicular to CA'
- BH is perpendicular to AC and AC is perpendicular to CA'. Thus, CA' and BH are parallel.

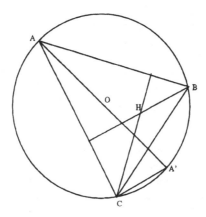

The following is a hybrid proof of this theorem.

Let $S = \{AO = OA'; OA \cdot OA = OC \cdot OC\}$ be a presentation of a domain of the two-dimensional vector space based on reals. This defines our algebraic context.

(1) $\neg Parallel(BH, A'C) : S$		negation of conclusion
(2) $\neg(x \perp y) \vee \neg(y \perp z) \vee Parallel(x, z) : \{\}$		lemma
(3) $\neg(BH \perp y) \vee \neg(y \perp A'C) : S \cap \{x = BH; z = A'C\}$		from (1) and (2) by (R)
(4) $BH \perp AC : S$		hypothesis
(5) $\neg(AC \perp A'C) : S \cap \{x = BH; y = AC; z = A'C\}$		from (3) and (4) by (R)
(6) $S \cap \{AC.A'C = 0\}$		by (DE)
(7)	$\square : \{\}$	by vector computation

Such a proof depends on the axiomatization of the geometry. It would also be possible to translate algebraically $Parallel$(for instance using the outer product) or to use the algebraic translation of the well-known Euclidean axiom: if D_1 is perpendicular to D_2 and D_2 perpendicular to D_3 then D_1 is parallel to D_3.

Example 4. Although naturalness of proofs is very important for teaching and this justifies the use of our method, a problem of the elementary Euclidean plane with an ordering for which there is no simple algebraic alternative is discussed below.

There exist some tricks to represent order relationships such as betweenness in algebra. These are sometimes used with the Wu's method for instance. However it is not possible to apply them when the order appears in the conclusion of the conjecture. Some methods such as Cylindrical Algebraic Decomposition [4] are able to solve these problems but they suffer from a lack of efficiency.

We propose to deal with these problems using a hybrid approach: the order relationship is represented by a predicate and other relationships are represented by algebraic equations. In [14], a first-order theory for the elementary plane geometry is presented. It is a clausal presentation of a system proposed by A.Tarski [17]. The only predicate involving ordering is the ternary B: $B(x, y, z)$ means y is on the line passing through x and z, and y is between them. The entire theory can be presented in our framework using only B and the equality predicate, which allows to reason about position of points within first-order logic and to deal with congruence algebraically. Some of the axioms, such as the reflexivity axiom for equidistance $\forall x, y \ xy \equiv yx$, can simply be removed while others are simplified. For instance, the upper dimension axiom is reduced to a contextual clause with one literal and an algebraic context instead of seven literals.

Example 5. Here is another possible application of our method: reasoning in the projective real plane. Consider an axiomatization of the projective real plane in first-order logic with equality and the usual algebraic context using three coordinates (x, y, z) for each point and such that $x^2 + y^2 + z^2 = 1$. Such an equation may be added for every point in the algebraic context.

While it is possible to introduce the analytic coordinates of points and make computations directly on them as other methods such as Wu's do, it is also possible to use the duality principle for swapping points and lines. So it is possible to exploit both the computable and the metatheoretic aspects by producing in parallel two deductions simultaneously with their algebraic context and to use the generated geometric information.

These examples show that hybrid geometry theorem proving provides new possibilities both for reasoning and computing in geometry.

5 New Possibilities

Some of the possibilities given by combining algebraic computations and logic inferences are presented in this section.

5.1 Strategies for deduction

Deletion of formulas when (LE) is applied may not preserve the completeness of our method. Elimination of every literal in a clausal part provides a contextual clause with a context only: this can prevent from applying the other rules. On one hand if a context is proven to be empty then the clause may be deleted while completeness is preserved. On the other hand if one can prove that a context is non empty then the set of clauses has been proved to be unsatisfiable. However, proving emptiness is a difficult task in general depending on the algebra.

The profiling strategy. Provided a decision procedure for testing emptiness of contexts and a complete strategy for the standard resolution calculus, one may apply the following strategy:

1. For all contextual clause $C : S$, translate as many literals as possible from the clausal part to the context part using the rules (LE), (A) and replace every non variable term which may be translated by a new variable according to the rule (N).
2. Saturate the set of contextual clauses using the (DE) rule.
3. If a contextual clause has an empty clausal part, then check whether the context is empty or not. If both parts are empty, it is the empty clause. Otherwise, add the context to the other clauses.
4. If no clause has an empty clausal part, apply one step of the inference rules (R), (P), (F) according to the complete strategy for resolution chosen as a parameter and go to step 1.

This strategy is called the *profiling strategy* and is complete provided that the strategy used for applying the standard inference rules (resolution, factorization, paramodulation) is complete. The elimination steps cannot necessarily be eliminated, depending on the axiomatization of the geometry and the geometry itself. If the theory is stated without contexts, then it is possible. Remark that replacing elimination steps by standard proofs means that one has some kinds of internal judgement for these rules in the considered theory.

The profiling strategy is also convenient for interaction with a user. This last one may be asked only for deciding whether one has to replace an elimination step or not instead of applying blindly tactics or asking for confirmation at every step. The decision of replacing an elimination step may depend on what the user expects from the proof. The role of contexts, algebraization and literal elimination is crucial in this strategy.

Contextual subsumption. It is possible to define a notion of subsumption by analogy with the standard case. A contextual clause $A : S$ is said to subsume a contextual clause $B : T$ if and only if there exists a substitution σ of variables in A by terms such that $A : S \cap \Sigma \subseteq B : T$ where Σ is the translation of the equational presentation of σ. This means that one intends to discard the clauses for which their algebraic interpretation is an instance of another one.

The subsumption test can be improved using literal elimination and naming; let $A' : S'$ be $A : S$ after elimination and naming, $B' : S'$ be $B : S$ after elimination and naming, if there is a substitution σ of variables in A such that $A' : S' \cap \Sigma \subseteq B' : T'$ then $A : S$ subsumes $B : T$. The inclusion must be understood as a mapping of algebraic domains.

The following algorithm checks whether $A' : S'$ subsumes $B' : T'$ or not.

1. Let $\mathcal{L} = \{\neg L : T'' \mid L$ non equational literal of $B'\}$ where $B' : T''$ is the clause obtained by applying the rule (A) with every equational literal in B'. Let $k = 0$ and $I_0 = \{A' : S'\}$;

2. If I_k contains $\square$: then the test succeeds else compute I_{k+1} the set of inferred contextual clauses applying once the rule (R) with a clause from I_k and the other one from $\mathcal{L}$, and the rule (N) at every position where a variable had been replaced by a term;

3. If $I_{k+1} = \emptyset$ then the test fails else set $k \leftarrow k + 1$ and go to step 2.

It is only the adaptation of a well-known algorithm for deciding whether a clause is subsumed or not. The only difficult point is that variables in the clausal part are not replaced by new distinct constants. Instead their translation has been added to the context.

If there exists a refutational proof for a set of contextual clauses, then there exists a proof involving only clauses not subsumed by previously generated ones. It suffices to use the subsuming contextual clause instead of the subsumed and to insert in the proof new steps for the algebraization, elimination and naming. However it is not clear whether applying the profiling strategy with contextual subsomption provides a complete strategy. This point has to be investigated.

Ignoring positive equational literals. Another interesting strategy consists in ignoring equational disjuncts, treating them like labels. That is, they are never involved in a unification. Thus it is not necessary to apply paramodulation which wastes a lot of computational time. It is clear that instead of getting an empty clause, a disjunction of equations between terms in an algebraic context may be inferred. Finally it suffices to check whether such a clause is satisfied in the algebraic context which can be done by a specific procedure. This strategy could probably improve the efficiency of the method.

5.2 Proof presentation

One can present a hybrid proof using a post-treatment on the generated formal proof. A presentation may be given in a pseudo-natural (English-like) language for instance. An application of the resolution rule such as $\frac{C_1:S_1 \quad C_2:S_2}{C_3:S_3}$ could be translated by:

From C_1 and C_2 we deduce C_3. As S_1 and S_2 are required, we get S_3.

Thus details are slightly hidden (especially boring sequence of algebraic computations) and the proof structure appears more naturally. It is also possible to discard algebraic computations not used in a future step. To make this possible, one can either compute again the algebraic systems and notice how the equations are derived, either annotate algebraic equations by the list of equations used to compute them during the search proof. Other steps can be presented similarly.

6 Comparison with Other Techniques

This section is devoted to a discussion about the relationship between our approach and five other ones: constraint satisfaction, logic programming with constraints, simultaneous search for proofs and counter-examples, semantic resolution and theory resolution.

6.1 Constraint satisfaction

Constraint satisfaction deals with the problem of finding values for variables in a set of constraints such that these constraints be satisfied. The problem of proving a geometry statement of the typical form $A_1 \wedge \ldots \wedge A_n \Rightarrow B$ may be considered as a constraint satisfaction problem, finding *values* for variables in B such that the constraints on these variables defined by $A_1, \ldots, A_n$ are satisfied. One can even consider an algebraic equivalent translation of these atoms. Solutions represent geometrically valid configurations of the conjecture.

The main advantage of this approach is to provide a more useful information than the validity. Unfortunately it is difficult to solve this kind of constraints symbolically: the problem is out of the scope of actual constraint satisfaction methods. Moreover even if one gets a finite union of domains, proving the conjecture is equivalent to proving that this union covers the entire domain of variables which is related to the so-called complement problem [18] and is undecidable in general. However it is possible that the complement problem be decidable in some geometries.

6.2 Constraint logic programming

Modern versions of logic programming languages such as Prolog allow to use constrained clauses on specific domains. However some of them only deal with a specific kind of variables whose type is part of the language. In the hybrid approach, no sort to represent the algebra is introduced as this point is managed by the translation, and literals may also be translated: we do not restrict ourselves to (dis,in)equations. Note also that the Prolog language is restricted to Horn clauses. The philosophy of these systems is different from ours: they are designed to solve efficiently problems expressed by logical and numerical constraints while ours is designed to give a proof of a geometry theorem by combining several languages. Finally these systems cannot simulate ours (except in using them as ordinary programming language) but could be interesting to implement prototypes. Their strategies could also be of a particular interest to improve standard strategies for theorem provers based on clauses.

6.3 Simultaneous search for proofs and counter-examples

A method for trying to prove or refute a theorem simultaneously has been developed by Bourely, Caferra, Peltier and Zabel [5, 1]. More generally it is a method of an emerging research field, namely automated model building. The language they use is an extension of clauses by systems of equations and disequations. Systems are reduced symbolically and lead to the definition of a model when the set of clauses is saturated and does not contain the empty clause.

This approach is very general but does not take into consideration the specificity of a particular theory. Although a lot of statements have been proved or refuted efficiently, significant geometry theorems are still out of the scope of this method. In integrating an efficient (but specific) treatment of algebraic equations it should be possible to overcome the difficulties of this approach and thus to fill the gap existing between theorem provers based on a logic and provers specialized to the theory. However using constrained clauses instead of clauses in our framework could be an interesting challenge for building counter-examples in geometry and capturing a larger class of models.

6.4 Semantic resolution

Semantic resolution essentially consists in using an interpretation to divide clauses into two sets and in preventing clauses within the same set to be resolved with each other. In our method an interpretation (actually it is a model) is used to in connection with an algebraic domain in which computations can be efficiently done. One may say that the interpretation is used locally to reduce the size of clauses and preventing the production of new clauses. With semantic resolution, the interpretation is used globally to prevent the application of the resolution rule. However, semantic resolution could be extended to hybrid reasoning as a strategy to reduce the search space.

6.5 Theory resolution

Theory-resolution [16] is a method created by M.Stickel to use specific reasonings in some theories and to incorporate them in the framework of resolution. It is also possible to use external procedures to simulate the reasoning in these theories. Such a procedure is used as a kind of black box to improve the efficiency. Many refinements of resolution may be simulated by this approach: it is the case of reasoning with sorts.

The hybrid approach also use external procedures but it is possible to control their use with the rules (A), (N), (DE) and (LE). This means that both parts interact. Moreover Stickel emphasizes the use of one theory T though the hybrid approach gives the possibility to use different algebras and is designed to connect logic reasonings and algebraic computations.

7 Conclusion and Further Research

We argue that the hybrid approach provides a convenient integration of logical and algebraic methods for proving theorems in geometry. Although only a calculus based on resolution has been presented, it is clear that this approach can be easily adapted to other extensively studied formalisms such that tableaux or sequents. However if one still makes the restriction of algebraic systems to equations, one has to take care about the definition of first-order formulas with contexts.

This presents a lot of interests. Obtained proofs are more readable and may be formatted and presented to look like human-style proofs in elementary geometry. Moreover this approach allows to improve logic-based methods by using the efficiency of algebraic computations. The profiling strategy allows to structure a proof, to reach a desired level of precision and to hide boring details. But it also gives the possibility to exploit powerful logic-based methods to improve them specifically in our framework. It also gives the possibility to structure a proof of a theorem using the logic part to provide a sketch and the algebraic part to provide details.

One must be conscious of some drawbacks of this approach. Of course the relationship between inferences and algebraic computations need to be improved to enable a true cooperation. This point is at the heart of the hybrid approach but is still entirely new. Intelligent strategies for deciding whether it is interesting to translate a literal or not would be interesting and even be one the key topic. Moreover algebraic methods are still used as black boxes though elimination methods provide a useful strategy for applying inference rules. This opinion is strengthened by some works on the generalization of algebraic methods to first-order logic [13].

This approach makes new things possible or easier to realize. However there are still a lot of possibilities to explore. In some cases it is possible to translate algebraic terms into their logical equivalent formulas. This possibility has not been exploited yet. It could also be interesting to mix the algebraic frameworks, for instance in associating one to each clause and giving relationships between them. It is still necessary to study how to adapt strategies for the resolution method or calculi based on a clausal language to our framework. A systematic method would be welcome. And it could be considered how to exploit geometric transformations to improve efficiency of provers or at least avoiding unnecessary operations. Last it could be interesting to characterize what are subsidiary conditions from a logical point of view: in some axiomatizations, equational literals correspond to such conditions. Thus resolution calculi could be considered without equality if we add algebraic contexts to clauses. This could make the cooperation with algebraic methods easier and simplify the task of provers.

An implementation of this system is currently in development. It is intended to use simultaneously several algebraic contexts to produce hybrid proofs based on the capabilities of external computer algebra systems. Production of interesting and pertinent proofs should be highly interesting for the discovery of new mathematical concepts especially in geometry. We think that this task relies more on qualitative aspects of proof theory than computation speed and hope that the hybrid approach is the first step towards the achievement of this task.

References

1. Bourely, C., Caferra, R. and Peltier, N.: A method for building models automatically: Experiments with an extension of OTTER. In: Proceedings of CADE-12, Springer Verlag, LNAI 814, 72–86, 1994.
2. Buchberger, B.: Gröbner bases: An algorithmic method in polynomial ideal theory. In: Multidimensional Systems Theory, Bose, N.K. ed., D. Reidel Publ. Comp., 184–232, 1985.

3. Caferra, R. and Herment, M.: A generic graphic framework for combining inference tools and editing proofs and formulae. Journal of Symbolic Computation, **19**, 217–243, 1995.

4. Collins, G.E.: Quantifier elimination for the elementary geometry. In: Proceedings of 2nd G.I. Conference on Automata Theory and Formal Languages, Brakhage, H. ed., LNCS 33, 134–183, 1975.

5. Caferra, R. and Zabel, N.: A method for simultaneous search for refutations and models by equational constraint solving. Journal of Symbolic Computation, **13**, 613–641, 1992.

6. Chang, C.-L. and Lee, R.C.-T.: Symbolic Logic and Mechanical Theorem Proving. Academic Press, New York-London, 1973.

7. Chou, S-C., Gao, X-S. and Zhang, J-Z.: Automated generation of readable proofs with geometric invariants: I. Multiple and shortest proof generation. Journal of Automated Reasoning. **17**, 325–347, 1996.

8. Chou, S-C., Gao, X-S. and Zhang, J-Z.: Automated generation of readable proofs with geometric invariants: II. Theorem proving with full-angles. Journal of Automated Reasoning, **17**, 350–370, 1996.

9. Coelho, P. and Pereira, L.M.: Automated reasoning in geometry theorem proving with Prolog. Journal of Automated Reasoning, **2**, 329–390, 1996.

10. McCune, W.W.: OTTER 2.0. In: Proceedings of CADE-10, Springer-Verlag, LNAI 449, 663–664, 1990.

11. Fèvre, S.: A hybrid method for proving theorems in elementary geometry. In: Proceedings of ASCM'95, Scientists Incorporated, 113–123, 1995.

12. Kapur, D., Mundy, J.L. eds.: Geometric Reasoning. MIT Press, Cambridge, 1989.

13. Liu, Z. and Wu, J.: The remainder method for the first-order theorem proving. In: Proceedings of ASCM'95, Scientists Incorporated, 91–97, 1995.

14. Quaife, A.: Automated development of Tarski's geometry. Journal of Automated Reasoning, **5**, 97–118, 1989.

15. Robinson, J.A.: A machine-oriented logic based on the resolution principle. Journal of the ACM, **12** 23–41, 1965.

16. Stickel, M.E.: Automated deduction by theory resolution. Journal of Automated Reasoning, **1** 333–355, 1985.

17. Tarski, A.: A Decision Method for Elementary Algebra and Geometry, 2nd edition. University of California Press, Berkeley and Los Angeles, 1951.

18. Trainen, R.: A new method for undecidability proofs of first-order theories. Journal of Symbolic Computation, **14**(5), 437–458, 1990.

19. Wang, D.: Reasoning about geometric problems using an elimination method. In: Automated Practical Reasoning: Algebraic Approaches, Pfalzgraf, J., Wang, D. eds., Springer-Verlag, 147–185, 1996.

20. Wang, D.: An elimination method for polynomial systems. Journal of Symbolic Computation, **16**, 83–114, 1993.

21. Wu, W.: Basic principles of mechanical theorem proving in elementary geometries. Journal of Automated Reasoning, **2**, 221–252, 1986.

22. Wu, W.: Mechanical Theorem Proving in Geometries: Basic Principles. Springer-Verlag, Wien-New York, 1994.

Author Index

Springer
and the
environment

At Springer we firmly believe that an international science publisher has a special obligation to the environment, and our corporate policies consistently reflect this conviction.
We also expect our business partners – paper mills, printers, packaging manufacturers, etc. – to commit themselves to using materials and production processes that do not harm the environment. The paper in this book is made from low- or no-chlorine pulp and is acid free, in conformance with international standards for paper permanency.

Lecture Notes in Artificial Intelligence (LNAI)

Lecture Notes in Computer Science